MW00786934

REPORT OF JOINT FIGHTER CONFERENCE

NAS Patuxent River, MD
16-23 October 1944

REPORT OF JOINT FIGHTER CONFERENCE

NAS Patuxent River, MD
16-23 October 1944

Technical Editing by Francis H. Dean

Schiffer Military History
Atglen, PA

Note: Slides and figures referenced in the text appear on pages 339-356.

Book Design by Robert Biondi.

Library of Congress Catalog Number: 97-67601.

Printed in the United States of America.
ISBN: 0-7643-0404-6

Published by Schiffer Publishing Ltd.
4880 Lower Valley Road
Atglen, PA 19310 USA
Phone: (610) 593-1777
FAX: (610) 593-2002
E-mail: Schifferbk@aol.com.
Please write for a free catalog.
This book may be purchased from the publisher.
Please include $3.95 postage.
Try your bookstore first.

REPORT OF JOINT FIGHTER CONFERENCE

U.S. NAVAL AIR STATION
Patuxent River, MD

To: Chief of the Bureau of Aeronautics.
Subj.: Joint Fighter Conference-Report on.

1. A Joint Fighter Conference sponsored by the bureau of Aeronautics, Navy Department, was held at NAS Patuxent River, MD., during the period 16-23 October 1944. Representatives of contractors for fighting type aircraft for the Army, Navy and Marine Corps participated as well as representatives from all the services, including the Royal Air Force, Royal Navy, and Royal Canadian Air Force.

2. From a review of the discussions and pilots' reports and questionnaires, the opinion of the participants relative to the basic requirements for fighter aircraft are expressed below. It is emphasized that the opinions expressed in this summary represent a cross-section opinion of the participants of the conference and do not necessarily represent the views of the Navy Department, nor of any of the individuals, services, or contractors.

I. A. *Stability.*
Weak positive stability (static and dynamic) about all three axes of flight is desired, although individual airplanes exhibiting stronger stability characteristics were favorably commented upon by most pilots.

B. *Controllability.*
Highly effective control about all axes of flight with light positive forces is desired. Pilots were reluctant to accept artificial means for attaining light control forces unless control "feel" was present. Dive recovery flaps are acceptable as a "stop-gap" safety means of recovery from the effects of compressibility. Contractors were urged to continue investigations of airplane behavior in the compressibility ranges.

C. *Performance.*
Although high speed and rate of climb performance is obviously desired, the problems of range and endurance are demanding greater emphasis. Selection of the most desirable airplane critical altitudes was left an open question, even though discussions indicated that the theatre of operation would be the deciding factor.

D. *Armament.*
.50-cal. machine guns in batteries of four, six, or eight are currently satisfactory, but the trend is definitely towards exclusive use of 20-mm. cannons. Cannons in combination with .50-cal. guns were viewed with disfavor. The 37-mm. and 75-mm appear to have no place

in current fighters. All fighters should be equipped with zero length rocket launchers with mountings for four rockets on each wing. Sufficient ammunition for 20 seconds of machine gun or cannon fire should be provided.

E. *Protection.*
The present protective armor is satisfactory except for the desire for additional protection underneath. Concern was expressed over the removal of miscellaneous gear and structure from behind the pilot to obtain the advantages of the bubble canopy. This equipment previously augmented pilot protection astern by causing projectile tumbling. It was concluded that adequate protection could be obtained without sacrifice of visibility astern through careful design.

F. *Cockpit.*
Pilot comfort and simple arrangement of instruments and controls become more important in consideration of increasing range. Provision must be made for adequate heating, ventilation, oxygen, pressurized cabins where necessary, the perfection of a single power control, the adoption of reliable automatic devices – all with the idea in view of increasing the efficiency of the pilot. Adequate visibility and absence of distracting reflections from the enclosure are mandatory.

G. *Power Plant.*
Increased power with due regard to economy is desired without sacrificing current dependability. The current air-cooled radial engines inspired the most confidence. The use of a conventional reciprocating engine with a supplementary jet or gas turbine in combination was viewed with disfavor. Pure jet propulsion engines were considered promising for future use.

II. Further opinion relative to the additional characteristics necessary for certain special classifications of fighters follows:

A. *Night Fighter.*
A night fighter should have an exceptionally high rate of climb in addition to high speed. It must be multiplaced to accommodate radar operator. Visibility ahead should be unrestricted. The pilot should not be placed behind an engine or propeller. Speed reduction devices must be incorporated. The heaviest armament consistent with the size of the airplane should be provided.

B. *Escort Fighter.*
The correct balance of protected and external unprotected fuel must be achieved. Extreme range must be stressed. Pilot comfort must be stressed. Critical altitude must be determined by the theatre of operation.

III. *Conclusion.*
It is considered that present service type fighting airplanes used by the United States and her allies are satisfactory, and comparable to conventional types used by our common enemies. However, emphasis should be placed on urgent consideration of design, production, and employment of fighter aircraft utilizing pure jet or pure rocket propulsion.

3. Members of the conference generally agree that the principal benefits derived from it are somewhat indefinable but none the less real and important. Aside from certain obvious conclusions, partly outlined above, there was wide diversity of opinion on the merits of individual characteristics of fighter aircraft. The greatest benefit of the conference is believed to be the opportunity afforded pilots and engineers representing many different companies and branches of the service as well as many different operational theaters to evaluate personally and exchange views on the merits of a number of the most advanced types of fighter aircraft under actual flight conditions. This collective experience, which will be translated into future airplane design work, is believed to be of far more value than any categorical generalizations approving or disapproving individual aircraft features on which no general agreement has been reached.

4. A detailed record of the conference proceedings, including analysis of flight reports and questionnaires, photographs of participating aircraft, and the agenda of the conference, which are appended, constitutes the final report.

A.P. Storms
Commanding Officer

ROSTER OF JOINT FIGHTER CONFERENCE MEMBERS

MANUFACTURERS

Aeroproducts Division, General Motors Corporation.

J. Talbott.

Allison Engineering Division, General Motors Corporation.

D. Gerdan.

D. G. Zimmerman.

Bell Aircraft Corporation.

E. P. Haughey.

M. M. McEuen.

Jack Woolams.

Boeing Aircraft Company.

Fred Collins.

Don W. Finlay.

Chance Vought Division, United Aircraft Corporation.

Paul S. Baker.

L. A Bullard.

R. H. Burroughs.

J. R. Clark.

F. N. Deckerman.

H. B. Gibbons.

E. J Greenwood.

B. Guyton.

W. N. Horan.

W. C. Schoolfield.

C. L. Sharp.

I. M. Shoemaker.

A. I. Sibila.

N. Siegal.

F. Taylor.

Ben N. Towles.

Curtiss Wright.

K. Boedecker.

Fred Brychta.

Fred Chamberlin.

Lloyd Child.

Allen Chilton.

G. B. Clark.

F. A. Fayerweather.

Lou Enos.

F. Foster.

Jack Hamilton.

W. K Houpt.

R. S. Huested.

R. McDonough.

V. F. Peterson.

G. E. Powers.

Fred Steele.

H. W. Thomas.

A. Von Valkenberg.

W. O. Webster.

J. T. Wetzel.

H. M. Wilkerson.

S. G. Winch.

Burdette Wright.

DeHavilland Aircraft of Canada, LTD.

E. G. Alley.

S. H. Burrell.

J. W. St. John.

Eastern Aircraft Division, General Motors Corporation.

R. W. Foot.

B. O. Malmberg.

A. Wescott.

J. R. Williams.

General Electric Company.

E. J. Specht.

Goodyear Aircraft Corporation.

Donald Armstrong.

Dr. K. Arnstein.

D. A. Beck.

W. F. Burdick.

E. M. Eichman.

R. C. Hager.

E. M. Hawley.

T. A. Knowles.

V. S. Kupelian.
A. R. Mueller.
C. J. Pierce.
E. L. Shaw.
S. Stroud.

Grumman Aircraft Engineering Corporation.

S. A. Converse.
H. W. Danklefs.
C. Engvaldsen.
Pat Gallo.
B. Allison Gillies.
R. L. Hall.
C. Meyer.
F. C. Rowley.

Hamilton Standard Propellers.

Carl Baker.
F. C. Macternan.
P. E. McCormack.

Lockheed Aircraft Corporation.

B. A. Martin.
P VonEssen.

McDonnell Aircraft Corporation.

Woodward Burke.
R. J. Baldwin.

North American Aviation, Incorporated.

J. F. Steppe.
E. W. Virgin.

Northrop Aircraft, Incorporated.

John Myers.
Fred Baum.

Packard.

W. M. Packer.
Col. J. G. Vincent.

Pratt and Whitney Aircraft Division, United Aircraft Corporation.

J. S. Conley.
P. R. Debruyn.
H. Gosselin.
T. Gurney.
H. Kuhn.
A. H. Marshall.
G. W. Pederson.
P. S. Riley.

Republic Aviation Corporation

R. G. Bowman.
C. H. Miller.
D. M. Parker, Jr.
J. F. B. Parker.
M. S. Wittner.

Ryan Aeronautical Company.

W. K. Balch.
E. O. Baumgarten.
M. G. McGuire.
E. D. Sly.

Sperry Gyroscope.

E. J. Petersen.

Wright Aeronautical Corporation.

P. R. Brown.
R. B. Miller.
W. G. Owens.
Mr. Sulzman.

United Aircraft Corporation.

Luke Hobbs.
J. J. Hospers.
D. J. Jordan.
C. A. Lindbergh.
C. J. McCarthy.
M. D. Raffo.
J. F. Sandrs.

BRITISH

Air Commodore N. R. Buckle, M. V. O., R.A.F.
Group Captain C. Clarkson, A. F. C., R.A.F.
Group Captain H. W. Dean, A. F. C., R.A.F.
Commander D. R. F. Cambell, D. S. C., R.N.
Lt. Comdr. G. Gunthrie, R.N.
Lt. Comdr. G. R. Callingham, R.N.
Sq. Ldr. A. W. Stockwall, R.A.F.
Lt. J. P. M. Reid, R.N.V.R.

Lt. D. B. Law, R.N.
Lt. J. Lawrence, R.N.
Lt. L. P. Twiss, D. S. C., R.N.V.R.
Fl. Lt. Singleton, R.A.F.
Fl. Lt. Bruorton, R.A.F.

ARMY

Col. J. A. Bertolero, Middletown Air Service Command.
Col. R. M. Caldwell, Headquarters, Army Air Forces.
Col. Lee B. Coats, AAFPGC, Eglin Field, Fla.
Col. R. S. Garman, Air Technical Service Command, Wright Field.
Col. J. B. Kirkendall, Middletown Air Service Command.
Col. H. Vicellio, AC/AS Operations, Commitments and Requirements.
Lt. Col. Jack Carter, Air Technical Service Command, Wright Field.
Lt. Col. J. S. Edgerton, Headquarters, Army Air Forces, St. Louis.
Lt. Col. L. B. Meng, AAFPGC, Eglin Field, Fla.
Lt. Col. M. A. Moore, AC/AS Operations, Commitments and Requirements.
Lt. Col. H. A. Pelton, Middletown Air Service Command.
Lt. Col. M. F. Reid, Middletown Air Service Command.
Lt. Col. J. H. Sams, Air Technical Service Command.
Lt. Col. C. H. Terhune, Air Technical Service Command.
Lt. Col. I. W. Toubman, AAFPGC, Eglin Field, Fla.
Lt. Col. K. A. Tyler, AAF Board, Orlando, Fla.
Lt. Col. W. Yancey, Director of Operations, Hammer Field.
Maj. C. W. Draper, Middletown Air Service Command.
Maj. T. Lamphier, AC/AS Operations, Commitments and Requirements.
Maj. A. G. McKay, Headquarters, Army Air Forces.
Maj. B. Muldoon, AAFPGC, Eglin Field, Fla.
Maj. Ralph Newman, Middletown Air Service Command.
Maj. W. E. Rhynard, AC/AS, Operations, Commitments and Requirements.
Maj. V. B. Schoenfeldt, AAPGC, Eglin Field, Fla.
Capt. J. C. Davis, AC/AS, Operations Commitments and Requirements.
Capt. V. A. Ford, Middletown Air Service Command.
Capt. J. Kessler, Headquarters, Army Air Forces.
Capt. R. A. McClung, Headquarters, Army Air Forces, OTI.
Capt B. E. Turner, AAFPGC, Eglin Field, Fla.
Capt. Webb, AAFPGC, Eglin Field, Fla.
Lt. Corey, Middletown Air Service Command.
Lt. S. D. Crawford, Hammer Field, Fresno, Calif.
Lt. R. Eluhow, Headquarters, Army Air Forces.
Oliver Brandt, Middletown Air Service Command.
G. S. Carroll, Middletown Air Service Command.
P. J. Hendrickson, Middletown Air Service Command
F. M. Kline, Middletown Air Service Command.
G. F. Koerting, Middletown Air Service Command.
J. W. Rush, Middletown Air Service Command.

R. E. Smink, Middletown Air Service Command.

MARINES

Col. C. L. Fike, Deputy Dir. Eng. Div., BuAer.
Lt. Col. D. E. Canavan, Flight Test, NAS Patuxent River, Md.
Lt. Col. J. O. Harshberger, VF Design, BuAer.
Lt. Col. C. J. Quilter, Service Test, NAS Patuxent River, Md.
Lt. Col. J. N. Renner, Military Requirements, BuAer.

NAVY

Rear Admiral L. B. Richardson, Assistant Chief, BuAer.
Capt. A. P. Storrs, Commanding Officer, NAS Patuxent River, Md.
Capt. T. C. Lonnquest, Director, Engineering Division, BuAuer.
Capt. S. B. Spangler, Power Plant Design, BuAer.
Commander L. H. Bauer, Armament Test, NAS Patuxent River, Md.
Commander C. T. Booth, Flight Test, NAS Patuxent River, Md.
Commander T. B. Clark, Executive Officer, NAS Patuxent River, Md.
Commander J. E. Clark, Armament Test Unit NAS Patuxent River, Md.
Commander E. W. Conlon, Structures Branch, BuAer.
Commander R. E. Dixon, Director Military Requirements, BuAer.
Commander G. M. Greene, Radio Test, NAS Patuxent River, Md.
Commander R. E. Harmer, VF Design, BuAer.
Commander J. P. Monroe, Armement Branch, BuAer.
Commander F. L . Palmer, Tactical Test, NAS Patuxent River, Md.
Commander P. H. Ramsey, Director of Tests, NAS Patuxent River, Md.
Commander S. S. Sherby, Flight Test, NAS Patuxent River, Md.
Commander J. G. Silney, NAS Sanford.
Commander W. E. Sweeney, VF Design, BuAer.
Commander M. H. Tuttle, Tactical Test, NAS Patuxent River, Md.
Commander T. D. Tyra, VF design, BuAer.
Lt. Comdr. J. H. Anderson, VF Design, BuAer.
Lt. Comdr. F. L. Bates, BuAer.
Lt. Comdr. R. A. Beveridge, Tactical Test, NAS Patuxent River, Md.
Lt. Comdr. I. W. Brown, Flight Test, NAS Patuxent River, Md.
Lt. Comdr. Cooke, BuAer.
Lt. Comdr. R. C. Merrick, VF Design, BuAer.
Lt. Comdr. R. M. Milner, Navy Liaison, AAFPGC, Eglin Field, Fla.
Lt. Comdr. E. M. Owen, Flight Test, NAS Patuxent River, Md.
Lt. Comdr. A. D. Pollock, VF Design, BuAer.
Lt. Comdr. J. F. Sutherland, VF Design, BuAer.
Lt. C. C. Andrews, Tactical Test NAS Patuxent River, Md.
Lt. R. L. Buell, BuAer.
Lt. A. G. Clarke, Flight Test, NAS Patuxent River, Md.
Lt. C. R. Clifford, BuAer.
Lt. W. A. Collier, BuAer.
Lt. J. G. Gavin, VF Design, BuAer.
Lt. W. C. Holmes, Tactical Test, NAS Patuxent River, Md.
Lt. A. Hyatt, BuAer.
Lt. J. B. Jorgenson, Service Test NAS Patuxent River, Md.

Lt. J. M. Kirchberg, Tactical Test NAS Patuxent River, Md.
Lt. J. E. Laurance, Flight Test, NAS Patuxent River, Md.
Lt. R. L. Loesch, BuAer.
Lt. C. L. Orphanides, NAAF Charlestown, R. I.
Lt. R. E. Reed, NAS Sanford.
Lt. D. Runyon, Flight Test NAS Patuxent River, Md.
Lt. A. H. Sallenger, Power Plant Design, BuAer.
Lt. W. M. Shehan, Jr., Operations, NAS Patuxent River, Md.
Lt. E. D. Soule, NAS Patuxent River, Md.
Lt. Walter Wilds, Air Intelligence Group.
Lt. B. R. Winborn, BuAer.
Lt. J. C. Weedon, BuAer.
Lt. (jg) L. W. Cabot, VF Design, BuAer.
Lt. (jg) E. E. Ford, BuAer.
Lt. (jg) S. E. Garrett, Tactical Test, NAS Patuxent River, Md.
Lt. (jg) N. E. Halaby, Flight Test, NAS Patuxent River Md.
Lt. (jg) D. L. Mandt, Armament Test, NAS Patuxent River, Md.
Lt. (jg) L. A. Ryan, W-V (S), Flight Test, NAS Patuxent River, Md.
Ensign M. Abzug, BuAer.
Ensign R. M. Head, BuAer.
Ensign P. L. Offield, VF Desigh, BuAer.
Dr. A. A. Brown, Air Intelligence Group.
R. H. Driggs, Aviation Design Research, BuAer.
W. Frisbie, Design Coordination, BuAer.
H. Hennington, Air Intelligence Group.
Dr. E. Lamar, Air Intelligence Group.
Dr. J. Wehausen, Air Intelligence Group.

NACA

Mr. Gilruth, Langley Field, Va.
Mel Gough, Langley Field, Va.
H. Hoover, Langley Field, Va.
W. H. Phillips, Langley Field, Va.

OPENING MEETING
16 October 1944
0930-1200

Commander Paul H. Ramsey, USN, Director of Test, NAS, Patuxent River, and Chairman of the Joint Fighter Conference: "The First Joint Fighter Conference at the Naval Air Station, Patuxent, will please come to order. First, I want to welcome you to the station, and tell you that we are very happy to have you aboard. We hope to offer you all the conveniences which our facilities will permit. We want you to feel at home and we hope our arrangements will make your visit happy while you are here.

"The previous fighter conference at Eglin Field was very successful and profitable, and we hope these meeting here will be even more successful. This is the first Navy-sponsored Joint Fighter Conference and we hope that it is only the beginning of many more to be sponsored at this station. We plan in the future to have a suitable building to take care of these conferences as you know, this station is still under construction and I think that in another year we should be well along the way to furnish a larger program.

"The purpose of the conference is to enable the services and the companies associated, with the fighter program to test and evaluate airplanes and to report on the different types on a comparative basis. Along this line I wish to state that we hope the contractors will not enter the meet in a competitive spirit.

"The first 6 days of the 8-day conference will be built on intensive flying and discussion: on the seventh and eighth days we hope to have final evaluations. The topic for discussion will be posted on the bulletin board outside and any criticism will be appreciated by myself and Commander Booth (Commander C. T. Booth, USN, Flight Test Officer, NAS, Patuxent, and officer in charge of conference arrangements). Flight schedules will be based each day on pilot preference. If you will be so kind as to enter on the preference list the particular types of plane you wish to fly, Commander Milner (Lt. Comdr. R. M. Milner, USN, Navy Liaison Officer, Eglin Field, and Conference Coordinator) will arrange for you to make as many flights as possible. Members will be able to talk on any topic they wish, in addition to those posted on the bulletin board. We will welcome any remarks on any subject you may discuss.

"In case of foul weather, which we are likely to have here this time of the year we have three foul-weather schedules to take up at that time and we hope to make it interesting enough even though you can't get in any flying. Alternate programs will include rocket demonstrations, a trip through the station, and various other items which may be of interest to you at the fighter conference.

"The conference operations, we hope, will be simplified so that you will not be annoyed by the details involved in flying your planes. Lieutenant (jg) L. A. Ryan, in the conference headquarters room, is pretty well briefed on all that is to happen and if you have any questions, please see her and she'll be glad to answer them. The operations officer is Lt. W. M. Shehan, Jr., USNR, who will be in the headquarters room also.

"All general information, including a chart of the station and surrounding territory, is printed in the conference bulletin which I hope you all have. The detailed daily program, flight schedules and transportation schedules will be posted in the conference headquarters room and any questions please take to the duty officer who will be on duty 24 hours a day. Airplanes available are listed in the program; to that list should be added the P-61.

"The contractors' representatives will give a 3- or 4-minute talk on the planes they brought down, the special qualities and restrictions, and general information relative to their use, speed, etc. Individual check-out pilots will be assigned to each aircraft and each pilot must have a check-out in order to insure safety. Special check-out flight line will be outside this building (Flight Test Landplane Hangar) and the spot will be shown on a chart which will be posted. Special calls to save you confusion with the control tower will name all planes Blue, for Example Blue 1, Blue 2, etc. Webster Field, as shown on the chart, is an auxiliary field and available for touch and go landings or for whatever landings are desired. I wish to emphasize the fact that if you are late taking off on the hop, please be back on time in order not to throw off the schedule. Try to take off on time, but by all means be back on time in order not to delay refueling and succeeding hops.

"I think we have sufficient local transportation available. There is a daily shuttle trip to Washington for those of you who wish to commute. I advise against that, however, because during October and November there is often trouble with weather. I think it is better, if you can possibly do it, to remain on the station. Communications, including incoming telegrams, telephone calls, letters, etc., will be handled by Miss Ryan. She will see that you get them all.

"Your attention is called to security. The printed programs are restricted and during the conference you may see everything on the station except secret and top secret material. Please do not divulge this information to others. Facilities of the ship's service, commissary store, officers' bar are all available to the members of the conference. Priority will be given you at the barber shop and for laundry and dry cleaning during the conference. Please wear the conference badge because special privileges are given to its wearers which are not given to other members of the station. The badge will give you special consideration from all departments on the station and will permit you to go into restricted areas. It will also permit you access to all information directly related to the conference, the files, technical library and the Director of Test's office and all matters of a confidential classification and below. It will also permit you ingress and egress to the Naval Air Station at Anacostia. Every member of the conference is requested to safeguard the information received. No secret matters will be brought out either directly or indirectly. No confidential discussions are at the dinner table, ship's service, etc.

"We will have an officers banquet at the BOQ Wardroom on Wednesday. On Friday we will have some good Patuxent River oysters for the oyster roast at Captain Wildman's barn and on Sunday an officers cocktail party. The conference will conclude Monday afternoon at 1730.

"I will now turn the meeting over to Lieutenant Commander Milner, who will cover some of the conference's operating details."

Lieutenant Commander MILNER – Conference Operations and Flight Procedures: "Just a word to the operating, the flight schedule, course rules, flight gear and other facilities to enable everybody to get into the airplanes and fly. Make left-hand turns at 1,000 feet around the Patuxent field; contact the tower in the usual conventional procedure. At Webster Field fly left-hand turns at 1,000 feet. Webster Field is available for touch and go landings but I want to caution you that the runways are 5,000 feet and 4,300 feet and there is a possibility of overshooting. At the Patuxent field runways are plenty long enough. One is almost 10,000 feet and the other is about 7,000 feet. As for the radius of operations, please fly within approximately 25 miles of the main field; we would like everybody to stay in that area and primarily to the north. On the bulletin board is posted a list of danger areas. Your knee-pad chart will also give the danger areas. We will have the spot posted on the bulletin board. Take an airplane out and bring it back to the same spot. Radio procedure there will be a few variations between available planes. All planes except the Seafire and the Firefly will be on VHF, with the Channel C for tower communicators and Channel B for inter-plane, if you wish, in the air. The Seafire and Firefly transmit on 3105, which is the cross-country frequency, and receive on 347. You will find on the instrument panel, probably near the sight, the radio instructions for each particular airplane. For flight gear, right below here we have a storeroom where you can draw headsets, oxygen masks and life jackets. Incidentally we are going to request that all hands wear life jackets because there is plenty of water around here.

"Now for the purpose of the record and keeping the logs up to date, we will use for the Navy planes the regular yellow sheet and for the AAF Form 1. On the main bulletin board are samples of both.

"We have also G-suits or zoot suits in six sizes and there will be a man standing by the airplanes to give you instructions. The zoot suit is just a common ordinary flight suit or coverall; on the back of the collar is given the waist measurement. G-suit operation is simple. All you have to do is put your suit on and connect up the fitting on the left side. If you don't wish to use the G-suit, don't connect up the fitting.

"One of the most important things for a fighter conference is getting data on the airplanes for discussion and record. We have a knee-pad questionnaire for each flight. The questionnaire will be explained later by Lieutenant (jg) Halaby. USNR, NAS, Patuxent).

"Now here is how we are going to work flight clearance. We will have the flight schedule posted on the main bulletin board and a copy in the operations office. If you are scheduled for a flight see the operations and he will hand you a kneepad which is your clearance. Show it to the crew chief or plane captain and all you have to do from then on is get radio clearance from the tower. When you return from the flight, please bring your kneepad back to the operations office. We want to get it ready for the next flight and we want the comment card. These cards will later be available for you to see. All flights are scheduled for 1 hour with 45 minutes between flights.

"Pilots scheduled for afternoon flights get their cockpit check-outs after this meeting. We would like all the Army pilots that are available to stand by their airplanes to assist us with cockpit check-outs, and we will have the Navy pilots stand by their airplanes. We will also appreciate having the contractors at their airplanes."

Capt. A.P. STORRS, USN, Commanding Officer, NAS, Patuxent, River, Md.: "Welcome to Patuxent River Air Station. We are glad that the Navy finally has an opportunity to have one of those conferences. This is the first time that we have had the honor to be able to put across one of these affairs. The one at Eglin Field last year, I understand, was very highly successful. You will find in your booklets a description of the Naval Air Station which is perhaps new to many of you. We want you to feel as much at home during the week while you are here as we can possibly make it; that is, during working hours and after working hours too. Any of the gentlemen who are interested in any of the other projects, considering security, of course will be able to go into matters which are cognizant to their companies or to their departments after contacting the Director of Test's office. I hope you enjoy your stay here and I hope we have good weather and can get in a lot of flying time, and I am sure the indications are pretty good at the present time. Gentlemen, I take great pleasure in introducing the Assistant Chief of the Bureau of Aeronautics, Rear Admiral L. B. Richardson."

Admiral L.B. RICHARDSON: "Gentlemen: On behalf of Admiral Ramsey and the Bureau of Aeronautics, I want to welcome you to the Fighter Conference. This conference is held primarily for the purpose of informal discussions on the suitability of airplanes from the standpoint of pilots. We want to place emphasis on the practical pilot considerations at all times. The basic idea of this meet is the comparison and evaluation of your own fighters and those of other people, both in our service and other services. The value of the results obtained here will be chiefly dependent upon what the pilots are able to report orally on their impressions as to what they can actually get out of the airplane, not only in speed, climb, and other numerical data, but what they can do with it when they meet the enemy. We want to emphasize also that this is not a competition. It is chiefly your chance for personal contacts and free, frank discussion and criticism and interchange of ideas on any and all developments. The more free and frank the discussion is, the better the results will be. Everything is completely informal in all respects. We would like to have you remember that nothing you say will be used against you, even if you start an argument just for the sake of an argument, and we would like to have everybody feel that they are free to say whatever comes in their mind, and not to worry about the consequences. The basic purpose, of course, is to get the best possible fighting airplanes for the combat unit of all services and branches. In the Navy, everybody realizes how vital the fighting planes are, and, of course, their primary job is destroying enemy airplanes.

"We frequently hear of only using two types of airplanes on carriers. You will notice that one of these two types is invariably the fighter. It is this type that seems to be paramount in everybody's thinking. They are used both as defensive and offensive weapons. As everybody knows, of course, they are used primarily for control of the air, with the new uses of the auxiliary tank as an incendiary, the firing of rockets, photographic missions, long-range scouting, and almost an all-purpose airplane. You all know of the four general classes of Navy fighters: The small-carrier type of about 8,000 to 9,000 pounds gross, the large-carrier type of about 13,000 pounds, the Marine fighter, and of course the night fighter which is a development of some of the others. A fighter has to have versatility, but not at the expense of performance or its primary suitability as a fighter. We have an excellent range of Navy fighters at this time. They cover the field extremely well as the scores of enemy losses contrasted to our own losses indicates. We have a great assortment of propellers, engines, and accessories which add up to the line of fighters which we now have to choose from. We have to keep that superiority of combined airplane, engine and propeller by looking forward and improving steadily. We hear that the Japs are beginning to introduce new fighters. We would expect that they would introduce these and it is up to us to keep as far ahead of them in the future as we have in the past. The Navy's entire fighter aircraft program does not change with the end of the European phase of this war. We will still need large numbers of fighters throughout the Pacific phase of this war. Quality is the paramount consideration. We believe and we hope that this conference will lead to improved rapid increases in quality which is the primary objective of all those present."

Commander P.H. RAMSEY: "We are going to make a complete record of the proceedings of this conference. We have Lieutenant Wilds here (Lt. Walter Wilds. USNR. Air Intelligence Group, Navy Department) who will be in charge of the record. Court reporters will take down everything and we will also cut records of what is said. To get the proper results of the conference, will you please take up the microphone when you are talking and please give your name first. It will aid materially the reporters and the air combat intelligence officers. If you have any notes prepared, please leave them with the reporters so they can write them into the proceedings.

"Also I didn't stress the informal aspects of this conference. We wish everybody to be informal. Please take off your blouses – this is an informal occasion.

"Now Lieutenant Halaby will explain what we want from the flight data cards."

Lt. (jg) N.E. HALABY: "I think I speak for all pilots when I say that they hate paper work more than anything, and so I am in the position of being a schoolmaster among a bunch of recalcitrant pupils. In explaining what we expect to get from you in the way of paper work, the engineers have only the flight data cards and reports to work with. We will try to make it as simple as possible here by giving you a kneepad, the data cards and pencils, and from there on we will have to rely on you to bring in as much information as possible in as short a time as possible so that the manufacturer's engineers and Army and Navy and British can work out some notes and actually get some good out of the reports. It will be very useful to all of us as individuals to fly the airplanes and to have a good time, but so far as the information is concerned, what we get on paper will help particularly the manufacturers to incorporate some of the points of their competitors. Another purpose concerns the Army and Navy and NACA handling requirements. The men who make those specifications want to know what is a nice airplane. I don't think anyone is in agreement yet as to just what is a nice airplane for all purposes, including armament, radio, long range, etc., so if you can cooperate with us on getting as much information as possible in filling out these questions and getting as much detail as possible, we will certainly appreciate it.

"We are going to make it as easy as possible to get in the air. You will have a kneepad and comment card and we will expect you to turn those in as soon after the flight as possible. In the airplane, in addition to whatever check-off sheet is there, you will find a general data card. For example I will take the F6F, which is pretty well known here-the card designates the airplane number, the take-off power at sea level, the other power ratings (normal military, and combat), as listed with the various blowers for level flight, rpm, the airplanes critical altitude, etc. Combat power is going to vary in some ships. In the P-38 there is no water. In the P-51 there is water and in the P-47M there is water. The next is the description of the propeller, and then the load on the airplane. Now our standard load for this meeting is as closely as possible the condition of the airplane the combat pilot takes into action. The weight empty is given. The gross weight shown will be what we have actually weighed this airplane at here on our scales. The fuel, oil, armament, etc. for which the ship has been ballasted, are listed. The radio equipment generally in the night fighters is of interest, and if there are any general features on the airplane on general restrictions, they are listed in a catch-all at the bottom of the second card. In the Army airplanes there is a diagram giving maximum speeds at quoted altitudes and we will attempt to have that data in each of the Navy ships.

"The pilot's comment card is a two-page affair fitted right in the kneepad and it asks for a lot. We want to get as much out of it as we possibly can. If you have an hour's flight, we don't want to spoil it for you, but we would like to get as much of your impressions on the following data as possible. The first heading on the comment card is about the cockpit, which is a particularly important point: what the best features or the worst features of the cockpit are, so that the engineer can be told to get a gear handle like the F6F, or flap lever as in the P-47, etc. If that seems to be the best flap handle or lever, he can go ahead and get the dope on it. The ground handling section is simple enough. Particularly is the Navy interested in how the brakes work and how well they hold take-off power, and how easy it would be to handle this plane on a carrier or a restricted field. The conference is particularly interested in things like manifold pressure regulators, single power controls, if we have anything like that this year, and the relation between rpm and manifold pressure. For example on a P-63 there is no rpm control and the throttle gives a set rpm. Comments on features of that nature are the object of this meeting. The take-off comments would be in general whether there is adequate rudder control, etc.

"In all this filling out of the cards, we wish to impress upon you that it is quite probable that the latest models of these airplanes are going to be used as much or more against the Japs as they are in Europe, and so if you will keep that in mind when filling out these cards, emphasis on any special features will be very useful. We place emphasis on the climbing ability and also in the Navy we require a great amount of instrument flying and over water navigation. Also on the carrier type we should give a little more thought to stability and control of the airplane, holding in mind that some of the Army airplanes might be used off a carrier. We also list force and effectiveness and by that we mean are the ailerons heavy and effective, light and effective, etc. For maneuverability, we have an A, B, C, D rating for comparing in a general way ability to move the airplane around in three different directions and get results. Stall speeds and recovery are especially important on Navy ships. Accelerated stall should be listed to produce any information for the engineers. With the installation of dive recovery flaps, we note it would be useful to have some information as to diving if you get clearance, and we would like to know the general feeling, how it is best controlled, and the timing effects. The wave-off is simply the "go around" to the British, or the "follow through take-off" to the Army. The Navy pilot is forced to make it in the most critical condition of stall. We are pretty interested in it and would like to sort of promote some interest and some comments on it. Night instrument flying is simple enough. Gun platform rating is one of the most important things that we work on here. How is it for gunnery, and that ties in with stability, particularly directional stability?

"These questions asked are to provoke comments, and the pilot comments they provoke will be used by the engineers and the manufacturers to improve the qualifications of the airplane. I stated one as the maneuverability rating. That was intended to provoke comments about how nice the ailerons were at high speed, how effective they would be in combat maneuvering. The night and instrument flying rating was intended to provoke notes about what features of the particular airplane you are flying which would make it nicer to fly on instruments, or at night. The cockpit evaluation, the landing qualifications, and the visibility etc., are all included. The gun platform rating is still the most important one in that the only reason the airplane is built is to carry some bullets to the enemy, or some bombs. Two more items on this comment card which require a little information are the estimated experience required to fight it, and the last is the frank opinion of the combat quality.

"I am sure that frequently the pilot comes down from flying and says, "It stinks" which may be true, but which is not much use to the engineer or to the other pilots who are about to fly it, and we would like a little detail. What we are anxious to do is to get some data on the card so that the good features may be incorporated and improved. We hope that you can fill out as much information as possible. After the last flight we will give you what is known as the general questionnaire and in the general questionnaire we will attempt to again incite or provoke your comments and criticism as to the whole project. Combat experience, if any, is asked for in the first part of the questionnaire. We will appreciate it if you will indicate what you have flown, a P-61A or P-61B because only if we know what you have flown can we properly evaluate your comments on the airplane. In the cockpit arrangements we ask for your opinion on the best cockpit, and we ask why and hope that it will provoke some comments about recommendations, accessibility of the controls with straps tight. We hope to have here in the next day or so a standard cockpit mockup which has been built by the Bureau of Aeronautics and we would like to get some remarks as to whether or not this would be useful not only to the Navy but for all services. If you have any suggestions or remarks here, set them down. In three ships, the Corsair, an FM2 and F6F, we have installed anti-blackout equipment using the G-suit to cover portions of the body in order to prevent, or at least to allow the pilot to stand, blackout. We are interested in the best cockpit arrangement, the nicest arrangement of engine control. Again, I want to emphasize the comparative best. By best we mean what did you like about it. Several questions in here are general questions. For example, "Do you believe general power control is practical?"

"I think we have a very good opportunity here to evaluate the bubble canopy and not only is the visibility important, but in some cases what effect does the bubble canopy have in flying an airplane. For example, it may be effective on the directional control. The most comfortable cockpit is important. The best visibility takes into account the optical qualifications of the canopy. The visibility around the arc and gunsight and downsight visibility are important. Also the best all around armor, the maximum protection and the minimum interference. One thing that we are concerned with is just how much of a pilot's body and head should be covered, and what that angle should be, and you will note that it varies with each airplane.

"In flying characteristics we have tried to take the NACA criteria and the Army's specification 1815, and the Navy's specification SR119, and provoke comments on each airplane which would permit further information; also the effect of overload take-off from a small area. In the basic qualifications that becomes an important question. We are also concerned with the maneuverability of the plane. I think when we say a nice airplane, we mean it is easy to operate. We are interested in the best aileron at 350 knots and the ability of the pilot to get sufficient aileron. The Navy is interested in comments on spring tabs. One ship, I believe, will have spring tabs on all three controls.

"The first question will allow the pilot to say that she is a nice airplane, flies nicely and that is what we are trying to do here, and if you feel there is duplication and would rather answer one question than another, please feel free to do so. Is it better on all all-around stability? Which fighter do you think has the best gun platform? Which fighter appears to have the best stability and control in diving? Which has the best night flying qualities? Which power plant inspires the most confidence? Which do you think the best night fighter and the best combination day and night fighter? This is intended to provoke some comment on the relative stability of single engine against two-engine night fighters, especially as respects their usefulness in a scramble. Which gun sight and armament control impressed you most? Which type of radio control do you prefer? Which is the best fighter, the best strafer, the best torpedo plane?"

Commander RAMSEY: "Are there any questions from anyone; or remarks? Night flying will be available on Tuesday, Thursday, and Saturday nights. The list of planes available for night flights is on the board. Please indicate to Mr. Milner or Lieutenant Shehan if you have preference for any particular night. I would like to call on Mr. B. Guyton of Chance-Vought."

Mr. GUYTON: "We brought down to the conference an F4U-4X which is simply an experimental airplane for test and does not have a revised cockpit, but it has a C engine. The engine is an R-2800-18 with water injection and a 70" Mil rating. We also have an F4U-4X airplane which is our latest airplane. The general qualifications are the same as the F4U-1. Flying qualities, stability, and control are essentially the same as in the F4U-1 airplane. The engine in this production F4U-4 airplane is an R-2800-18W. The present rating is 60" MP at

2,800 revolutions per minute which gives greatly increased performance at all altitudes. The critical altitude of the airplane engine in the F4U-4 is approximately 5,500 feet higher than the F4U-1 with the R-2800-18 engine. The cockpit of the F4U-4 airplane is completely redesigned and revised. Pilot comfort has been accentuated by such measures as the bucket seat which tilts as it raises, the revised instrument panel for better vision, the bulged canopy which will be incorporated in production shortly, the raised rudder pedal and foot position, the simplified radio, and switch panel, the cockpit deck installed, the new ventilator which keeps all direct draft off of the pilot's person and many other items such as an accessible miscellaneous box for sandwiches, etc., a properly stowed chartboard and map case. I am not sure this airplane is on the flight schedule as yet but it may be later.

"The F4U-1D airplanes that you have available to fly have the same characteristics as the F4U-1. They incorporate several changes. The F4U-1D has twin pylon and center line tank installations, an antishimmy device in the tail wheel. It has the same airplane controls and the same flight characteristics as the F4U-1. The cockpit is exactly the same. That airplane, the F4U-1D, has a bubble canopy with a flat front windshield. It is here and I believe available for flight. The ailerons are normal in all respects, and if you fly it you will find that the landing characteristics in any F4U airplane are now such that you will find it very nice for any type of landing. You will find no trouble landing with a 50-degree flap at all times. You can take off or land with flaps. There is no directional instability on the ground. If there are any questions we will be glad to help you. If you will see me I will be glad to go down to the cockpit and check you out with any equipment that might seem new. Any Vought pilot here will be glad to do the same."

Commander RAMSEY: "Thank you very much Mr. Guyton. The Grumman representatives have not arrived as yet. Lieutenant Commander Owen will give a talk on what they have available here."

Lieutenant Commander OWEN: "The Grumman planes which are available for general flying in the conference are the F7F which is the new day and night fighter, carrier based, and equipped with the droppable radar installation. The other Grumman airplane, the F6F5 which is the modification of the F6F3, is equipped with spring tabs on the ailerons and the cowling is drawn in thereby giving a little more speed. The F8F is due to arrive today; I believe it will be for display only. That is the small R-2800-22W Grumman fighter, the C engine and an entirely new airplane. Those four models will consist of the Grumman airplanes available for flying and for display."

Commander RAMSEY: "Mr. Johnny Myers, Northrop pilot, will please give a small discussion on the Northrop plane."

Mr. MEYERS: "There are probably a number of us who are familiar with the P-61's. Therefore I will give just a brief discussion. It is entirely conventional with one exception. It has a wing area of 662 feet and weighs with the turret 28,600 pounds; without the turret, around

27,300. The ship was designed with a four gun turret, remote control. Unfortunately the B-29 decided to use the same turret and it had priority over the P-61's, so our airplane has come out without gun turrets. The rest of the armament is four 20-millimeter cannons in the belly. The only unconventional thing about the airplane is the lateral control. Instead of conventional ailerons, it has a special type. The object of this was to use full span wing flaps on field landings at night. I don't think you will find the airplane will be quite as unusual in flight as you would expect. There will be a tendency to use too much rudder at first. And consequently you make fairly decent coordinated turns with almost no rudder. It has top speed of approximately 375 MPH with the turret. Its controllability and stability I think you will find quite unusual. And I wish those of you who fly it will try a single engine stall. You'll find that you can safely and conveniently make continuous accelerated stalls with a dead engine. You have to snap roll it into a spin. I think that's the outstanding point about it. I hope you'll like it. We're short on performance. Unfortunately we could not present the turbo supercharger with the R-2800C engine which will give added performance and altitude."

Commander RAMSEY: "Lt.Col. I.W. Toubman will give us a short discussion on the Bell aircraft."

Lieutenant Colonel TOUBMAN: "I have not looked at the P-63 here; however, it is quite similar to the P-39 as far as cockpit lay-out. It has the Allison V1710 engine and it is rated at 60." I haven't checked on it to see if this P-63 has water. They will have water and then the rating will be 75." I believe this model down here is without water. The tricycle landing characteristics are exactly the same as conventional gear. In making a good landing try to keep the nose wheel off and try to drag the tail; just let the nose wheel fall down of its own accord. The 63 is still restricted in quite a few maneuvers and I will give you more information when I check on this plane. At present it will be available for just level flight and chandelles."

Commander RAMSEY "The P-47 airplane – by Mr. J.F.B. Parker, Republics pilot."

Mr. PARKER: "I brought an M airplane down here. It has the C engine. The military power rating will be 54" and 2,800 revolutions per minute. The handling qualities are the same as for the rest of the P-47's and we have dive flaps on the plane which lift the restriction on the dive speed. You will get compressibility without the use of dive recovery flaps, but they are to be used in recovery. At high altitudes you will recover at approximately 4G" without any trouble to the pilot. At low altitudes it will be approximately 6G" so you actually have no restrictions on the plane as far as dive and compressibility are concerned. I suggest when you fly them you use 400 miles per hour to get used to the compressibility recovery flaps. The bubble canopy has caused some comments on the P-47, and on the M airplane we have a dorsal fin which eliminates the unsatisfactory yaw condition. However, on the D-30 you don't have the dorsal fin and tail and there is a possibility by rough uncoordinated flight for an outside snap roll. Landing characteristics are about the same. We have used flaps for take off and there are some comments about the flaps being unequal when they are down. However, you should arrange your flaps to get the approximate down angle, and as you gain air speed, the aileron flaps will equalize themselves. You may have mechanical difficulties like a leaking selector valve, but that can be tested on the ground. On this M airplane, the carburetor air temp is critical especially in WEP, either in a climb or level flight, especially at high altitude; you have to watch it pretty closely. The combat weight is 14,500 pounds, and the dry weight is 11,330 pounds. The CG location landing wheels down is 27.93 percent, and 27.97 percent wheels up. The war emergency climb is 72" Hg and 2,800 revolutions per minute. We have automatic coolers and we have a power condition where you can use 1,400 revolutions per minute for maximum economy, but that is low at altitude to obtain the best mileage per gallon under the best conditions. One thing that I want to point out is that it is very important to have mixture control all the way forward to get in the automatic rich position. I will be glad to answer any further questions."

Commander RAMSEY: "I will call on Mr. V. S. Kupelian, Goodyear engineer, for comments on his airplanes."

Mr. KUPELIAN: "Goodyear will have two airplanes, an FG-1 and an F2G-1. The FG-1, which is now here and available for flight evaluation, is equipped with a free-blown bubble canopy and a flat front windshield. The canopy is manually operated, using a sprocket-chain and incorporating a single release jettisoning system. Installation of the bubble canopy has resulted in a slight increase in the speed and no appreciable destabilizing effects have been noted.

"Later in the week we expect to have available for inspection an F2G-1 airplane which is a medium altitude fighter. This airplane is a development of the FG-1, in which the power plant is a Pratt & Whitney R-1360-4 Wasp Major engine. It has a manually operated bubble canopy and a removable overturn structure.

"While the engine installed in this particular airplane has a single stage-single speed supercharger, subsequent airplanes are being equipped with a single stage variable speed blower giving an airplane critical in excess of 17,000 feet. In this installation of the R-4360 engine, we have made an attempt to wrap around it the smallest possible diameter cowling. This interchange between the R-2800-S and the R-1360-4 engines in the same airplane has resulted in no increase in cowling diameter and only 4 1/2 inches in total power plant length.

"Your attention is further invited to the following power plant features: A top cowling induction air scoop incorporating a boundary layer removal duct. Twin siamezed jet exhaust stacks on all 28 cylinders. Automatic oil cooler flap controls. Twin oil coolers using a parallel flow system and combination oil dilution and diverter valve system. Fuel transfer system in which fuel from either of two drop tanks or a fuselage auxiliary tank may be pumped into the main fuselage tank from which fuel is discharged through a submerged centrifugal booster and deaerating pump.

"The F2G-1 has an entirely new cockpit arrangement with a new night instrument panel and new grouping of controls. The seat which is adjustable offers rear, bottom and partial side armor protection. Fresh air ventilation is offered both at pilot's feet and at the oxygen regulator. The loaded airplane, 300 gallons of protected internal fuel, weighs 13,182 pounds. The airplane which is to be exhibited here in its partially loaded condition weighs about 12,000 pounds with C.G. at 26.7 percent MAC (W.D.) and 27.9 percent MAC (W.U.).

"At these weights and powers, this airplane offers no marked operational differences from the FG-1 airplane.

Commander RAMSEY: "We have several British airplanes here, and I will ask for a discussion first from G/Capt. C. Clarkson, RAF, of the British Air Commission, and then Commander D.R.F. Cambell, R.N., also of the British Air Commission. Group Captain Clarkson."

Group Captain CLARKSON: "As two of the aircraft that are here belong to the Navy, at a later point I am going to get Commander Cambell to talk about the Seafire and the Firefly because he knows much more about them than I do. The other aircraft we will have down here is the Mosquito which is coming down from Canada and should be here today. They are sending a representative with it who has been with DeHavilland since the start of the Mosquito development, and who is in a position to tell you far more technically about the Mosquito development than I am. As you know, the Mosquito was designed so that it had a large number of different roles as a fighter, fighter bomber, and reconnaissance aircraft. Each of those variations differed, but differed only slightly. The chief difference (and I think I am right on this, although Commodore Buckle will call me if I am wrong) is in the Mark B16, with the bomb bay underneath which is now being so modified that it will take a 4,000-pound bomb. The fighter has 20 mm. cannons and .303" machine guns in the nose. The reconnaissance aircraft has no armament at all, but has long range tanks, to help its performance at high altitude and high speed. The fighter bomber has limited bomb-bay capacity with the same guns in the nose, and you will see that one, I hope today, when it arrives. If anybody is anxious to get more details, I would like him to get them from the DeHavilland representative who is coming down with the aircraft. As far as the flying characteristics are concerned, I hope you will like it. It is not perfect. I don't know of any airplane that is. Its chief advantage when it came out was that it had very light lateral stick force, great maneuverability and is not 100 percent stable. Perhaps in a fighter that might be an advantage. You will find that on the approach there is a certain amount of sloppiness in the ailerons which gives a slightly unstable feel on the approach. There is nothing alarming about the landing of the airplane, provided you stick to an accurate air speed, but it does require accurate approach speed, one of the reasons being that the position error correction is subtractive. In other words, the indicated speed is a little bit higher than the true speed. You will realize that for every 5 mph that is dropped in indicated speed, you may be dropping perhaps 7 mph so that it leaves you much less margin for error on approach. Thus, it does need an accurate approach speed and if you will do that and just hold it off and do not bring it down too hard on the wheels, you can do it very nicely on three points. On the take-off, don't try to open the throttle on the brake, just open the throttle and let her run ahead. But if you open the throttle with a hell of a bang, she will drag the tail around. Apart from that, I don't think there is much more I can tell you as far as the flying is concerned, But Lieutenant Singleton has been flying these on fighter operations and he will be glad to answer any questions in regard to flying."

Commander CAMBELL: "The two Royal Navy planes here for the conference are the Firefly and the Seafire, L2C. There are actually two Seafires here. One is the L2C and one is a 3. The L means low and the L2C has a Rolls Royce Merlin 32 engine, which I believe, is special to the Navy. It is a low rated engine, around 5,000 feet. In all other respects the airplane is very similar to the standard Spitfire 5B, now obsolete in the Royal Navy. It is replaced by the Seafire 15 and right now the version in service is the Seafire L-3. The difference between the 2 and 3 is the folding wings. This one does not have folding wings. There is nothing unusual about it and otherwise it is just a standard Spitfire fitted with a hook and catapult spools. The Firefly Mark 4 is an airplane built to a specification I think especially for the Royal Navy-a two-seater fighter with no rear defense. Its mission was primarily intended to be long-distance scouting when radar aids to getting out and coming back were not visualized and it has fairly luxurious provisions for a navigator in the back. The performance is nothing to write home about. The Firefly has an unusual use of the maneuvering flap of a type known as the Youngman flap. It has a flap which extends rather in a trailing position first of all, thus giving added wing area for maneuvering. Two other positions of the flap are take-off and landing. There are three positions: maneuvering, take off, and landing. I recommend pilots to try this flap in the maneuvering position. Just as a matter of interest, it enables us to tighten up the turning circle quite a bit. The engine is a Mark II, Rolls Royce Griffon, 1,725 H. P. 54. Performance is even more contemptible than I thought at 300 mph top speed and I think that is about all on that."

Commander RAMSEY: "I will ask Colonel Toubman to discuss the North American and the Lockheed fighters the Army Air Force has here at the conference."

Lieutenant Colonel TOUBMAN: "They have the two P-51D-5's, with bubble canopy. The only restriction I can think of is with 85 gallons of fuel in the fuselage tank. I believe you will fly with 25 gallons of gas in the rear tank and use it only in the case of an emergency. It has no other restriction. The D-5 has a dorsal fin and 6-gun installation. The cockpit is cleaned up a good bit from the B airplane and has one of the nicest bubbles of any of the bubble installations. It was restricted for power stalls at one time, but now it has the dorsal fin and the restriction is lifted. It has a tail wheel lock on the stick. If you want to unlock the tail wheel you have to push the stick forward and neutralize the rudder.

(The following statement was added by Mr. J.F. Steppe of North American.)

"The P-51D-15NA is equipped with a Packard built Merlin V-1650-7 engine, rated at approximately 1,500 horsepower at 61" Hg and 3,000 rpm for the take-off. The war emergency rating is 67" Hg. There is no water injection installation.

"The two airplanes available here are the long range escort fighter version of the P-51. Internal protected fuel totals 265 gallon; 75 gallon droppable tanks on each bomb rack give a combat radius of slightly less than 1,000 miles. It is equipped with six .50-caliber free-firing machine guns with approximately 1,900 rounds of ammunition. The wing bomb racks will carry all bombs up to and including 500 pounds or 75 to 110 gallon droppable fuel tanks.

"In addition to the increase in armament and fuel capacity, the most notable changes over the previous models are the new windshield and bubble canopy and the dorsal fin. The cockpit has been cleaned up considerably and the seat installation modified to increase pilot comfort.

"Automatic controls are provided on both the oil and coolant system. The engine blower speed shift is also automatic. Manual selection on all automatic controls is also provided.

"Despite the rearward CG's resulting from the addition of the fuselage fuel tank flight characteristics similar to the P-51B have been obtained by the addition of the dorsal fin and a bob weight on the elevator control system."

Lieutenant Colonel TOUBMAN: "They have a J and an L-5 model P-38 airplane available for the conference. I just looked at it this morning and the J-25 has dive flaps and the L-5 also has that. The Allison 1710 engine is in there and they are not fitted up with water as yet. We have manifold pressure regulation on both the J-25 and L-5 and that is the Minneapolis Honeywell control system. One point to notice on that boost aileron is that you will not turn your boost aileron on or off in flight; we had a little trouble with the P-38 hydraulic system so, if you contemplate using the boost ailerons, turn them on the ground and leave them on in flight. The dive flaps control is right on the wheel and it is quite a novel sensation which is like compressibility. You can overpower it very easily by pushing forward and you can practically come down in a saw tooth dive. I have not tried the 47 dive flaps, but the 38's are quite a novelty. On the turbo control you have a Minneapolis Honeywell system. You cannot overspeed the turbos and there will be no turbo indicator in the cockpit; I think it has worked out fairly nice in the P-38 and a note might be taken of this for turbo installation of some of the planes contemplated. This tricycle landing gear is the same as the P-63 and lands very nicely and there should be no trouble with that phase."

(The following statement was submitted by Mr. Jack Woolams, Bell.)

"The Bell XP-59A airplane is the first jet-propelled type flown in this country. It is powered by two General Electric 1-16 Whittle type jet power plants rated at 1,650 static thrust each. The engines, which burn kerosene, are located side by side at the juncture of the wing roots with the fuselage so as to present as little drag as possible and to keep at a minimum the amount of directional trim change occasioned by the reduction in power of one engine.

"The airplane is equipped with a pressure cabin and tricycle landing gear, but is otherwise conventional with the exception of the power plants. Flight characteristics, including aerobatic maneuvers, are normal and the airplane is unrestricted except for outside loops and dives in excess of 350 miles per hour observed air speed. High mach number dives have been executed at high altitudes with adverse compressibility effects.

"Engine operation, although involving the use of two new temperature gages measuring tailpipe and rear bearing temperatures, is extremely simple, being entirely governed by two throttles which control the amount of fuel delivered to the engine. The engines increase rpm as altitude is gained, and since the automatic constant rpm governors are inadequate at the present time, it is necessary for the pilot to throttle back as altitude is gained to avoid overspeeding. Since the static thrust of the engine is low, and since the power of such an installation is figured as the thrust times the velocity, acceleration is poor at low speeds. Therefore, the pilot should be careful not to approach the stalling point prematurely during landing, since there is not enough power available to catch the airplane in the manner of conventional fighters once it starts to drag. Aside from the two characteristics noted above, the operation of this airplane is extremely simple and safe."

Commander RAMSEY: "We will now hear from Commander Bauer (Commander L.H. Bauer, USN, Armament Test, NAS, Patuxent River)."

Commander BAUER: I would like to enter a word of caution to the pilots in regard to the Armament Test gunnery area. The Armament Test unit is located at the southeast corner of the station. The total area south of Armament Test (extending south of the Patuxent River) is the area which we use for gunnery. Particularly those pilots who check out with the lead-computing sight in the F6F5 will probably jump at various times other planes in the area to get an impression on how that sight works. But do not jump any planes in that particular area. If pilots should jump an airplane there they may not be seen by the people operating turrets and the possibility exists of running into some complication along that line. This danger area is drawn up in the sketches which are posted on the bulletin board and there are many additional copies available in the Board Room for the pilots to study."

Commander RAMSEY: "I would like to call the attention of the factory representatives, the Army and Navy people and those of the Bureau of Aeronautics who have not registered please do so. We would like to keep a record of all those who have been issued a badge so you can get the privileges due you.

"We would like to reiterate: Please get off on time on your hops and please get back on time so as not to botch up the schedule. The rest of the morning will be taken up by a general shake down.

"Again, I want to wish you a pleasant time during these 8 days.

"I wish to thank Admiral Richardson for coming down to give the opening speech. We will now secure until the late afternoon conference.

SECOND MEETING
16 October 1944
1530-1715

Commander RAMSEY: "Several items have come up from the first day's operations. First is the F7F – the mockup of a cockpit which follows the Navy specifications as to what the Navy thinks a cockpit should be is down on the lower deck and we would appreciate any comments you have to offer.

"We had some trouble getting the second hop off and it was due to the pilots on the first hop taking time out before dropping back to the operations desk. If you will report immediately to the operations desk after the hop, then take time out to fill in the comment card, the second hop can go off without any delay even though your data card has not been turned in.

"There is a locker room now available for dressing, two doors down from here on the second deck. There are lockers and you can use the room to put on flying suits, etc.

"An SNJ will be available to accompany the F6F with the lead-computing sight, up to 6,000 feet. If you put your name on the board, that SNJ will be available for you to take runs on.

"The flight gear situation is slightly critical and if those pilots who have more than one hour during the day retain their gear it might be difficult for others. If a pilot has the first hop and the fourth hop of the day, it is requested that he turn in the gear so that others can use it. If he has the first and second, he should naturally retain it.

"The Army planes are using headsets instead of earphones and those desiring earphones or headsets can draw them from the flight gear headquarters below this room.

"For those who do not know about the parachute storage, there are bins for storing in the flight gear issue room on the lower deck. "There are planes for those desiring transportation to Washington and return in the morning. If you will express your desires along that line to Miss Ryan, we will know what type of plane is needed and be able to schedule the flight. The trips will be posted on the bulletin board outside this room. Those who wish to take the hop tonight, please tell Miss Ryan and she will see that you get to Washington. Those who miss the plane in the morning due to weather or some other reason, will be able to get a bus at Anacostia that will leave 15 minutes later than the scheduled plane. Your badge gives you ingress and egress to the Naval Air Station at Anacostia. "The first conference flight in the mornings, in order not to delay its take-off at 8:15, will be posted at the BOQ where the officer status people are quartered and also announced at this last meeting in the afternoon.

"The F6F5N is available but without scope hood. You may take flights but you won't get much out of it unless you take a hop at night. Comments on this plane are requested.

"The bar is available between 5 and 7 in the afternoon and from 8 to 10 in the evening. The bus will leave the front door of BOQ #1 at 10:15 in the evening to take the contractors' representatives back to quarters.

"For those suffering any injuries, slight or major, there is a small dispensary in the operations building on the lower floor. There is a flight surgeon in this building at all hours of the day. The main station dispensary is also available for all members of the conference.

"The Zeke-52 will be put on the schedule tomorrow and it will be subject to limited use. Anybody wishing to fly the Zeke please contact Mr. Milner and check out with Commander Palmer or Commander Andrews. There are a lot of ramifications to flying that plane and if you can check out to one of those officers you will be free to fly.

"All civilian pilots have not signed waivers. Please sign these with Miss Ryan in order to cover the legalities. All those who have not registered, please see Miss Ryan this evening to make sure that we have billeting space and eating space available.

"We have had very few requests for transportation. We have a lot of transportation available – according to Navy Standards. If you have occasion to make a trip around the station or to Washington, please see Miss Ryan.

"We have an NH plane out here with the Hoover Horizon and it will be very interesting to those concerned with night fighters and night fighter development. It will be available for those who desire to go out in it. I will ask Mr. Milner to give a dissertation on the Hoover Horizon. Before that, however, I would like to call on Commander Owen (Lt.Comdr. E.M. Owen, USN,NAS, Patuxent) on the comment cards. As we said this morning, there will be a lot of comment about the cards. There is too much on the card to fill out in an hour's hop and it would take 6 or 7 hours of actual flying to get the data complete, but we want to get as much data as you can give us on the hops."

Lieutenant Commander OWENS: "We realize that we're asking for quite a lot on this comment card. I wish you would look at it this way; We hope that the card will help you plan the flight. On some airplanes it may be easy, for example, to get a new idea of stability in a short time and in that connection we should like to have you cover as much on the card as you can. Some pilots feel it's too much to ask in an hour's flight. Some others said that they were able to check every item on the card and give some particular characteristic of the airplane. So please give as much as you can and, if possible, try to cover each item. You will notice the next to the last item says "Estimated experience required to fight it." We didn't clarify that enough. We would like to have your opinion as to how long it would take for an operationally trained pilot, as the Navy refers to it, or the flight trained pilot, as the Army refers to a pilot with 350 to 500 hours single engine total flight experience; in other words, estimate as to how long it would take him to use that airplane as a fighter. The other items on the card we have tried to put down so you can take some specific sequence and if you can't get them all, we would appreciate just as much as you can get. If you desire to take one particular phase, such as lateral stability, and dwell on that, fine. We do feel that considerable information can be gathered from impressions that you have from one flight in the plane, answering as many of the items as possible. Also consider a second flight in the plane to cover some items you might have missed before. I believe Commander Ramsey would like to have some discussion regarding these data cards."

Commander RAMSEY: "Boone Guyton, would you give your opinion on the data cards?"

Mr. GUYTON: "I tried this card on the first flight this morning and, to me, if I pick out something peculiar on the plane, it is easier to carry it through instead of covering the entire card. To get the most out of the conference it seems that some planes have peculiarities and different characteristics from others, and it is a lot easier to pick those out and it seems it is a better way to go about it."

Commander RAMSEY: "Johnny Myer, would you like to say anything?"

Mr. MYER: "I agree that it is a pretty big order to try to carry the entire card through. Lots of us are fairly familiar with the planes we're flying. In this case, I know what you're trying to get and I would like to suggest that we use our own discretion on those planes we are fairly familiar with and if we can cover the items, fine, and if we get in a plane we are not familiar with, we can follow through on particular characteristics. I think that is the best way to do it. On our plane we would like to have comment on any unusual characteristics about it."

Commander RAMSEY: "Joe Parker?"

Mr. PARKER: "No"

Commander RAMSEY: "Commodore Buckle, have you anything on the data cards?"

Commander BUCKLE: "The general story is that you don't have long to go through a rather big program. I broached the detailed point with Lieutenant Halaby and it has been taken care of."

Colonel MENG: "On the first flight, if you write across the card "Familiarization" and put the general impression of the plane, and on the second hop go into detail, it seems to me that would be a better way of doing it."

Commander RAMSEY: "That would work if we had more than 1 hop and we could guarantee each pilot's getting more than 1 hop. But there are 60 some pilots to get into 20 some planes. If we use our discretion and fill out as much of the card as possible, agreeing that we cannot possibly do it all thoroughly. If you do just get familiarization, put down the remarks on the general evaluation of the plane from just 1 hop. Is everyone in agreement for that general trend on the cards? Hearing no other comments, I assume it is satisfactory and everyone will fill out the data cards to the best of his judgment.

"We would like to have Lieutenant Commander Milner from Eglin Field tell us about the Hoover Horizon."

Lieutenant Commander MILNER: "The Hoover Horizon is an instrument which combines the gyro and the artificial horizon into one. There are six of them now in an experimental class and people interested in flying by instruments should take a flight in this plane. The

Navy knows this plane as an NH. It is a single engine Howard. It is not necessary for you to check out. The plane is equipped for instrument flying. We will have a pilot and it will be available for the second flight tomorrow. We will have it listed as the Hoover Horizon on the bulletin board.

"Briefly, as you look at it, it has a conventional artificial horizon and the silhouette of the plane is an actual miniature. On the horizon line marked off to about 60 degrees to the right and to the left you have the directional gyro part which is very clear; and above the horizon you have an artificial sky with clouds and below the horizon you have the panoramic view of the ground with roads and trees, etc. You set the directional part of the horizon in the normal position, match it with the magnetic compass. When you change course or turn to the right 90 degrees the scenery changes. If you put your nose down you can see a little picture of the ground. It is a definite attempt to break down the un-naturalness of instrument flying. Mr. Hoover, who is the inventor of this horizon, will be here tomorrow."

Commander RAMSEY: "If no one has any comments to make, we will adjourn this meeting."

THIRD MEETING
17 October 1944
1530-1745

Commander RAMSEY: "The session will commence for this afternoon. I have a few items before the speaker of the day takes over the microphone. The Zeke 52 which is parked on the line has very fragile wings and a lot of people have been stepping on the wing where the sign says "do not step." There has been a lot of maintenance work on that plane. We would like to have a number of contractors fly it, so please be careful not to step on places that are fragile. The plane has certain weaknesses and the plane captain will advise you where to step.

"In case anyone flies it, it is not advisable to pump the Zero throttle because it acts as a primer and will start a fire. Since it is the only Zeke we have here right now, we would rather not have a fire.

"The plane data cards on the bulletin board seem to have caught the eye of some people and many more copies are being made up. If you desire some of those cards, see Miss Ryan or Commander Milner. "Yesterday, for information's sake, we had 44 hops, which is not as many as we hoped. The BAC got 8 hops, the Army 5, contractors 20, BuAer pilots 5, Patuxent pilots 6. Some of those may not be in accordance with the pro rata arrangements we desired in the conference. Because of colds and other conditions, the pro rata arrangement will not hold for each day. We'll try to make it even and hope that today's record will prove that. Today we have had 80 hops. The weather for tomorrow looks promising – clear, some haze and a good stiff breeze form the northeast. Everybody be on your toes tomorrow morning at 0815. It is mandatory that we get the hops out on time and back on time.

"Yesterday's recording, according to the reporters, was very good. Everyone announced themselves before speaking and spoke clearly and slowly. I ask all of you today who do speak to announce your name, your job, and the company or service you represent before talking. "It is my pleasure to introduce Mr. Mel Gough of NACA, who will give a short talk on stability and aircraft handling characteristics. Mr. Mel Gough."

Stability and Handling Characteristics

Mr. GOUGH: "Thank you, Commander Ramsey. Gentlemen, this to me is a most unusual occasion. I'm sure that though I have met many of you personally at various times, I have never had the opportunity to be with a group so closely associated with the same type of work as I have been engaged in. As you know, this question of stability and control is a most complicated business and I am awed by the presence here of so many who are specialists in some one of the many branches of this problem, and who know so much more about them than I. But I hope to be able to provoke some discussion and so engage in a meeting which is of value to us all. We have assembled here designers and builders, the people who have the different tasks of fabricating and producing the airplane, the test pilots who have the first crack at it, the acceptance pilots who must pass on it, the evaluating engineers who must take the information from the test pilots and decide what can be done about their recommendations, and then we have the user of this product – the man who must say what he can tolerate if it is to do the job at hand, the man who tells us what degree of control and stability he can put up with to do the job he knows far better than any of us who haven't been there.

"The purpose of this meeting is really for all of us to get together, become better acquainted and share our problems. We all have generally the same problem, but we see it from a different part of the sphere. We all, I hope, will have the opportunity of making our contributions and I am sure that many of the things I'll say will be controversial and I, too, will learn the other fellow's viewpoint. By so doing we will be united then at this fighter conference which has been arranged for our mutual benefit, and we will be able to produce better aircraft and stay out in front as we have so ably shown in our armed forces.

"It was a surprise to me to call this a lecture because that is definitely out of my line, but I would like to present to you a general picture of a test pilot's viewpoint on the problems of stability and control in the hope that it may convey to you some new aspects of these problems.

"At the end of the last war we had much the same situation as we have today in that we had many qualified pilots, each of whom was high in his praise for the particular machine which he had been flying, and each of whom had a machine which he considered as his ideal or his standard. I had the opportunity to listen to many of those pilots suggest at meetings just such as this that we get together and measure in some way what those characteristics were that made up the good airplane that they liked and see if we could combine the good features in one machine. It was evident even then that relative opinions on stability and control characteristics were practically valueless to the engineers. Lots of quantitative data were collected to help evaluate various phases of the operation of aircraft, but it remained, to my knowledge, to E. P. Warner to list for the first time the handling requirement specifications of an airplane when he wrote the specifications for the DC-4. He attempted to establish quantitatively and systematically and to break down the various characteristics of an airplane that would be produced on contract and which was a major venture. At the same time that was done, it was immediately recognized that there were many characteristics of aircraft that had not been investigated and on which fundamental information was needed, so the NACA embarked on a series of tests in which they attempted to measure the characteristics of existing aircraft with the hope that in the future these quantitative data could be pieced together with the opinions of individuals so that they might be able to establish in quantitative terms what they thought of an airplane.

"In the past several years, the NACA has by means of special instrumentation accumulated a wealth of material with regard to existing airplanes. In many cases it has been able to take specific problems of certain airplanes, measure quantitatively the conditions existing, vary the geometry of the machine and again measure the characteristics of the airplane; and by so doing it has been able better to tie down specifically the characteristics that are associated with stability and control. As yet those data are not thoroughly evaluated to the place where an entirely satisfactory specification can be written as to what stability and control should be, but the Army and Navy have undertaken to establish their own specifications indicating what they thought stability and control should mean and I refer you again to the Navy specification SR-119, which has to do with control and stability, and the Army specification C-1815. Now in those specifications the services have systematically listed what they supposed to be the desirable characteristics that make up a good flying airplane. They have established quantitative values for certain characteristics and, as I have said, those values are not secret-they are the best that were known at the time they were established, they are still subject to change and they are characteristics which are known to be possible of achievement.

"Now in the present war the airplane is far more complicated than it was in the past war. We have entered new flight regimes. We are going at speeds where the laws of airflow no longer follow those we knew in the past and where we are entering new types of flow that bring with them problems that had not been touched before the war.

"In this age our aircraft industry is moving fast. We have little or no time to measure quantitatively all the characteristics that we would like to know about our airplanes before they get into service, and again we are back, it seems to me, to the place where we have to rely on relative, qualitative opinion. Now that is a real challenge to the test pilot because he must evaluate the machine and he must tell, in terms which the engineer who has already been on his way to interpreting quantitative data can understand. The test pilot must now tell the engineer in qualitative terms his relative opinion of the airplane. And that is one reason why we are here today. We have the privilege of flying each and every type of airplane and we must be careful that we do not evaluate it only as it compares to others we are accustomed to, but that we are very inquiring and systematic, that we understand thoroughly the condition in which the airplane happens to be that we are flying, and the necessity for it to be in that condition to do a job.

"I have in mind a two-engine bomber which I flew recently and which I felt that I knew about because I had read a quantitative, instrumented, complete, systematic report of its flying and handling qualities and was able to interpret those characteristics into the feeling of being in the machine and flying it. When I flew it for the first time, at a far removed place, I was very much impressed by the poor characteristics of the airplane, as were many others who also flew it. I regret that none of the others were sufficiently inquiring to find that there were 750 pounds of sand in the tail end of the airplane which should not have been there, that the airplane had four turrets for guns which people in combat thought that the airplane needed and which were not on the airplane in the original handling qualities tests, and that in general the airplane was purposely removed from being a good flying machine because of the objective for combat. To me that was tolerable, but the airplane was surely barely flyable. I had the impression that the people who left that flight in that airplane were quite happy that their particular machines were far better than that one – the ride they had experienced made them feel quite elated. We must be very careful in knowing exactly what we have and the purpose for which that arrangement is intended.

"I would like to digress for just a moment. I would like to talk to the test pilots. I would like to indicate that the test pilot of today is physically the same as the test pilot of years ago and he will be the same tomorrow regardless of the size, shape, and type of aircraft we are called upon to fly. Physically we can tolerate large temperature changes and large pressure changes, so it doesn't make much difference about performance range – how high, how fast, how far. We are all subjected to the same acceleration limitations and we all have in general the same strength as far as our ability to apply the forces on the controls of an airplane are concerned. It is interesting to note that an aileron which is heavy might be one which requires a force of but 50 to 75 pounds, because of the positions from which the force has to be applied, whereas the rudder which is heavy might require 400 to 500 pounds. Specific investigations have been made of those points and we find that we can relatively assert on the 3 controls of the airplane forces in the ratio of something like 1-3-10: 1 on the ailerons, 3 times that much on the elevators, 10 times that much on the rudder; so we are immediately faced with something I hope to touch on in more detail – the problem of stating what we mean when we say heavy or light. Which control are we discussing? About which axis are we applying it?

"In addition, physically we are the same in that our hands are relatively sensitive to small force changes and small position changes. Our feet are far less so. We are quite capable of estimating forces to a very high degree as far as magnitude is concerned if the force is in a reasonable limit range; and I think of force in terms of 30 to 40 pounds on the elevators. Lighter force we are apt to underestimate and heavier force we are apt to overestimate and we should learn more of our own limitations. Physically we are also extremely sensitive to motions. We can discern very, very small angles visually, we can feel extremely small accelerations and we are sensitive to quite small variations of both position in space and acceleration. Now all the airplanes that we have to discuss are ones which must be controllable within a very small range of human capabilities as far as force is concerned and as far as acceleration.

"So much for the pilot. This is to me a sort of build-up toward what we can expect qualitatively from pilots.

"Now about the airplanes of today. I have no intention of touching on performance – those measurable qualities of speed and climb and those capabilities which lend themselves to the slide rule and more exacting calculations.

"One of the old textbooks said that the elements for flight were sustentation, stability, and control. I will dispense with sustentation, because I will assume that the plane will fly.
"Stability and control to me have been a controversial subject that seems to become more confused as time goes on, but I feel that they are almost inseparable. The real problem is: Can we place an airplane, and we're thinking more now of fighter types, in any desired position in space, at will, easily, and will it stay there? Now this airplane that we have, even though it is a fighter type, is really many, many airplanes because there is a large range of conditions in which this particular airplane can be placed. And again I will state that which bears repeating, that as test pilots flying a great many airplanes of different models and makes, we must be very careful to qualify exactly the condition in which we had the airplane and to which our statements apply. We have a large range of loads and a large range of distribution of load. We have a large range of external configuration and I think of cowl flaps, scoop outlets, gear, flaps, and power. With all these conditions possible with one airplane, the engineer has a terrific job trying to make of all these airplanes ones which, when combined, will please the average pilot. We must help him.

"Now in trying to tell him what we think of these airplanes, do we have fixed in our minds a reasonable definition of such terms as stability? Do we all understand stability to be the tendency to return to the position from which it was disturbed? Do we use synonymous terms in discussing stability or any of these other characteristics? When we think of control, do we consider it the ability to move the airplane about a given axis and do we think in terms of the force required to do it and the motion on the control as regards magnitude and direction of those forces and positions? Are we exact in our definitions or can we improve them so that the engineer will understand? When we say that a control is mushy, do we mean that we can move the control column a long way and get no response? When we say it is sloppy, do we mean that there is no force required to move it even though we do get response, or do we qualify it? When we think of maneuverability, do we think of the combination of the ability to control the airplane at a given speed about all three axes and, in addition, moving the center of gravity of that airplane through space, combining performance and control to obtain maneuverability? Couldn't we have a very small, poor performance airplane that was highly controllable and not maneuverable? And couldn't we have a poorly controllable airplane which, due to its performance, was considered maneuverable? The point is do we qualify what we say in this way when we do not have an opportunity to measure quantitatively the conditions upon which we report? And do we investigate the characteristics upon which we report systematically?

"I think we might have done something worthwhile if here we were to leave with the feeling that in flying airplanes we would make use of every minute in the air and we would attempt to make use of the quantitative reports which exist and have been published for the benefit of various manufacturers, in studying those reports which tell the engineer in his terms the characteristics of machines; and we in turn, who have flown those same machines, might look at those reports and see what our feelings look like in the form of graphs or figures, and thus when we are flying and reporting in the future we can try to duplicate those figures in spirit if not in magnitude.

"Now I would like to talk a bit more specifically about the manner in which we might be able to present and evaluate the control characteristics of an airplane in more understandable terms, and which you may or may not agree is easily performed. For instance, we talk about friction and say that the control has too much friction. Where, when and under what conditions does it have too much friction? Couldn't we, without being too highly specialized engineers, think of friction as being something like this (see fig. 1). For a given control we would plot forces either to right or left or forward or aft. We would plot control force against control displacement on the ground with no air loads. Then for an aileron we might get something of the sort shown in figure 1. We get a variation of the force required to move the aileron on the ground with no air load, with displacement. And there was an initial force required to move the airplane from zero, so I can plot the direction in which the force would be applied and which could be indicated to an engineer as a measure of the friction. Now if we left the control there, it would probably stay and if we wished to reverse the control motion, we would find that we can plot that sort of thing too. In other words we present in graph form the magnitude of the friction forces – that's all. Similarly we can think of other characteristics of the airplane in the air and by being closer in contact with existing quantitative data we can begin to show our qualitative thoughts.

"I have in mind another example. I am thinking now of ailerons. We have many reports with regard to ailerons. They seem to be a most important control at the present time. We usually hear that one plane has better ailerons, or the ailerons are sweet, or the ailerons are very lousy. What do we mean? Under what conditions are we telling the engineers that these ailerons are good, or in talking as one pilot to another how would we qualify these ailerons? I would like to present a possible simple method of presenting those characteristics. For instance: Suppose we just think of speed increasing in this direction (fig.2). We know something about the possible speed range of the airplane, we know that it has a stalling speed and for this particular condition, this particular airplane (of the many kinds that the same one can be, depending upon the flaps, the loads, the power, etc.) has a limiting speed. Now, with regard to ailerons, we're interested in two things. We are interested in the effort that we have to apply to the ailerons, and the result that we get from it. We can express those in engineering terms without getting too mathematical. First, before we fly the machine we must learn from the engineer what the limiting conditions are under which the ailerons can be used. He would tell us something of this sort: He says that you have available a certain deflection range that is possible plotted from zero the ailerons can move 20 degrees. So that is a possible range of deflection whether you consider it the motion of the stick or the motion of the aileron itself. Now he specifies a certain speed because of structural limitations, and usually the pilot has to request what that limitation is or where that speed is, because beyond a certain speed you do not dare to deflect those ailerons fully because the wing strength or something else will interfere. Thus you will immediately get a picture of the allowable deflection range.

"At the same time we have our limitations as pilots, so we're interested in the force we're going to apply even though we have that deflection range available, and again we are interested through the speed range. Now we know that standards have been set, and our physical limitations almost impose them anyway, so that at somewhere around 60 or 75 pounds we have a maximum force that we can exert. As we fly the machine and go through the speed range, at the low end or at the stall, it requires hardly any force, but it has some value. The ailerons advance along some such curve as that shown in figure 2, but the forces increase with increasing speed. Beyond the point where they reach the maximum allowable speed we have to consider the force as constant; so the resultant is the force that the pilot will be called upon to apply. And what does he get from this force range and that deflection range? It might well be that the force allowable is less than or occurs at a deflection less than or greater than that which the structural man said could be used for deflection. And it may well be that the dotted curve (fig 2) is the deflection curve that corresponds to the force applied for full deflection. Now the reason we move the aileron is to get roll. Rolling velocity is what the pilot is interested in – how fast does the airplane roll due to full deflection? So, we're interested in another quantity which we will call roll. But how does the roll vary for the deflection that is allowed throughout the speed range? Now we get to the crux of the situation. Does the pilot who talks about a plane having good ailerons talk about that point? Is he interested in that point? Has he been engaged in combat, or what is his purpose in life? Unless he qualifies it on some condition the engineer is at a great loss?

"Now in addition, the engineer applies another factor which he calls pb/2v or an efficiency rating if you will, the Helix angle through which the airplane goes. He has felt that regardless of the size of the machine, if the pilot says the aileron is good it will give some value of this factor which is somewhere around .7 or .8. As pilots we do not have to worry about it too much. It has to do with the semispan, the rolling velocity and the forward speed, so it describes the Helix Angle which the wing tip gives in making a roll. It has been found that regardless of the size of the machine when the pilot gets a Helix Angle of this value, he thinks he has a pretty good rate of roll even though on a large machine the roll is low and on a small machine it is very high. Whatever these actual rolling velocities that are obtained, the pb/2v standard of excellence may be all over the lot and will depend upon how much of the allowable wing the engineer has made into the ailerons. I don't think we need worry about that too much, but I think we should try to qualify in some way how we used the controls; we should state that we are interested in the controls at a certain range.

"Similarly there are many things that we are interested in, at any one speed for these ailerons, to help to describe to the pilot why he likes or dislikes them. And they do lend themselves to feel and are not too dependent upon measurements. For instance, let us take at any one speed the characteristics of the ailerons. At any speed we can plot two other things that are of extreme interest. We will plot against deflection from neutral to full deflection either side the force required to operate the control and the response we get (fig 3). It will be noted on figure 2 that the force and response occurs only at the full deflection end, and that does not completely describe the aileron. If we move the aileron but a short distance, a small portion of its total allowable range, do the forces progress linearly with displacement and if we release the control would it snap back to zero? Does it have a self-centering characteristic? Is it light only through neutral and then with forces increasing through neutral, does it snatch by reversing direction of the force necessary? Do we move the control a little way and get little roll, and then reach small amounts of motion giving large rates of roll? So we examine at these speeds the control characteristics of that particular condition.

"Now similarly we could investigate elevators and rudders. The specifications systematically arrange for you the many characteristics that we demand of these various controls and intimates to you the interaction between stability and control and between the motions of the airplane about its various axes that are not the ones about which control was required. I would like to touch just a few of those.

"In the case of the elevator, we are again interested that this angle-of-attack controlling device is capable of not only producing any angle of attack that we desire, but maintaining it. Here I would suggest that an airplane which has good stability and poor control may not have good stability to the pilot, and similarly an airplane that has poor stability but good control really does not have good control, because if it is necessary to apply a control force in the opposite to the logical direction, even though the control may have the capabilities of maintaining the angle of attack which is desired, it is not a good control since the application of the force has been reversed. It might have been a good control; it surely should not be blamed for the lack of stability. Yet the control is not good. In the case of the elevator and the rudder, I can run through it fairly quickly and again present by this method the fact that we could describe through the speed range the characteristics which we obtain, those we think we would like to have had, and the engineer would be better able to interpret them. For instance, in take off we are interested in and we should obtain from these airplanes that we fly, information as to whether the rudder power is adequate to maintain the heading, and is it necessary to use only a reasonable portion of our physical strength to displace the rudder to the position to hold the heading, and if we desire to change the heading slightly, do minor motions of the rudder accompany minor changes in force so that we have what is termed "feel"? Do we get that sort of force variation with displacement of the rudder as with the ailerons if we plot again displacement against the change of heading (fig. 4)? Would we get a variation in the change of heading without variation of displacement or would we have a rudder which is flat through the center and one which over-balances toward large angles of yaw? We do not need figures and we do not need great instrumentation that only few people are capable of obtaining and which is in very great demand to indicate clearly what it is that we want.

"Now this had to do with the rudder characteristic at a given speed. Similarly we could test at other speeds and under other conditions and get a variation of this rudder linear force with displacement and for other conditions at higher or lower speeds, and thus we would define at many different speed conditions the characteristics of the rudder through its entire range. In addition, we would be interested in the ability of the rudder to maintain trim through a speed range. Through what angles must the rudder be displaced to maintain a constant heading through the speed range? Does it have a steep gradient from the trim position (fig. 5)? Do we have to continuously change trim? Do we go through only a small speed range before we have completely used up our force capabilities to maintain straight flight or can we go through a large speed range with small changes in trim?

"These are the things that we should have in mind when we describe our pleasure in having an airplane ride. Are we happy to return, or did we enjoy the motions we applied and the flight we obtained? "Captain Diehl of the Navy told me a story that impressed me. Two very fine pilots were talking on their relative merits and they disagreed as to who was the better; the one who could land in 5 minutes or the one who took 20 minutes to land. The one who took 20 minutes to land was sure the one who had landed in 5 minutes was far the better pilot because it took him 15 minutes less to find out how to do it.

"Now I have touched on the rudder and ailerons. The are usually inseparable controls. I haven't talked about dihedral effect nor the effect of aileron yaw on the rolling power of the ailerons. There are two or three high spots that I would like to touch on and then I am sure that I will subject to your best criticism.

"There has been a tendency for our airplanes to exhibit less and less stability. This has complicated the problem by making varying demands on our controls. The tendency toward rear CGs, the tendency toward increased power, and the extension of the nose have given rise to decreased stability which has brought with it problems that we as pilots may not be entirely aware of. I have in mind that much work is being done on the problem of tail failure – tail failure which results from no fault of the tail as such, but the fault of a maneuver which produces it. With small directional stability and an increased desire for greater rates of roll and larger aileron deflections, it is possible to roll out of a turn on one side violently into a turn on the other side and obtain yawing angles far greater than the angle which was assumed in the design load conditions of the vertical tail. This characteristic seems to become more important as the normal acceleration is increased. The higher the G turn, the more violent the use of the aileron, the greater the tendency for the load on the tail to become more critical than that for which it was designed. As test pilots, I think we should be interested in such things, because the engineer might forget to tell us and we should be in a position not to exceed conditions beyond the design of the airplane.

"Another characteristic that we could discuss at length is the combination of directional and lateral stability, which is an old, old problem: also the combination of directional stability and its dihedral effect and its effect on our opinions of ailerons with adverse yaw, and a rudder sufficiently powerful to take care of the aileron adverse yaw. When the dihedral angle was changed markedly the adverse yaw of the ailerons produced sufficient yaw so that the roll from the dihedral was greater than the roll that was produced by the aileron. Thus the aileron effectiveness reversed. Moving the stick to the left caused roll to the right. There is a chance that that characteristic could have been improved by an addition to the vertical tail area. I am not trying to diagnose why this occurred. I am saying that any variation, any liberty that is taken with an existing arrangement, must be borne in mind by the pilot so that he is aware of the forces and characteristics about other axes, and the possibility of changes in the characteristics about other axes. In this war we are forced to take liberties with aircraft and make changes that we can make, but we are not necessarily getting to the root of our problems. It becomes more necessary for the pilot to subject the airplane to every known test before assessing the value of a modification.

"Now, there is one other characteristic that I would like to mention that has to do with the elevators, I am hitting the high spots on this thing because it would take far longer to go systematically. If there is interest about any one specific thing, I am sure there is someone here who will discuss it. We think of the elevators as the angle of attack mechanism and we say that it has done its job if it has the capability of trimming airplanes through the speed range and if it has the capability of producing motion and acceleration about the lateral axis to the designed strength capabilities of the wing. In other words, the elevator can put the plane up to CL Max as long as that is allowable, and it can pull it to the allowable acceleration. Hence we think of the VG diagram (fig 6).

"Now we are reaching the place where our elevators are no longer capable of reaching the terminal velocity or the limit speed and acceleration because of other characteristics that we associate with compressibility, so we can't any longer get in this corner (see fig 6). In addition we are learning by continued instrumentation that we must be extremely careful of touching CL Max at the left upper corner of figure 6. There are indications that in making pull ups to that corner the buffeting loads can double the magnitude of the normal load achieved and we have to be careful of tail failure, and that is a tip from one pilot to another. Now, in addition, we expect of elevators the ability to fly a machine through a large CG range and still not get overbalance – and still not get high force per Gs. I want to touch lightly on two tests that we must make that have arisen because of the attempt to spread the CG range. One is in turns. The pilot can tell the engineer what he thinks of the plane or he can measure these forces, but still he can plot the variations. It has been decided by consensus that force per Gs under 3 pounds is extremely light and for fighters 8 pounds I think is the upper limit, although I would personally prefer 8 to 10. That is the force range in which we would like to have the elevator for combat maneuvering. Now, characteristics of tails and elevators are such that they do not remain constant throughout the CG range. If we had an allowable CG range such as shown in figure 7, it would be extremely difficult to get an elevator with sufficient characteristics varying with the position of the center of gravity which would look like that diagram. Now, that box indicated in figure 7 is the area in which the control forces must lie to contribute to you a good feeling airplane, and obviously if you push the CG back still further (they are relative figures), the force per G moves back to zero and we say we have the maneuvering neutral point. It takes nothing to disturb the airplane and we do not like it. In order to get the CG range larger it is necessary so to balance the elevator that its hinge moments are less, lighter. This will have a flatter curve and we still stay in the new area of increased CG range. Now, when we do that, we prey on another characteristic which we call C11 Delta – the hinge moment due to just deflecting the elevator itself we get almost at zero. We get an airplane in which, if we fly it in straight level flight and we hit a bump, the stick begins to pump. If we just tap it we can pull several Gs because the elevator is so light. As such a gradient shows, there is no change in the force per G, and quick motions produce large accelerations. There is another test then that we should make to define our elevators, and it is becoming more important. We should look for oscillations in the controls, short period ones that are produced either by the air or by ourselves. They should be noticeable to us immediately – pumping of the stick, and we should so report them.

"At the same time we get another characteristic that I am sure we do not like. I would like to show you something about the variation of force per G at a given speed. We pull 3 G's at 200 MPH and it takes 10 pounds, so we say we have put on those extra G's at 10 to 5 pounds per G. We hold it steady and that is the force per G that we have been talking about pretty much. It occurs out at some point in time. The control that we have been used to recently is one which, if we made abrupt pull ups thus deflecting the surface before the plane gets a chance to move, gives us a larger force per G. So in small periods of time we would get a small force per G (fig.8). It is surprising that we can tolerate two or three times the magnitude of force per G if we do it quickly and don't sustain it long, as in abrupt pull ups from level flight. So we have, I think, been accustomed to a force per G variation with time that looks like the diagram in figure 8. Now the other type elevator – the one which is very highly balanced and which allows us to have a small variation of force per G through the speed range – is one which takes little or no force to deflect it abruptly before the airplane moves and we get a control which is very light and we get a tendency for the force per G to vary in the direction shown. But we aren't always turning. We are mostly in straight flight and this new one detracts from a gun platform. I think that this is a reasonable place to conclude that if you wish a good gun platform, you cannot have an airplane that has a very large CG range because the military load which you wish to put in it alters the characteristics the designer had in the original airplane.

"I have taken a good portion of the time that was allotted me – probably more – and I would like to conclude without mentioning anything about stalls or aerobatics. I would like to conclude with a statement to the pilots that there is in existence lots of information about which you may not know which describes the possible arrangements of airplanes – airplanes that you have flown which have been specifically measured. These measurements may lend weight to our opinions, which are undoubtedly correct, but which vary from the other fellow's because you are not talking about the same thing or you are not thinking in terms of the same maneuver, the same condition of the airplane, the same position in the speed range or the same detailed peculiarities which caused you trouble. Flight testing is fast becoming an exact science and the systematic measurement of what the pilot feels is progressing rapidly. We are thus becoming better able to tell how the pilot got this impression that he had and why he got it. As I say, the information exists for him to take advantage of and thus gain more by flying those many types of planes by sticking to the facts, by reporting exactly what he saw and felt and by qualifying it with every characteristic that was changed between the previous and the last condition, and lastly reporting it with no attempt to tell why it should be so, necessarily, and thus possibly influencing the designer, but with a sincere attempt to investigate the entire range of characteristics of the airplane properly qualified.

Commander RAMSEY: "Thank you very much, Mel. That was a very nice discussion on stability and control and I am sure it will provoke some interesting comments. I would like to hear from a pilot or engineer on this subject. As a matter of fact, I talked to an engineer this morning who stated that in his opinion an engineer should go up more often with the pilot to see what the pilot means and in order to talk his language more intelligently. I would like very much to have some engineer discuss this. How about a pilot giving a short talk on his point of view as to stability and control?"

Mr. Paul S. BAKER (Engineer, Chance-Vought): "Well, I will be glad to speak from an engineer's point of view and tell Mel that I heartily second his remarks on pilots qualifying their criticisms and comments on planes. It does an engineer no good and, as a matter of fact, I know of nothing that is more infuriating to an engineer than to have a pilot say he doesn't like it, but fails to give his reasons clearly. I think that what Mel had to say and especially the diagrams that he drew on the board are excellent. It would help all engineers if pilots in discussing flying qualities of planes would resort to sketches and diagrams. Even if they cannot put the exact values on them, they can at least indicate the tendencies which would be very helpful to an engineer who generally knows the direction that the tendencies may or may
not take."

Commander RAMSEY: "I think those remarks suggest that the cards which each of us takes up on the hop might well be filled out with some little sketch to aid in further discussions which we will have during this conference. I think that most engineers are handicapped quite a bit by mere words such as O.K., good, lousy, etc., whereas if we drew a sketch to describe our point, it might be very helpful to them, and also helpful in our discussions. If we all would take a little time in filling out the back of our comment cards, it would be very useful. We would be very happy to hear from a pilot along that line. Dick, haven't you something to say?"

Mr. R.H. BURROUGHS (Chance-Vought pilot): "I would like to say that while Commander Ramsey has mentioned that the engineers are at quite a disadvantage because pilots don't say what they mean, I think the whole situation could be helped a lot if the engineers would make more effort to educate the pilots into finding out what they are particularly looking for. You seldom see an engineer get up at a blackboard in front of some pilots and show exactly what the pilots have been looking for for 2 or 3 years, but didn't know it. I think in general education of pilots by engineers is indicated just as much as education of engineers by pilots."

Commander RAMSEY: "Thank you, Dick. Lieutenant Colonel Toubman of AAF Proving Ground Command will have something to say."

Lieutenant Colonel TOUBMAN: "I would like to start a discussion on this hydraulic boost aileron, which I think is wide open. I believe the manufacturers have the feeling that they are going to have to go to boost ailerons for high speed to get what they want. We have such an example in the P-38 and F7F boost rudders. They are giving us really a roll, but it is not the roll that the pilot wants. I would like an open discussion on that. If this is the last alternative, can we get a natural aileron that we can feel and still gives rate of roll? I am skeptical of the artificial means of getting better ailerons and better rudders, and kind of fear for it. I know that we have airplanes planned (and pretty high-speed planes) and I think that they are going to have the same disadvantages that the present ailerons have."

Group Captain CLARKSON: "I would like to go back just for a minute, if I might, to the question as to how we describe these controls. I think that there is a little more to it than coming down and perhaps drawing out something for the technical man on the ground. I believe it would be possible to have an airplane which had all the good points of the type of curves that have been drawn there by Mr. Gough, but that the majority of the pilots would turn around and say they did not like flying it. I think there have been cases like that where according to the book the plane ought to be perfect to fly, but when you got it up in the air it wasn't perfect and the pilots didn't like it.

"I am absolutely certain that finally we are going to have to come down to describing the controls technically in that way as we have had them described to us this afternoon. I am not sure that we ought not now, and perhaps before we reach the stage when we can come down and write about it, try and find a common terminology for let's say the aileron control. I believe we are going to need that as well as the curves, and I would like to suggest perhaps one or two nomenclatures for these things. I was talking to Mr. Gough about this at lunch. I split my requirements up into three, and I know he doesn't agree with that. I thought we ought to have standard nomenclatures for stick force for the ailerons, for the rate of roll, and possibly for the response. I think that he was of the opinion that the rate of roll and response should come under one heading, and I'm not sure but that having thought about it he's not right; but I would suggest that when the pilot comes down and is asked by the designer what the stick force is like, he should be able to say if the stick forces are light, or if stick forces are heavy, or overbalanced. If there were some written description of what was meant by heavy, we might get a little further. Now to get that we might have to take an airplane that is well known to all pilots and compare others with that and make that a sort of criterion. I cannot think of an aileron force at the moment without making it a bit ambiguous, but I think that most people here will agree that the stick forces of the old original P-51 were about as nice as anyone ever had, from the pilot's point of view. Perhaps we could find an airplane which is known to have good stick force, wasn't too light, not too heavy, no sign of overbalance, and make that the criterion; then if necessary get our curves from that, which would give us something to work on. But I believe you could show someone the curve like that and they still wouldn't know exactly what the feel of the plane was. As far as rates of rolling are concerned, you could divide those into very high rate of roll, rapid rate of roll, perhaps something like acceptable or normal, although that's a rather difficult thing, and something like poor. I believe if we divided stick forces and rates of rolls which would be coupled with responses and tried to get some standard description for each of those we might fill up an interim period between what we are doing now, which is scrambling around in the dark rather, and the time when we can come down and write everything down in the form of a curve. I would be very anxious to hear what other people think about that."

Commander RAMSEY: "Thank you, Group Captain Clarkson; I think perhaps Mr. Mel Gough might like to refute."

Mr. GOUGH: "No, I am in hearty agreement. I would call attention to the fact, though, in connection with his remarks, that the P-51 with its original ailerons has been very finely evaluated quantitatively. There is a confidential report to the Army Air Force. I am not sure of the status of it, but I think it could be obtained for those who are in a position to receive that kind of information. It would do for Captain Clarkson exactly what I had hoped we would be able to do in the future for all. He had in mind that particular airplane with that particular control. Now that particular control is very carefully defined. At many speeds the variations of the force required to displace the control are shown, the deflection in the system between the pilot's hand and the control surface itself is shown, and the response obtained at any position of the control and for any force is shown.

"Now as far as the magnitudes are concerned, I think that the limits have been pretty well established. I think I am right in saying the Navy would like to have 30 pounds as a maximum limit, indicative of a proportion of the possible physical effort that is acceptable. The Army has I think a standard of 50 pounds, and I know of tests made at 75. Personally I can't make them at that force to the left with my right hand, but I can go to the right with my left hand. Those magnitudes would then be shown so that if the captain would get that particular report, he would see quantitatively what his relative opinion is. I think that similarly when it is possible to release all these reports you would be able to establish those values for yourself.

"Now, I would like to add just a little to what I said about the ailerons, and that is it is surprising how you can either measure quantitatively or tell qualitatively what you think of a control, or in the case of the ailerons if you were at a given high speed and you moved the control from neutral you noticed that it took a force to move it, and suddenly that force lightened a little and then as you moved further that force remained constant. It is immediately apparent our physical limitations are such that we can judge very finely between forces. If there is any nonlinearity of the control force with motion or of the response with motion you can tell it immediately. We have some ailerons in which airplane rolls fast at first with deflection, then stops abruptly as side-slip builds up. It has been necessary to standardize the types of tests that were made in order to reduce the number of variables and in making aileron studies. For instance, we have subscribed to the thought that if we trim the airplane about all three axes at a particular speed, or lock the rudder with our feet by jamming the heels to the floor so that in the ensuing motions the rudder will not move, then all the roll that is obtained is due to the motion of the ailerons, all the side-slip that is obtained is due to the adverse yaw and the directional stability characteristics of the machine. The result is, however, that if you got lots of side-slip and had lots of dihedral, though the first displacement of the control would roll the airplane rapidly, as the side-slip builds up the dihedral may stop the rolling motion entirely.

"Without benefit of figures we could show that the ailerons were deflected and required a linear force to deflect it. When we get to the place where we've stopped the deflection, the roll which has resulted due to deflection comes back to zero; so we define what we mean by the motion of the control. Similarly if there is any nonlinearity in the force of deflection we'll tell it right off the bat. Take an airplane which has lots of friction at its neutral position or which binds at the neutral position, or in which after we start the motion, the control forces are light. We define it this way; We would get a large force for the large deflection. We feel that we were riding on the high spot of a cam and it is surprising that when you measure it, there will be only 2 or 3 pounds difference between that and one which is perfectly linear.

"Now I would like to answer briefly the colonel's remarks about boost controls. I have often wondered the same thing, but I have adopted a philosophy which permits me to accept them. In thinking of flying the plane from the pilot's viewpoint, we apply a force and motion to the stick and the only thing we are conscious of is the kind of response we get. It seems to me it makes little difference whether the control is aerodynamically balanced, spring assisted, or power boosted as long as we satisfy that requirement which meets with our own physiques, and that is that we must have the control change increase as we displace the control from neutral. I am not sure that I would know what was in a system – whether it was a boost or not, it the boost were designed to give me the characteristics of linear force with displacement and equivalent response in roll with deflection. In the case of elevators for instance, we have had to change the gradient of force per G with CG by changes in balance, by spring tabs, by gear tabs, and by bob weights, and the peculiar part of it is that when they are properly designed we do not know they are there. In the original airplanes which had elevators hinged on their leading edge, we had in effect a bob weight; when we make a turn the inertia moment of the control itself tended to make it hang down. We got stability by the fact that the control was not aerodynamically balanced. We had a bob weight and we didn't know it."

Commander RAMSEY: "Thank you, Mel. Someone wanted to ask the Group Captain if he liked the first aileron on the P-51?"

Group Captain CLARKSON: "Yes, I did. That was with the original Allison engine in the 51. I would make this qualification, that I am not sure that when that plane went into service with the United States Army they did not alter the control because they felt that there was insufficient control at the slow end of the speed range. I think it was an alteration in gearing. So the ones you got may not have been the same as ours."

Mr. STEPPE: "The original P-51 which we built for the Army had the same as for the British and I want to make a point in referring to pilots' opinion of ailerons. The group captain said he liked the ailerons, but the Army universally didn't like them. I don't mean to start an argument, but to illustrate: A boy from a commando outfit had a P-51 with a lot more supercharge. I don't know if he had the bubble canopy, but the thing that made the whole difference was the fact that it had sealed ailerons and that he could get rapid response at high speed. Nothing meant as much to him as the aileron change it had. The other might be better but the boys out in the service know what they like to fly."

Mr. C.H. MILNER (Republic pilot): "I would like to ask Mr. Gough what his qualitative analysis of the ailerons showed."

Mr. GOUGH: "I recall quite well the original ailerons on the P-51, and the restricted throw which we thought they had because they were of the internal balance type. We have with us Mr. Hoover, NACA test pilot, who made all the flight tests of the P-51 and who was quite enthused with the rolling characteristics of those ailerons at high speed. We all agreed and measurements showed that they were quite low in rolling response at low speeds though they were light. In general they could be liked because they didn't take large forces to move them, so we had to be very careful in stating our opinion as to what they were. I would like to have Mr. Hoover comment on these P-51 ailerons. It is difficult to disassociate yourself from the light forces, but Mr. Hoover went further and worked out every known method to improve the rolling response by changing their shape. An aileron was developed which, as I recall, produced considerably more of a roll at low speeds and gave the same response at high. As I recall, and I would like to have him check me, they gave considerably more roll at high speeds, but the reports that we had were that at the speed range at which the airplane was capable of flying, particularly at high speeds, other planes had ailerons that stiffened up and you couldn't move them. Perhaps Mr. Hoover could add to that."

Mr. HOOVER: "The original P-51 ailerons were good at high speeds but at low speeds they stank. The P-51B with the internal sealed and balanced ailerons is about the same at high speeds, but is much superior at low speeds. I think the reason the British liked the ailerons at the beginning was that they were using the plane for high speed flight, whereas our Air Force were interested in low speeds and they didn't get the performance. But to answer Mr. Miller's question, the original ailerons were about as good at high speeds as the P-51B's. They had the pb/2v around .04 at 450 MPH."

Lieutenant HALABY: "We have talked a lot about ailerons and we have just now had a specific case which makes it a little easier to work with, although perhaps it is taking a little advantage of the North American representatives. Ailerons mean lateral control to most of us. It seems to me that we have it tied in with directional control and stability, and the P-51B is a good example. Whether the addition to the fin was caused by the installation of the bubble canopy, or the higher rates of roll at high speeds, which Mr. Gough pointed out, induced greater adverse yaw angles, is not important. It brings us back to his original point that any modification that is made on the planes should result in a thorough-going test. There are a lot of pilots who object to charts and curves and I am one of them, but if we draw up those graphs and charts in company and service test checks, we run across these things. For example, the P-51B would not have come out with ailerons that could induce those high yaw angles and shed tails and kill pilots. The point to me that has come up in this discussion is that when we change the ailerons, we change the whole airplane. We can't stop there and we have to check directional stability control along with the bubble canopy, along with superior ailerons at high speed.

"We have neglected directional stability, I feel, a great deal, not only in the Army planes, but in the Navy planes as well. For example, the P-51 has more fin, and almost every plane on the field has another chunk of tail added on at some point-dorsal, ventral or spoiler, or something that indicates that the designers' calculations were a little awry. So I would like to emphasize that I think Mr. Gough's ideas on checking each modification throughout the chart or curve or pilot's viewpoint is most important."

Commander RAMSEY: "You may have stepped on some contractors' toes. Maybe they would like to refute."

Mr. BAKER: "I would like to say one thing and that is in connection with any of the stability or control tests. I remarked before of the value to the engineer of diagram sketches that the pilot could draw. I might go a little further and say that I strongly feel that an engineer must have, in order to make intelligent modifications of the plane, to effect a desired improvement, the quantitative values on curves or charts or sketches such as Mel put on the board. He simply can't get along without that. Now it has been my experience that necessary and valuable information of that sort can be obtained by relatively crude instruments. It tends to be burdensome to the pilot, who might much prefer to come down and say, "I think it's so and so, and so and so' with possibly a few sketches, rather than to make a long flight, obtaining points of stick force against stick displacement or time to roll 90 degrees against stick displacement at various speeds. Such testing does tend to be tedious and time consuming, but it has been my experience that the information obtained by such tests in the end, in arriving at a satisfactory airplane, more than pays its way."

Commander RAMSEY: "Does North American wish to say a few words?"

Mr. STEPPE: "I just want to clarify the lieutenant's question on the P-51. The addition of the extra fin was brought about by the modification which added the fuselage gas tank and I think that could be directly pinned down to the addition of area rather than the action of the aileron. I agree that not too much investigation in directional characteristics on the roll with that modification was made. In line with Mr. Gough's thoughts those modifications are usually hurry-up jobs and do not allow time to fully investigate the conditions that result. Most of them are dictated by theatre conditions and the time can't be allowed."

Mr. BURROUGHS: "I am not presuming to refute what Captain Clarkson had to say about the undesirability of obtaining a lot of data, but I think a lot of discussion has pointed to the fact that in almost every case you cannot get away from it. For example, we had brought up the case of the P-51 and certain pilots thought it was good, other groups disagreed with them; and the reason that they disagreed was that they had not separated all of the variables. You never get to first base in this question, particularly of stability and control, until you do separate the variables. For instance, suppose we wanted to evaluate ailerons with no instrumentation. We could not even start up because there is no pilot who can tell whether he is going 150 or 225 miles per hour. He has to have an instrument – the air-speed indicator. If he does not, he is not going to be able to evaluate the ailerons. Mr. Gough has shown how you can separate the variables, and how from a design standpoint you can get a better airplane when you're through despite the fact that it was very tedious and caused discomfort to the test pilot. I would like to repeat again what Mr. Gilruth said at Langley Field, which is almost immortal; "There's nothing like a helluva lot of data."

Mr. HOOVER: "To get back to what you've got to do to have the pilot and engineers get together, with the NACA, we work in terms of pb/2V. For instance in connection with ailerons, you might say we are calibrated to the fact that when we take up a plane without instrumentation, if we say the ailerons are good you can bet that the pb/2V will be about 0.07. If we say they are very good, then the pb/2V will be somewhat higher than that. If we say the aileron stinks, we mean that it is something less than 0.07, but in the services you calibrate the pilots to acceleration by putting the accelerometer in the plane. We have to calibrate the pilot for making any qualitative statements. We know how the plane felt. We know what the measured data was so we can give you a pretty good qualitative analysis of its handling characteristics. I think that that would be the way for the manufacturer and engineer to get together. Let him instrument the airplane, have the pilot fly it and show him the results of his test."

Mr. GOUGH: "I would like to amplify that a little bit. That is in direct line with my premise that we have tested many of the planes which you have flown. I am sure that most pilots of the various companies representing those planes have not had the opportunity to see the measured quantitative data which has been obtained on the plane they are familiar with and if the could, they would immediately interpret in terms of graphs and figures these things which they have felt on the airplane which they know. I think that Hoover is personally correct, if you could instrument your own airplane if it has not been done and let the pilot get his impressions and compare those with the engineer's data. If the data exists, it is the right of the company, it is its duty to show it to the pilot, the man who flies the plane, so that he can become familiar with the magnitude of the thing he is handling. Now, in general, if you have the instrument, and you have qualified personnel, it takes 20 to 25 flight hours in the average single-engine pursuit plane to thoroughly evaluate those standard things we now measure. After practice in going through those systematic routines, which are the possible characteristics of the plane that we should have, you would get a relative opinion which would describe in great detail the characteristics of the machine and immediately bring to your attention those characteristics which might deserve more minute investigation."

Mr. I.H. DRIGGS (Bureau of Aeronautics): "I have very little to add except to congratulate Mel on a very excellent presentation of the subject. Frankly, I fail to see how the engineer can possibly develop a good plane unless the pilot interprets for him the flying qualifications in terms that he can understand and view toward making an improvement. It seems to me it is a matter, as Paul Baker suggested, of mutual education. At the Bureau, we are concerned at the present time with the setting up of flying quality specifications. We are working with Army Air Forces, CAA, NACA, and the Flight Test branch here, trying to set up what is desired and what is necessary. We are also endeavoring to rationalize the subject, trying to collect all the data together in one place that will, we hope, help the contractors and help ourselves to understand the design in its preliminary stages. We want to eliminate as much as possible the changes that now occur from the time the plane starts on the drawing board to the time it is in service. In other words, we want to shoot closer to the mark. It is necessary that flight test personnel supply us with the information to enable us to do that. We are also concerned with consultation of course with the various manufacturers, trying to assist them in obtaining the flying qualities required. And another branch of our work is concerned with the interpretation of flight test results in connection with wind tunnels. We are trying to present our wind tunnel flight results so that we can compare them with flight tests. That is absolutely a much more quantitative rather than qualitative matter and the test pilot has got to help us in that way.

"This whole discussion smacks very loud of rather poor research and test flying. I would like to point out, having traveled around the country for the last 2 or 3 years, that practically all flight test units of the manufacturers are not involved in prewar research flying. They have more or less been restricted by the circumstances of war to testing for the pilot's line. We get a change either from the Navy or from our own engineering department and we are forced to put that on our resident test airplane and evaluate that change as rapidly as possible so the change can be made in the line as rapidly as possible. Many of us here have a genuine interest. It seems to me that for that landing power research problem, it will be several years before many of us are allowed to get too deeply into it. I envy Mel Gough his position in that he can indulge himself to the limit in this particular work.

"I have one other comment. I have nothing against aerodynamists as such. We have a lot to thank them for. But it has been my impression as a test pilot that in years gone by they have been primarily interested in performance – how fast can my next plane go. In reaching for an extra 2 miles per hour, we have been prone to overlook details of an airplane which directly affect the performance which our customer measures. That performance is not the performance of an airplane. It is the performance of an airplane plus a pilot. It has been very difficult in our particular outfit at Lockheed and in many others to convince the engineering department that a change should be made. A great many test pilots that I have talked to have expressed the desire to have some feature on their airplane changed and it so happens that we spend 90 percent of out time proving that the change should be made and 10 percent making the change. I think the time is here or it soon will be when for a given H. P. and a given gross weight one aerodynamist will be able to make his article go just as fast as the other aerodynamists'. It will then go back to the vital problem of stability and control. The situation to me is very interesting and very important and it will, I think, be a long time before the manufacturers get back to that type of testing."

Mr. GOUGH: "May I discuss that just a minute? I am very sympathetic of the manufacturers' problem. I am not as sympathetic with the desire to obtain a quick answer, hoping you have made the right guess. I am also sure that in the long run it will pay dividends to obtain the research viewpoint, go directly to the problem and fully evaluate it before making a change. When you see what has to be done – do it. The manufacturers' problem, it seems to me, is complicated. It has never seemed to me to be desirable to spend months and years making quick changes when, if you had stopped and evaluated what you had and then corrected it, it would have paid off. One day you stick two more fins on a plane, the next day putting in twice as much power. The present tail has already been found adequate and you pass it if just enough to get by, knowing that you're going to put a supercharger in tomorrow. I suggest that it might be worth the manufacturers' consideration to stay closer to the research attitude. Do not try to deny the fundamentals under which we must work. Go to them immediately. Get the data, make the changes and don't waste three times the effort trying to tell why the fundamentals are not desirable. The fundamentals which have been laid down are based on obtained performance, as far as stability and control are obtainable. If they cannot be met, it is a shortcoming of the designer. If a plane crashes frequently, it is a design error. We might as well face that situation now. Let's don't tear down the standards, let's not depart from the standards because individually we might not be able to meet them easily."

Commander T.D. TYRA (V F Design, Bureau of Aeronautics): "One point we have more or less overlooked is what you do after you find out what is wrong with the airplane – or what you do when you get these operational crashes that indicate you have overbalance, high force at low speed and not enough stability in yaw. That to me seems to be one of the principal problems confronting the service at the present time. We have measurements on a lot of our planes and we know what they are supposed to do and the problem is how to correct them. It takes 6 months to 2 years to get a good set of ailerons and the same is true of the rudder. I would like to suggest that whenever a contractor starts out to design a new airplane he lay down on the drafting board three sets of tail surfaces, different configurations of the aerodynamics balance, and throw in one tail that he thinks would be adequate and one that it is felt the tactical testers would like, and one tail about twice as big as the last one, that will satisfy the flight test department of the Navy. I think that if we go at it that way, although it is not an exact science, we would get better flying airplanes a lot quicker."

Commander RAMSEY: "Gentlemen, I must stop your discussion because the time is drawing short. I wish you would save your comments on this particular subject for the agenda of Sunday and Monday.

"A lot of people have not signed the register and ensuing events during the week depend on who has registered. Please register with Miss Ryan in the outer office. The banquet Wednesday at 8 o'clock promises to be good entertainment as well as good food in Wardroom No. 1. The oyster roast in Captain Wildman's barn on Friday will be a good time for you to eat a lot of Chincoteague oysters."

FOURTH MEETING
18 October 1944
1530-1815

Commander RAMSEY: "We will call the meeting to order for this afternoon. In order to get through by 1815, we will start out with a few items of general interest and then come to the discussion of night fighters. The white data cards posted on the bulletin board outside are subject to quite a bit of criticism. Some of the manufacturers think we have not put the performance on as they saw it. The performance we have on the cards is either that recommended by the manufacturer or found out by the Flight Test Section here for that particular airplane. If there are any great differences of opinion will the manufacturer concerned please see Lieutenant Harman immediately after the conference at table 11.

"We would like to have everybody take advantage of the good weather by getting into their airplanes promptly and getting off at the scheduled time.

"I wish to call your attention to the cocktails and banquet tonight at B.O.Q. No.1. There will also be some entertainment. How good it is, you will find out for yourselves; I have never seen it myself."

Night Fighters

Commander RAMSEY: "This afternoon the topic for discussion is night fighters. As you all know there are many open questions on their use in both the European and Pacific theaters. The opening remarks will be made by Commander Harmer. He has had quite a little night-fighting experience and I am sure he has some ideas of what a night fighter should be."

Commander R.E. HARMER, USN (VF Design, Bureau of Aeronautics): "My night-fighter experience has been in single-seater night fighters and operating from a carrier, so if you will give that consideration to all my remarks it might explain any attitude you might not consider normal. I have had about 2 years altogether in night fighters and I have formed some ideas of what I would like. So I will just go right ahead listing specifications and characteristics that I think a night fighter ought to have and then go over some of the compromises which will have to be made to cut down this airplane from the ideal. The first requirement is that it have two seats at least, two seats side by side if possible, otherwise in tandem close together, one for the pilot and one for the radar operator. The second requirement is that the pilot's seat should not be located behind an engine but have a free unrestricted forward view. In most conventional types of aircraft this calls for a twin engine airplane and there may be other ways to solve the problem, but we will call a twin-engine airplane the second requirement. The airplane must have sufficient and effective air brakes – a means of rapidly cutting down air speed from high to low. That is the third requirement. The fourth obviously is the best visibility in all directions, especially forward. It may be necessary to install bullet proof glass that can be moved in and out of position in order to obtain desired visibility when it is most needed. That is important indeed and should be given much consideration in a night-fighter airplane. Instrument panel is something that is never gone into deeply enough in a night-fighter airplane. There is discussion as to whether indirect red is the best lighting, but nevertheless it can be decided and put in the night-fighter airplane to give the best instrument light possible. Other requirements are that the instruments should be arranged symmetrically around the pilot in convenient locations. Such instruments are gyro horizon, altimeter and rate of climb, air-speed indicators. Your flight instruments and engine instruments should be on separate switches and controlled by separate rheostats. There are times when you want that cockpit dark and you might want to cut down the instruments to the minimum brilliance for a short time. The next is the heaviest possible volume of armament you can put in there consistent with the performance you need. These guns should be located around the center line of the fuselage. I consider that very important because your ranges are extremely short in shooting at an enemy plane, it might be shorter than 100 feet and sometimes that cuts down your hitting power by the guns of one wing which you cannot bring to bear. They should be cannon. I am not satisfied with .50 calibre guns and if you can add any more guns toss them on – we can use them. The guns should have effective flash dampeners and the muzzle so located in relation to the pilot that his vision will not be affected by the muzzle blast. The radar installation should be such that the following conditions are met. (In other words, I think the night-fighter airplane requirements should be built around a good set of radar equipment.) Here are the conditions that the radar equipment should meet. The antenna should be unrestricted throughout its coverage by any part of the airplane. That is pretty obvious. The essential parts of the equipment should be accessible to the operator so that he can make minor repairs or do tuning while in flight and the radar equipment must also be easily accessible to be removed or installed for maintenance purposes. Another requirement concerns flight characteristics; this airplane should be able to fly pretty slowly and be controllable as slow as 90 knots and while at that speed flown at a pretty flat angle of attack. The only practical way I can think of to accomplish this is by having flaps that you can drop at a low angle of attack so that if you are following a plane at a low altitude and at low speed you won't have to be steaming around with your nose up in the air.

"These items I have mentioned so far affect the basic design of the airplane, and there are some compromises that have to be made to meet the requirements. I am not too worried about performance. So far I have not had to use full performance of my F4U in any night interception. I am not saying that definitely you won't, but I think as far as the single-seater goes, it has plenty of performance in speed, rate of climb, and maneuverability. In order to give this ideal plane of mine a definite standard of performance, I will put it like this. I would like a rate of climb and speed just about the same as the F6F. I don't know how hard that is to meet in a twin-engine aircraft. Maneuverability should be comparable to the P-38. It has about everything you need, and you have to follow the evasive action of any bomber you intercept. On the landing characteristics it should be easy as the F7F; or, in other words, good enough for carrier operations. I think it ought to have a 4-hour cruising endurance, including 1 hour of military power. The cruising speed should be roughly about 160 to 180 knots. It seems to take a lot of gas.

"I realize that all of these things cannot be had in one airplane, particularly a carrier airplane, so I will mention some of these items I would be willing to part with in the carrier airplane. I will list them in the order I would least hate to part with them; I'd hate to lose any. I'd have to give up the twin engines for loading aboard a carrier. I'd have to give up the two seats close together, perhaps, as long as I can keep two seats. The last one which I think it is necessary to remove for a carrier airplane is that the radar equipment be accessible to the operator in flight. I'd hate to see that go. What I want to keep is all the other things that I mentioned, especially the two seats. I think we did a pretty good job with the single-seat night fighter, but it is not going to be enough to do the job in the future when the going gets a little bit tougher. I mentioned this and put it in reports and I have not heard it being considered very much, but I think I would like to leave this group with the thought that in the future the Navy might suddenly and very loudly call for a two-seater, carrier-based night fighter, and if some of the requirements have been given consideration we might be able to get one."

Commander RAMSEY: "I would like to hear from Lieutenant Colonel Harshberger who has had experience as far as the Marines are concerned with night fighter work in the Fleet."

Lt.Col. J.O. HARSHBERGER, USMC (VF Design, Bureau of Aeronautics): "My sentiments are pretty close to Commander Harmer's on quite a few things. I hate to waste too much time but there are some of you that may not know exactly what is necessary for night fighters. A night fighter can be used in two ways. It can be used as an intruder or it can be used with ground control equipment. The G. C. I. is the standard ground control equipment and is far superior for several reasons. Its accuracy in directing other planes is infinitely better. A long-range radar can be anywhere from 5-15 miles in error. The radar itself is not in error, it is the slowness of the people using it in plotting and interpreting plots. The G. C. I. controller works directly from the tube. Aboard ship I suspect the G. C. I. or radar coverage is good. In the South Pacific, ground control radars in their present form are greatly hindered by land return and by siting problems. You have to take the land from the Japs before you can place your radar and then you are apt to find a mountain behind you. Our problem brought out several features in the night-fighter requirements that were not outstanding aboard ship. For instance, before long-range radars could be set up we had no early warning. Therefore we were orbited from aboard ship. Our problem was to cope with a signal coming in 30 to 40 miles from the night fighter. Sometimes he dropped his bombs before we intercepted, so naturally I am looking for a little bit faster plane. We had PVs the Navy wished on us. We had a top speed of 285 statute miles with the modification we had to put on the nose. However, the gear we had was good if we were up high enough – 2 1/4 miles on our gear. It is obsolete gear so that figure does not mean too much. If we had to be as low as 5,000 feet, we could see 1 mile. That meant that the controller had to make an accurate, ideal interception.

"Now if we could get a long-range radar out there on the same plane, the same controller could vector you in a little more carelessly, and at the same time he could vector another plane around. So the importance of a long-range radar, especially under poor ground control, stands out. In the patrol neighborhood where your sole problem is to cruise along and pick out stray targets, a set with twice the range will give you four times the coverage. The countermeasures against ground control radar seem to be more effective than against airborne radars. The effectiveness of countermeasures is to either make the set impossible to use or to make the control less accurate. Ground control can put you only in the general area. The long-range radar makes the plane possible to be used where countermeasure is taken against the enemy.

"Now, I too, have some requirements for a night fighter. Mine are not quite as detailed. We need high speed, to get from where we are to the vicinity of an interception. My second requirement for night fighter is visibility. I think that is of paramount importance. You see a faint dark target. We have a screen that is shuttered up, letting practically 95 percent of the light through which means that the tail gunner will see you first and you are going to meet interception down and to the side in evasive action. You want to be able to see well up so you don't have to twist your neck like in the P-61 out here. We should have a windshield as close as possible to the pilot. The closer you get to that glass the better you can see through it. In other words, make the pilot comfortable. In a twin-seater fighter the prop should not obstruct vision. Don't make us look through those things too much ahead. If you can get them clear behind the pilot so much the better. I am for having the cockpit of a twin-engine night fighter smack in the nose, right behind the radar where he can see down at a 30 degree angle if that is possible. We have one particular interception where it is hazardous to come in under the guy when there is a full moon and a white solid bank of clouds; if you are flying above that and making interception, the best spot to see a guy is down in this particular instance. If you come in 500 feet above the guy you can see 1 1/2 miles ahead in some planes. When strafing surface targets (naturally the fighter in its present form is used for other missions than shooting down enemy aircraft, though that is the prime consideration), when we find one or two barges and make strafing runs, we should see well over the nose. Third I put rate of climb. Commander Harmer had a good airplane and I had a PV1; my rate of climb was only 1,000 feet a minute at a low altitude so naturally I put rate of climb as important both to correct the error in G. C. I. interception and on intruder missions where you don't know how you are going to intercept the enemy so may have to climb up to him. It is fine if you have a lot of time to climb, but you are going to do your intruding near the enemy and you don't have too much time before his information, his radar, warns him. So rate of climb is important to us for that mission. Next I put maneuverability, enough to follow violent evasive action of bombers or targets of the torpedo plane type. I can see a night fighter, both ship-borne and ground type, going after torpedo planes.

"A suitable radar installation with the thought to new design and ease of maintenance is very critical and I think radar has to be maintained every second or third day. For a suitable radar installation, however, we need longer range. The longer range radar we can get, the greater the force that can put out to oppose the enemy. The plane should be well instrumented, and you can take your hat off to the P-61 on that score. The red illumination from the Navy standpoint is all we will accept because the doctors tell us that red illumination hurts the eyes less than any other color. I wish we could get some beef behind instrument people to put indirect illumination inside the instruments. I would like everybody to get behind and push the manufacturers to give us indirect illumination from a red or white source inside the instrument.

"A good gun sight is important. There is no point in going out trying to see the guy and losing him in the gunsight. Using an extra glass there means that you have to get that much more light from the guy to see him. If you can reflect the light on your bulletproof screen, I'd say that is the best. I probably will be in disagreement with the commander on this. And if there are a few engine instruments on the forward panel, I don't think they will hurt me seriously. I don't want a lot of lights all around the cockpit. I have a flashlight and I should know the cockpit layout very well; if necessary I think I can use a flashlight. The green fluorescent and green radio-active illumination is out because, as I said, red is all we'll accept. I am dead set against the pilot, in a twin seater with an operator, having a pilot's repeater scope. He is paying attention to that when he should be looking for the guy. The commander and I will gave our biggest difference right there and I wish people who have definite ideas on that would get behind one or the other of us. Flame dampening is something that practically all manufacturers pay little attention to. Somebody mentions flame dampeners and they say, "Oh boy, we'll add that." At the rate of 10 to 20 miles per hour you add it. I think you manufacturers ought to get together, pool your resources and get something definite on this thing. With night-fighter missions coming to the fore, flame dampening should be something that you are working on constantly, to improve the technique and cut down weight. The PV lost 10 miles an hour on a rough check. It had the ideal flame dampening. You could never see any light from the PV-1. "Frankly we cannot blame the manufacturers for the condition of our night fighting in the Army and Navy. I don't think we pushed it enough. I don't think we take an active enough interest while you are building the airplane. I think you could give us a good two-seater airplane, in fact I am all for it. I am afraid we will have to throw out the turret to get enough speed. If the turret is a success when they direct it with automatic radar, perhaps we can throw out some of the fixed guns; then possibly we can put up with a turret in a big airplane."

Lt.Col. William YANCEY, AAF (Night Fighter Training, Hammer Field): "In the work that we are doing in the night-fighter program at Fresno, Calif., we at first definitely decided on two or three points that are very close to the ones brought up by Commander Harmer and Colonel Harshberger. We definitely decided after a series of tests that we did want a two-seater plane. There was some discussion because of naval operational work with the single-seater fighter, and they asked us to run a series of tests with a P-38 equipped with Ash and that we did. In all of our interceptions we assumed that, figuring closely, we would lose some 80 percent of our interceptions because of the fact that it was a single-seater and not manned by a two-man crew. Going further with the same proposition of the same aircraft and the same equipment but with two men in it we put a temporary piggy-back deal in our P-38 and placed an RO in there with the pilot. With this very temporary installation of both radar and radar observer, our success was pushed up to close to 90 to 100 percent of the contacts made which we completed successfully. So I think that the Army is very definite in their idea on the aircraft being a two-seater. I don't know of too many points on which I disagree with either of the individuals. We do not have a definite answer yet on the question of the pilot's scope. Some individuals like it and some do not. I am a little prone not to say one way or the other because I feel that maybe after the pilot scope is in one individual may find it profitable and the next man not care for it, perhaps could not use it, and do his work in the best of his ability. Our field is the only field that is using the P-61, and we have not used them in our training until recently. In fact the P-61 is not in our RTU program except in the transitional stage which is 10 hours straight piloting work and no combat night-fighter team work is as yet worked out. We think we have enough airplanes now so that November 10 we expect to put the P-61 in our last month of training where the crew will get some 60 hours work in a ship before going overseas. There is some thought that we are not too pleased with the ship, as you well know. We think that it should go higher, and we think that it will not go quite fast enough. We think this will be cured soon with the R2800C engine with turbine. We would like a little more altitude because of the fact that, although we are having success in the Pacific where the ship has just gone into operation and to the best of my knowledge all good contacts that have been made have been followed with accuracy and success, we feel the Japs will soon go up again. We are in agreement with the thought of the manufacturers themselves that the instrument panel is too far from the pilot, that the visibility is not good because of the cluttered-up windshield and because of the fact that the pilot's face is too far from the windshield. We very definitely like the landing and slow-flying characteristics of the P-61. Naturally another very important factor at the present time in the P-61 is that of its range and that we think can be cured with droppable tanks. As to the radar equipment itself, I think that we have been a little more fortunate in our work with radar equipment, and most of the work has gone ahead with our 720. When Colonel

Harshberger came to our field some 3 or 4 months ago and talked of their work in the Pacific he was using 540 equipment which was completely obsolete to us. We were surprised that they had any victories at all. We feel that the equipment which came through with the new Spitfire that is here on the field now has even better capabilities.

"I think that is about all that I have. I think that you might ask for some comments, if you have not already planned to, from Singleton and his RO who were out there and have the experience."

Capt. J.C. DAVIS, AAF (AA/AS, OCR): "Gentlemen: I am from Headquarters in Washington. Requirements Division. We set up requirements for military night-fighter aircraft. I think you may be interested in the night-fighter aircraft and its military characteristics. Incidentally, this paper is dated 23-August and is the latest we have. I'll go through it very rapidly.

"The primary mission is the destruction at night of hostile aircraft in flight. The secondary mission is to attack hostile ground or water-borne installations at night by employment of gunfire and other destructive agents, and through the use of radar and other detection devices. In the performance, we have asked for high speed. We require 375 miles per hour at sea level and 475 miles per hour at 35,000 feet. That is the required figure. In the climb, the requirement is 11 minutes to reach 35,000 feet. The above requirements are to be met with full internal fuel, maximum armament, and other internal installations. War emergency powers may be used to reach the above requirements.

"Combat radius of action required is 150 miles. I am not going to go through the conditions of all that. I have it if anybody cares to see it and you will be perfectly welcome to look at it. For ferry range, we require 3,000 miles.

"Again, with the Marine colonel, we have requested that if, after all performance requirements have been met, additional performance can be realized, that the additional performance be used to increase the speed in the day fighter and increase the maximum combat radius of action in the night fighter.

"Our flying characteristics are more or less general. A high degree of maneuverability and controllability is mandatory. It is desired that night-fighter aircraft be as small as possible, consistent with the mission, and that attributes such as rate of roll, acceleration, "zoom" ability, vertical and horizontal loads, vision and simplification of control, and construction be given maximum consideration.

"On the guns we're going to disagree. We want a minimum of six free-firing machine guns of not less than .50-caliber installed in either the nose or wings, or divided between them. If .50-caliber guns are installed, we want space provision for future installations of larger caliber guns.

"For ammunition, we want that amount of ammunition which will give approximately 20 seconds of fire per gun.

"On the gun sight, we want a computing type.

"On bomb installations, we want provisions for at least three bomb suspension points (shackles normally provided for drop tanks may be used). Installation shall be such that bomb shackles and sway braces are recessed into the wings or under the surface of the fuselage. Shackles shall be so arranged that standard size bombs (from 100 to 1,000 pounds) or small frag bombclusters (up to 500 pounds) can be carried on any shackle.

"On rockets, we want provisions for at least eight rockets.

"Armor: The crew shall be protected from .50-caliber armor-piercing fire from the front at all angles not greater than 10 degrees to the longitudinal axis of the airplane, except as follows: With the airplane in a normal attitude of horizontal flight, windshield armor glass need only extend upward to a position where the average pilot's head will be protected from forward horizontal fire when sitting in the normal position against the back of the seat with the shoulder straps fastened. The crew shall be protected from .50-caliber armor-piercing fire from the rear at all angles not greater than 20 degrees to the longitudinal axis of the airplane.

"For fuel, all internal fuel, regardless of flash point, will be in self-sealing tanks. External fuel will be the total fuel required externally for maximum combat radius of action problems over that amount required to climb to altitude from point of take-off and cruise to initial arrival over target.

"On equipment, we have requested cabin pressurization. We want automatic pressurization provided and maintained at a safe differential of pressure.

"We put in a definite requirement for an automatic pilot in our night fighter. We are interested in this pilot that, I believe, is put out by Grumman and weighs about 25 pounds. Of course we want pilot relief.

"We want a pressure demand oxygen system. We have required that there be sufficient oxygen for 100 percent of the mission, which will consist of the time from point of taxi-out for take-off until return to the line after the mission is completed. That's assuming a nonpressurized operation.

"We want VHF equipment, having not less than eight channels.

"We want a gun camera.

"On the instruments and crew we see eye to eye with the Navy. We want the pilot and the radar observer and we want them to sit as closely together as possible.

"There is no point in going through the radar-detection unit. I have it if anybody wants to look at it. We don't disagree with anybody on that.

"We also agree with the Navy on the flight brake. A means of rapid deceleration of aircraft upon closure with the enemy will be provided. Particular emphasis will be placed on this feature in the case of jet propulsion.

"We have asked for escape hatches to be on top side.

"We want a taxi light, of course, and if it is a tricycle we would like to have it on the nose or the nose wheel.

"We have asked that consideration be given to the provision of a fire extinguisher for each engine of multi-engine aircraft. "Visibility: The forward downward visibility over the nose, when the pilot's eye is aligned with the gun sight, shall be not less than the following depressed angles from the line of sight:

9 degrees for aircraft with a maximum speed less than 500 miles per hour.

10 degrees for aircraft with a maximum speed of 500 to 550 miles per hour.
10.5 degrees for aircraft with a maximum speed of 550 to 600 milers per hour.
11 degrees for aircraft with a maximum speed in excess of 600 miles per hour.

"We have given over an entire paragraph in this report to visibility. We agree with everyone on that. It's here if you want to see it.

"We also require that the time required for an eight-man crew to remove an engine should not exceed 20 minutes and the time required for an eight-man crew to install an engine should not exceed 30 minutes.

"This aircraft will fulfill all fundamentals of a night fighter, including night-interception missions, low-level attack, bombing, and rocket missions.

"These military characteristics have been approved and they are the latest ones in effect now."

Commander RAMSEY: "Thank you very much, Captain Davis; It kind of scares me, taking off and putting on an engine that fast, but I guess it can be done."

Flight Lieutenant SINGLETON (British Air Commission): "Gentlemen: There is not a great deal that I can add to Colonel Harshberger's remarks. In the first place, what we have found is that for the performance in Europe where we are dealing with short distance, where we have only 20 minutes to complete an interception, we needed very good overtaking speed. We need at least 20 percent command of speed over the bomber. In the Pacific, conditions might be rather different where you can get longer runs. We found that a lot of the interceptions took place when the bomber was on its home run and we had to have speed to catch him. That is where we have always had our trouble. Second, the visibility of the pilot is most important, and I think this has been stressed before. The seat should be near enough so that he can get right up to the clear vision panel and have good visibility. The seats side by side is again, I think, most important. We have found that very useful in making identification. I believe the P-61 has binoculars now fitted so that the pilot can use them. But we have found that the pilots have been able to make the visual on a dark night at about 1,000 feet; just get a silhouette and the observer immediately comes off his scope and takes a good look to make the preliminary identification. It is essential that the observer should also have good visibility. Regarding instruments, it is most important that they do not reflect and throw light all over the windshield. Air brakes again are very useful. The Japs, I understand, are coming in fairly slowly. In Europe we have had to contend with very fast aircraft and we have on occasion been troubled by not having maneuverability at slow speeds in the case of a bomber which has suddenly climbed very steeply and very slowly. As regarding turrets, the turret is, I think, very useful if you can keep the command of speed over the bomber, but if you have to sacrifice speed for a turret, then your front guns are far more useful. Most of the interceptions and kills have been made, I should think about 90 percent, from dead astern, and front guns can be used a lot, if not more effectively than a turret. In the cases of the bomber that sees you and the dog fight at night and deflection shooting, then the turret comes into its own. I think an observer for the radar is most important, and regarding the pilot's scope, this has cut down and destroyed the pilot's night vision. The greatest use for a pilot scope is in following evasive action and the pilot can do far more effective work if he has no scope to watch, but at the same time he will not make his visual quite so soon. The other consideration regarding the aircraft, as Colonel Harshberger mentioned, is that it should be easy to enter for a quick take-off. In the case of a crew standing by it is essential to get into her as quickly as possible and you should have an aircraft you can check readily, have everything laid out so that all you must do is climb into the aircraft cockpit, put on your harness and get into the air. You don't want a lot of instruments that you have to check at the last moment. Flame dampening is most important and enough stress has been given to it here. The other idea regarding the guns is that they should be easily equipped for quick rearm. All my other points I think have been covered. Thank you."

Commander RAMSEY: "Thank you very much, Flight Lieutenant Singleton."

Lieutenant HAHLET (Radio Test NAS, Patuxent): "The limitation on the installation of radar in night fighters has been in the case of carrier-borne planes pretty important. The space required by the manufacturer building the set so large and the airplane manufacturer making the plane so small creates quite a problem. The maintenance crews are pressed and they are insufficiently trained to do the jobs capably or I would say perfectly in the present set-ups. If they had more accessibility and the installation were more stable, that is, the sets themselves were made up more stable, this would improve. As a comparison the 720 the Army and the British are now using, is a stable set. With X Band radar we lost a lot of the stability and the installation requires more service. Those points should be considered. Is there any question that I can answer on that, Commander?"

Commander RAMSEY: "Thank you. We have many items which are controversial as far as the contractor and the designer are concerned. Would someone like to talk from a manufacturer's viewpoint?"

Mr. MYERS: "I would like to say that my company is pretty much in agreement right down the line with the requirements that have been stated today. I wish we could build an airplane that has all the qualifications that Captain Davis wants, it would be quite an airplane. There are a few points which I thought would interest everybody in this discussion. In the first place the question of pilot-repeater scope. In the Pacific where I had the opportunity to spend some time a short time ago we found that, as Flight Lieutenant Singleton has pointed out and as we all know, the pilot scope does destroy night vision. Consequently we built a little door that could be put over it so that you can leave it turned on and yet cover up the light if you are not using it. A pocket handkerchief worked very well stuffed in the box. The advantage of using it following evasive maneuvering and also for navigation under conditions of poor visibility when you are trying to get back to your home base are great enough to justify having it. You don't have to use it under conditions when you don't want it and you can very easily build that door. We talked about two-seaters; we have not talked about three-seaters. In the Pacific the squadrons that were unfortunate enough to get P-61's without turrets, fortunately most of them had it there when I left, sent an extra man along to do nothing but sit in the seat behind the pilot and observe and make visual contacts. There was a great deal of discussion as to whether the radar observer should be up forward with the pilot and see what is going on. Tests were run on the thing and we found in most cases the radar observer is almost completely disqualified from making first visual contacts because his eyes are not dark-adapted. Also we found the radar observer should keep on his scope because if you get in close and the bogie makes an evasive maneuver and the R.O. has to go back to his scope at that point, he has wasted quite a bit of time. We should give thought to the idea that a man can be well used who has nothing to do but look preferably through a pair of night binoculars. Incidentally we found out there that the man in the gunner position of a P-61 makes something like 99 percent of the first visual contacts which, of course, is very worthwhile. The question of flexible guns that was brought up might be interesting. My information is most unofficial but I am getting letters daily from the men I was fortunate enough to work with out in the Pacific and approximately 50 percent of their kills have been made with a flexible shot – a shot off from dead astern. It may be that conditions were different from those encountered by Flight Lieutenant Singleton or maybe the radar observers are not as sharp. It is a fact that none of the boys had ever seen the 720 equipment before and they went into combat immediately with it. In the letters I get back about 50 percent describe a chase having a close-in and then almost invariably he suddenly found himself flying formation with the bogie off to his port side or off to his starboard side. He would tell his right gunner to take a shot or something like that and the right engine was set on fire. It is true, of course, that the turret is a hell of a bulky thing and heavy and probably it would not be as necessary if you had the fighter brakes as has been indicated we're going to on the P-61. But I think it has definite advantages. I know in certain cases we have had Zekes accompanied by Bettys and when the Zeke gets on the P-61 tail it is pretty nice to have something to turn back there. That is something to consider. Also we have had cases of an interception where for some reason the radar operator picked up a leading Betty and when they closed in for the kill they found another Betty on their tail so I think the flexible gun from the experience I had is a very useful thing. If it costs speed, then it is not satisfactory. Unfortunately on the P-61, the turret costs us 4 miles an hour. It does cost in rate of climb because it is heavy, and I expect the only reason it costs us 4 miles an hour is that there are so many bulges and humps on that thing that a few more don't make any difference. Incidentally, Colonel Harshberger underestimates the amount of time we spend on flame dampening. I would like him to come out and see two or three rooms stacked with dampeners that were not satisfactory. We know they are not good but we have not seen anyone yet who can get really adequate flame dampening on a short stack plane. On the turbo airplane we are using a heat exchanger which we hope will get rid of every bit of flame. In the P-61 we have an installation of night binoculars for the pilot. Much of my experience leads me to believe that that installation would be better placed in the gunner's position now that we know more about it. Right now we are putting a monocular in the gunner's position. Well, I think I have talked much too much. Thank you."

Flight Lieutenant SINGLETON: "I would just like to add a bit more to my previous remarks. I think Mr. Myers mentioned first of all the turret as opposed to the front gun. Well, I'd made one statement that 90 percent of the kills were from stern shots. That is, you have four cannon and if you hit a fellow with four guns he goes down. That is quite naval and I don't think the turret can have such heavy guns as you can have with front fixed guns. The other point was the night binoculars. Our great problem in Europe always has been identification. We have not had a satisfactory method of identifying friend from foe. There are different methods, but none of them is fool-proof. The only identification has always rested with the pilot. With a pilot scope you are able to get 1,000 to 1,500 feet on a dark night. I believe the single-seater fighters can get to about 300 to 400 feet. If you are going to use a pilot scope you are going to come across the same difficulty they have and if you have a rear turret gunner in the bomber you are not going to have a hope of getting into range. You must make the last approach as quickly as possible and get in and fire before he sees you. There is no good hanging along behind a bomber and there is no point doing a deflection shot when you can do a front-gun shot without deflection. My observer would like to mention the pilot scope, for navigation, some time during the conference. Thank you."

Commander RAMSEY: "Thank you, Flight Lieutenant Singleton."

Lieutenant Colonel YANCEY: "I think that in our situation in the night fighter training program, we have definitely taken a stand on the turret in the P-61. We want it. I personally think that has two angles. From the experience of individuals in the South Pacific, all of them have noted in their letters to us as a final word, "Please send us turrets." That is the first point. The second point is that as Johnny mentioned the boys that we have in the South Pacific are individuals untrained in the 720. If they had any 720 experience it was a mere smattering before they left and certainly no actual work. So maybe the turret situation is now in demand because of this inexperience and if the interceptions were properly made as Flight Lieutenant Singleton points out, it is quite true that your stern chase and your forward firing guns from the nose are much more effective. Probably if our boys get a little better, they will take the same stand. The third seating arrangement that was brought up has its definite advantages in, if not the turret, the second pair of eyes. As Flight Lieutenant Singleton told us, immediately upon their visual contact in actual combat, Hazlin the R.O. dropped his set and grabbed the night binoculars for two reasons, I believe: First to identify the plane in case of any evasive activity that he must go through if the enemy is doing evasive action. So very likely the third place in any P-61 or in any night fighter is definitely an aid in that the R.O. should stay on his scope at all times and the pilot can very definitely use a second pair of eyes."

Flight Lieutenant HAZLIN: "I think I am quite correct in saying that we all want the turret if we can have it without sacrificing speed or maneuverability, but with us in the European theater, I think Flight Lieutenant Singleton will agree, it has been a question of pilot indicators raised by Mr. Myers is one on which I think a little more detail should be given. If a pilot indicator is to be at all useful I think it must give a continuous and accurate presentation to the pilot. With 720 the method of presenting to the observer and to the pilot is such that you get an intermittent picture and the question of following evasive action I think with a picture, that is with its flashing on and off, is too much to expect of a pilot. So the pilot indicator, I think, is good but not if it is intermittent. It is more disturbing than advantageous even in the following of evasive action. The question of mapping can, I think, be quite easily handled by an observer. He can sit with a map on his knee and the picture laid out in front of him in greater detail than can be given the pilot on the pilot indicator. I cannot really see why there is any requirement for the pilot to read; if you are in bad weather or really worried about where you are it is better for the pilot to fly the plane and the observer to do his natural job of navigating. There is one requirement which nobody has seemed to touch on, for a night fighter, and it is a very definite requirement for a British night fighter and especially in its second aspect, and that is a tail warning device. I think that in the Pacific Mr. Myers has mentioned the question of aircraft being shot down and then the fighter finds somebody on his tail. That is very unpleasant and I think too that it would be covered by a tail device and for low down intruder work there is no radar set which gives you more than about 180 degree coverage. If you have a normal azimuth range presentation covering you backward 60 degrees either side of your tail then you should be enabled to sweep the sky pretty well as you fly low but not essentially to enable you to get out of the way of somebody else, but to enable you to get around to intercept the fellow, who is trying to intercept you. I think that is all I have to say."

Commander RAMSEY: "I think now is a good time for Commander Tyra to tell us the requirements of the Navy night fighter as opposed to Army requirements which Captain Davis gave us a few minutes ago."

Commander TYRA: "At the present time the Navy does not have any airplane which was laid down primarily as a night fighter, and I think that is one of the biggest handicaps in our present operation. We have been using adaptations of planes which were already in existence. The first of those was the F4U: some two dozen were built and sent out to the Pacific. They were single seaters with pilot scope and they turned in a good record for themselves considering the limitations of the aircraft and the fact that they were the old, low cabin Corsairs and the visibility was not too good even for day work. The second aircraft was the F6F which is now the standard Navy night fighter and is carrier-borne. The third one coming up right now is the F7F, and they look like they have the best possibilities of any plane we have at present. However, as you know, it was not designed as a night fighter to begin with. The second seat has been added and the radar observer put in but it is not comfortable or complete and the radar operator does not have the visibility he should have. I think as a result of this discussion here that maybe the Navy may lay down some actual requirements and try to get a real good night fighter."

Commander RAMSEY: "I would like to call on Commander Harmer to add to or refute previous remarks."

Commander HARMER: "We have covered pretty well the requirements of the night fighter for the Army and the Marine Corps, and I might add that the F7F although it is in the Navy program is not a Navy night fighter. As far as the program has been outlined it is a Marine Corps fighter, and the Navy is not going to get a look-in. We still have to develop a good two-seater that we can take aboard a carrier and I think we are going to need it pretty soon. So far we have not talked over any of the answers that might come up. I will agree with Lieutenant Singleton that the pilot scope does ruin the night adaptation and I will say that as far as picking up a bogie after coming over the scope that the maximum range on dark nights as far as I can judge is 150 feet and it is too close. I am in favor of having a second man that can be looking at the scope and let the pilot take advantage of that 1,000 to 1,500 feet that he claims he can make a visual at if he is night adapted."

Commander RAMSEY: "Thank you, Commander Harmer. I would like to call on Lieutenant Halaby who can give the flight test viewpoint."

Lieutenant HALABY: "I happened to have flown all of the night fighters here on the field and I think that most of the men who have spoken today have flown them. It seems to me in view of the continuing trouble of whether it should be a single-seater fighter or what kind of characteristics it should have it might be appropriate to ask in turn, Commander Harmer, Colonel Yancey, Colonel Harshberger, and Flight Lieutenant Singleton three questions. It is a bit presumptuous but I think they are not too difficult to answer. One, in terms of results what was the percentage of kills per contact in the theatre in which each operated or about which each knows? Second, what percentage of the time was the airplane and the radar used in that plane in commission? Finally, a question about what kind of stability and control they would like based on the planes on this field now. For example, would they rather have the extremely stiff directional stability of the P-61 or the rather loose directional stability of the Mosquito or perhaps the P-38 or F7F? Would it be appropriate to ask those questions, Commander Ramsey?"

Commander RAMSEY: "I think while we are giving the officers named a chance to think out their answers I would like to hear from Colonel Meng."

Lieutenant Colonel MENG: "I want to ask questions, and one was when you get into a twin-engine fighter, there are a lot of gadgets there. I am wondering about automatic controls on these instruments at night. I believe that when the P-61 came out it had about 8 different levers which you needed to pull and twist to control it. I was talking to someone, I don't know to whom, the first part of the week down here who said that sometime ago he was talking about developing little lights that went on when the temperature went overboard or the pressure the wrong way. I was wondering what these night-fighter pilots would think, whether they would know if everything was all right if they just saw the light. Also about automatic controls."

Commander RAMSEY: "I think that is a good question of the percentage of kills which we made you must divide up into two – what percentage of contacts did we get on approach, and what percentage of kills after contact? They run about 50 percent in both cases. In about 50 percent of the interceptions we got to the point where we could get the contact and make about 50 percent kills after contact. I must qualify in that the rate of kills after contact was about 25 percent due to pilot error and 75 percent due to radar failure after contact. On the second question, in carrier-based work I could almost say about 95 percent, because out of 6 months on a carrier we were sent off on actual interception nine times and spread over a period of 6 months that is not a great deal of activity. The rest of the time we spent flying around with those planes and testing them every night and we actually had only two complete radar failures after the plane was air-borne. When you think of the small number of flights that we made the two complete radar failures builds up the percentage quite far. And the third question is hard to answer. Being a single-seat man and a fighter, I got about all the maneuverability I wanted. I have been short of it. It seems to me I want the loose directional control because you must be able to kick that plane around. I would take the maneuverability and directional control. I am all in favor of any automatic controls you can have in an airplane, and the use of lights as long as they are red and dim is very good. They show up clearly in a blacked-out cockpit. The only warning light I have in mind is a radio altimeter light which will flash at 59 feet or whatever altitude you have it set for. It wakes you up and I am sure that system would work all right. If you can depend on automatic controls I am all for them."

Lieutenant Colonel YANCEY: "Gentlemen: The Army activity in the night fighter situation has been somewhat limited. We sent a number of squadrons out to the English theater and most of them were immediately enveloped and more or less lost to us and absorbed by the English situation. They were given Beaufighters and put in complete control of the English commands and were using Mark IV equipment. As to the first question on the percentage of kills per contact we can only base our opinions on the work of our squadron since about July, at which time the P-61 made their first appearance in both England in the 422d Squadron and throughout the Pacific. Since that time for instance the 419 squadron in the South Pacific on Guadalcanal, etc., have made 48 percent of their over-all contacts that were reported in their area which takes into consideration both the GCI and the actual AI contacts. Now of the AI contacts that were attempted and were found on their screen I think I would be safe in saying that these squadrons in the Pacific with 61's and 720 equipment have made almost 95 percent kills. Once they get the bogie in their screens they have been successful in knocking him down. I will say that the big difficulty in the Army situation has been ground control and also I think that from talking with Colonel Harshberger it is his opinion, I believe, that the success of the Marines with their 540 equipment could almost be laid to their excellent GCI control. I think the Army has been 100 percent in the other direction. Colonel Harshberger was working with one or two of the GCI stations in New Guinea and he told them he would rather not work with them until they improved because they would surely get him lost. So the percentage of kills of the contact with our present equipment in combat has been exceedingly high. We feel with the present equipment we have, this situation will not exist too long because of the fact that we cannot get the altitude we need and the Jap will soon learn this and go too high for the equipment. On the second question, the percentage of time that these aircraft and the radar were in commission, once again all of the equipment has been new but the big problem that we find in the night fighter situation is getting all of this equipment in our aircraft operating at the same time and effectively. And it is a fact that we have had trouble in our training program keeping the radar equipment in commission for our training purposes. With that fact in mind and the fact that we have the best maintenance that there is in our program and the only complete base set-up and without bombs being dropped we can only visualize that squadrons which are in the field for any length of time will definitely begin to have trouble with their radar maintenance. I think we all agree that the maneuvering and looseness of control is something you need to follow the evasive action once you make the interception. I do think that the aircraft should be such that it is an excellent instrument flying aircraft. As we all know night fighter work is such that 90 percent of it is on instruments. The automatic control ideas are very definitely needed. Flight Lieutenant Singleton's statements are true about your quick take-offs and easy controls and the fact that you do not want many controls to turn on when you take off as well as the fact that you are on instrument work quite a large percentage of the time while actually making the contact and the pilot is doing two things. He is definitely flying the aircraft carefully and at the same time being controlled by another man's directions so I do think that the automatic controls are very definitely necessary."

Flight Lieutenant SINGLETON: "In answer to the first question the percentage of kills to contacts with 720 gear which we have been using, the kills have been 90 to 95 percent. And that has been ever since we started using it – at the beginning of the year. Before that we were using Mark IV, that's your 540 and the Mark VIII and, of course, the percentage was lower, depending upon the two radar equipments used. The Mark VIII kills were in the region of 8 to 5, and the Mark IV, much lower, of course. The second question – the percentage of time which the radar and aircraft were in commission. The radar with 720 is in commission about 97 percent of the time and the Mark VIII a little higher – about 98 percent. That is now, this year, when they have Mark VIII really on the top line and 720 has got over its teething troubles. As to whether the instruments should be automatically recording – definitely yes. In fact, if any of you have seen the Mosquito will see that we have cut out the fuel pressure gage and substituted fuel pressure warning lights and there is a question of cutting out further instruments – oil pressure gages, etc., to simplify the cockpit. Stability of controls – we want an aircraft that is stable, that has weak positive stability at all speeds and also one other thing, an aircraft which can be flown with one hand, preferably a stick – not a wheel unless you have an aircraft with a wheel that you can fly when the bomber is taking evasive action, with one hand just as easily. There is also the question of inter-com, which we have in our aircraft and I think would be generally accepted, maybe somebody will correct me if I am wrong, but you want an inter-com system with which you can speak to your RO and he can speak to you just as you do in ordinary conversations without having any buttons to press."

Lieutenant Colonel HARSHBERGER: "I want to briefly touch on a few things that I am still a little bit lost about. I notice that the Army has preference for a lead computing sight, I believe you call it. I think Flight Lieutenant Singleton is probably the outstanding night fighter present and I wish you would call on him about the lead computing sight. If it requires looking through an extra piece of glass or something, no. If it automatically computes the lead, I am all for it. On the inter-com, I believe Flight Lieutenant Singleton is quite right. Some of our boys failed to find the enemy even with an AI contact because the pilot was listening to the GCI and the radar operator to something else and they couldn't get together. In addition, that inter-com should be like a telephone. When one man is talking, the other man should be able to talk back. The fact that you don't have to push a button will do it. In the F7F we have to push a button, but the radar operator can now push the button and talk at the same time we are talking and we will hear him. In regard to tail warning, we are considering it and especially on intruder missions. It is expensive in size, weight, sideways draw drag, so it is not considered on planes solely for GCI, but on intruder missions I would want it. On the questions, I am going to have to go begging on those. The percentage of kills per contact. You can't compare it unless you consider the radars. Our radar was such that a plane could pass close aboard behind us and we could see it and it was a contact. The radar is such that the plane can pass immediately underneath and they couldn't see it and whenever the plane was ahead of them they had a better chance to get it than you would have if the plane were behind you. Our percentage of kills per contact were in the neighborhood of 10 percent. Personally, I have seen four planes I did not get near due to performance of the PV-1. They were too high above me. Yet we had AI contacts on them. My radar gear would work two or three nights and then it would go out. I would say 60 to 75 percent for our gear being in commission. We made extensive tests, too, while we were on operational patrols and I might put that figure at about 85 percent.

"On stability, I feel like Flight Lieutenant Singleton that I want it to be stable so I can fly instruments and if I let go the stick, it does not swing off and go into a spiral. I want it controllable with one hand and with not too much force. You are looking around and at the same time you are flying the plane and checking instruments and you don't want to have to beef it around. It will hurt in making tight turns if you have to pull hard. Also, the addition of an automatic pilot, which I think we all feel is essential, if it can be put in does mean that to and from the area on intruder missions or home again after a combat patrol, we can switch to automatic pilot. I am all for automatic controls. If you mean lights versus automatic features, no. I want lights to warn when the automatic features go out, but I am all for the automatic features."

Captain DAVIS: "The question was brought up a while ago about the American night fighter units in the Mediterranean. I have definite figures if you are interested. These figures I have here are for only 3 of the squadrons and for a 1 year period ending September 15, this year. One squadron has performed 84-day-and-dusk patrols over enemy waters; 89 intruder missions; 1260 convoy and harbor duties; 88 scrambles after bandits and bogies; 32 air-sea rescues; for a total of 4,442 missions and total hours at 9,948. That is total for 3 squadrons. The ships they have run into are the Ju 87, Ju 88, Ju 188, He 177, He 111, Do 217, Me 210, Me 410, Me 110, and SM-79 – a torpedo bomber.

"The claims for the 3 squadrons, and I again want to point out that these may be incomplete, at least 18 definitely shot down, 6 probables, and 12 damaged.

"They haven't met enough enemy aircraft on intruder missions to even talk about. The equipment they are using is the Mark VIII, AI, 10 centimeter type. It is very unsatisfactory. They have much difficulty maintaining it. There is a limited supply. They class the range as bad. They use also a SCR-695-A I. F. F., and an AN/APN-1 – a radio altimeter.

"I would like to say one more thing, that is on this I. F. F. I didn't cover the military characteristics for night fighters. It is more or less listed in general under instruments. We have approved the minimum instruments for fighter aircraft to be installed, plus any additional operational instruments needed. That is a broad statement. I don't know how the Navy works. In the Army we don't sit down here and decide what the theater commanders will get. We leave the question up to the theater commanders; if he wants I. F. F. he gets it, and the rest don't have to take it if they don't want it. I believe Flight Lieutenant Singleton has said that an identification device is necessary. We agree absolutely. And there is no doubt it will be in our night fighter aircraft. One of our boys in the Mediterranean came up in the night and picked on a B-25's tail gunner. He made two passes and didn't shoot him down and he came back for the third and then the tail gunner got tired of it and clipped him. There have been numerous cases of just such things."

Commander RAMSEY: "Flight Lieutenant Singleton, can you tell us something about the lead computing sight?"

Flight Lieutenant SINGLETON: "I will have to apologize because I am afraid I have no first-hand knowledge of the sight. I only have second-hand knowledge somewhat as I have picked up for myself. I feel that it would be ideal so long as it doesn't interfere with the clear vision of the pilot. The sight has got to be small. If you can get a gyro gun sight that works satisfactorily, O. K., so long as you don't interfere with front vision there. We have been trying it, and you have it in your F7F; the gun sight projected right on to the front wind screen. We had difficulty with our screens not being at the right angle to start with. We have corrected it now. Shooting from a gun sight directly projected on to the windscreen is far easier than if you have two pieces of glass to look through. I am sorry I can't answer the question fully."

Commander RAMSEY: "Gentlemen, I think we have had enough meat to think over before Sunday and Monday when this subject will again come up for discussion. If any person cares to take a few minutes more to discuss it, we'll hold the status quo. Otherwise, we will call the meeting adjourned so we can get off the Washington plane and the rest of us can get ready for the banquet.

"The first hop for tomorrow is posted on the bulletin board. Please try to get off at 0815. If you will all be so kind as to check the schedule before leaving tonight and if you cannot make that 0815 schedule, please let Lieutenant Commander Milner know."

The meeting is adjourned.

FIFTH MEETING
19 October 1944
1600-1745

Commander RAMSEY: "In order to get things rolling, the meeting is called to order. We will probably have a lot of people who must depart prior to Monday afternoon at 1800. If any choose to leave before that time, which we hope they will not, please check out with Miss Ryan. Also, we would like to have the confidential questionnaire filled out and I assure you that your names will not be taken in vain, but we merely want to get some data for our record of the conference. Please fill out the questionnaire and leave it with Miss Ryan before departing. If you have any hesitancy in putting down some remarks which are called for in the questionnaire, it can be mailed back to Fighter Conference Headquarters, or to Lieutenant Walter Wilds in Op 16-V. Navy Department, Washington.

"There will be no night flying for tonight due to probable inclement weather. The P-63 is now available for slow rolls. I just got word from Wright Field.

"Gentlemen, it is my pleasure to present Capt. S.B. Spangler, Power Plant Section of the Bureau of Aeronautics."

Captain SPANGLER, Aircraft Power Plants: "The field of power plants is a little bit too big to be covered in a discussion like this so I will restrict my remarks to what might be termed a military engine for fighters. Even then there are too many experts in this room for me to commit myself too far, so I will just give you some of our ideas and let it go at that. Obviously the pilot in a fighter aircraft would like to have the highest performance airplane. He is not specifically interested in what the engine will produce, provided he gets the maximum performance. Well, that covers a multitude of sins, but in general it defines the engine that you'd like as having the highest power at the lowest weight that you can get, and that means the highest useful power to the airplane so that it also includes the propeller and intercoolers, oil coolers, etc., which are some of the things that must be considered. If we define the lighter engine as the highest power at the lowest weight, obviously the engine that best fits that category is the jet propulsion engine.

"Now I am not prepared myself to say that the jet propulsion engine is the final power plant for fighter aircraft. I don't believe that it will be necessary. There are a few catches which must be taken into consideration. There is no question about the fact that the jet engine is probably the simplest, easiest to build, cheapest and perhaps easiest to maintain. All that is subject to some proof. However, there are other things that must be looked for in a fighter airplane, and I am not quite so certain that high speed as such is exactly what you want. There is no question about the fact that you will get the highest speed at the moment with jet propelled aircraft and one advantage of a jet propulsion engine which you cannot eliminate, of course, is the fact that a jet propelled aircraft increases its efficiency with speed and you can't sell short any device for aircraft which does that. However, there are some catches to the thing. One, and I believe it is the principal one, is the fact that there is high fuel consumption. Not that high fuel consumption is almost necessary when you are talking of a purely military engine, and for several reasons. In the first place you are trying to get a lot of power out of the engine. You don't specifically care whether the engine lasts a 1,000 hours or not, and while you definitely want reliability, it need not necessarily follow that you want durability. How to decide between those two in the average case is beyond my knowledge. However, I think it is fair enough to say that that is normally the case. Now that isn't strictly true with naval aircraft because we, of course, of necessity as you can well realize, need a definite time between overhauls on our engines because when we go to sea on a carrier, every aircraft lost for any other reason than enemy action is just as surely lost. And in general a purely military engine leads to an engine which has fairly low life. Now being operated at high power in the general case, your fuel consumption is high. In the jet propelled engine we do not at present know any way of improving the fuel consumption to a point where, if you were just interested in going from here to there, it can compete with a normal engine. Obviously, if you must go from here to there at 400 miles per hour, the jet propulsion airplane has an advantage. However, if you just want to go from here to there, and you don't care how fast you go, you must consider that the jet propulsion engine doesn't quite fit that picture.

"Now for certain types of fighter use, particularly interceptor fighters, the jet propulsion engine is highly useful because you don't have to go from here to there – all you do is go up and come down. Another thing as applied to the normal engine and when you start considering the purely military engine is that, due to the fact that you want bursts of power, call it what you will, war emergency power, combat power, in general, the engine weight based on its normal power is likely to increase and increase somewhat beyond reason if you like, when considering the same engine as applicable to long life and good cruising range.

"Another thing that enters into this same picture which is more pronounced probably in the air-cooled engine than in the liquid-cooled engine has what you might call a reserve of cooling power, is the fact that your cooling drag goes up. That is necessary for several reasons. The principle of this is that to obtain these high powers you must have more supercharging and the temperatures go up with the increase in pressure so that you must have inter-cooling or after-cooling of some kind. In an air-cooled engine this is a little difficult to take care of because the temperature rise is, I won't say instantaneous, but somewhat near that and therefore you need your increased cooling immediately as you start using excess power. In a liquid-cooled engine you do have some reserve cooling so that you can take a short burst of power, provided it is a reasonably short burst of power, without having the increase of the total cooling drag.

"Another thing that enters into this particular picture, which few people consider, in fact the pilots are likely to overlook it completely, is that increasing the manifold pressure of an engine does not of necessity, in the air, give you any improvement of performance. In fact, I can quote several examples where the exact opposite is true. That is due to several things, one of which is that the propeller, normally being selected for other purposes than high speed, tends to lower efficiency so rapidly that the increased manifold pressure in effect only increase the loss in the propeller and you don't gain very much. In general the going to higher powers, even for short periods, has a very marked effect on the operation of the carburetor for a couple of reasons and in general it takes a special treatment in each case to get a solution. The general control problems of engines for high power operation are fairly well known. I think we have reasonable assurance that we can operate as we wish at the so-called higher powers. There are a number of ways of going about increasing the specific output of an engine and particularly in military engines. If you look over the field, the engines which might be considered primarily military engines normally have what is considered a reasonably low compression ratio. The reason for that again is over-all temperature rise. To quote a specific example, a given engine when operated at a compression ratio of about 6.7 took 77" manifold pressure to deliver a given brake horsepower. By reducing the compression ratio to 6 the engine at 88" delivered the same power. However, on both of these figures, the 77" figure was a detonation limited, and thereby actually get somewhat of an increase in power. Another way of improving the specific output of an engine, of course, is to increase the supercharge. That has its defects and drawbacks. If you are able to keep the mixture temperature down to a reasonable number you can get quite a bit of power out of most of our engines. In fact one of the arguments presented in favor of the turbo-supercharger is that you can considerably overboost the engine for short periods. That may or may not be an advantage. It's a little risky to put in the hands of the average pilot, in my opinion, the possibility of considerably overboosting engines. I recall the old saying we used to have, "Put your faith in the Lord and keep your powder dry." It is not longer quite so true. Some of the pilots put their faith in the Lord all right, but they depend on somebody else to keep their powder dry for them. As a consequence we must put in all kinds of safety devices to prevent just that kind of abuse of the engine. High supercharging however can be taken care of providing it is a value to a particular engine by using intercooling, which of course is what you can do on the two stage and turbo-supercharged engine and also in some cases, more effectively although quite hard to apply to an air-cooled engine, after cooling. After-cooling has some defects, some theoretical, some practical. In most cases where it has actually been used, however, considerable increases in power have been possible; because in general with present day fuels the engine is limited more by temperature than it is by pressure. There are methods of obtaining better horsepowers at part throttle. For example on a highly supercharged, mechanically supercharged engine, it is better to vary the inlet to the supercharger to obtain considerably more power at sea level. That is a rather complicated process and a little difficult to apply to an engine, but it has been applied successfully.

"Another method of obtaining increased output from the engine is by the use of internal cooling. You are familiar I am sure with water injection as a method of improving power output. There has been a large amount of work done on water injection by both the engine companies and also by the National Advisory Committee for Aeronautics. I think it is possible that we know a considerable amount now about what are the possibilities of water injection. As you are all aware the first service use on water injection in the Navy was on the 2800 engine, the Pratt-Whitney 2800 engine in F4U airplanes, that at the moment we have not found the actual physical limit of how much internal cooling can be done by water injection. However, the limit is probably one of the engine itself, of its carburetor and mixture temperatures and fuel to air ratio, that the engine will function under high ram conditions. One of the interesting experiments carried out by the National Advisory Committee for Aeronautics has been to inject water directly into the cylinder in the last part of the charge. That is at the moment hardly a practical method and undoubtedly would lead to a complication which may not be justifiable. Another way of improving the specific output of a given engine at particularly the attitudes above the critical is by means of boosting the amount of air going into the engine or the amount of oxygen by other means. The Germans have made use of a variety of methods in doing this. We ourselves in the Navy have done a considerable amount of experimentation with liquid oxygen boost. The tests so far have been conducted, that is the significant test, in F4U airplanes. The results are somewhat disappointing, although here we seem to have an emergency method for obtaining high-altitude performance at a quite reasonable cost, provided of course we can depend on the pilot to do what he is expected to do and we don't have a few accidents like blowing off the rear end of the engine and burning up superchargers and such, which we have had. But as an example of what can be obtained from this kind of procedure I might quote some actual figures. I would like to point out that these figures should be taken as qualitative and not quantitative. With an F4U-4 at 28,000 feet with a 2800 B-engine we have been able to increase the horsepower from 1,112 to 1,521 horsepower at the same revolutions per minute, which increased the speed of the airplane somewhere around 40 miles an hour at that altitude. That is a typical figure. In high blower we were able to increase the speed at that same amount – about 41 miles per hour. The time to climb from 28,000 to 36,000 feet was more than cut in half. That is from 24 minutes to 10 minutes. However, the catch to this particular method of doing these things lies in the cylinder head temperatures which were obtained under these conditions. The cylinder head temperatures went up from 200 degrees at no small speed to 255 degrees with liquid oxygen boost. I am afraid that this is more or less of a general theory. Incidentally we used water injection with this particular series of tests and we had an even worse condition than this before we started using water. That, as I said, is something that we now consider an emergency method of using boosted altitude provided we find out that we need it at some time. We're afraid that putting this thing into average service use is likely to lead to a considerable number of engine failures. A method of improving the power output throughout the whole range is of course to get continuously variable turbo supercharger drives. Now I have already pointed out that turbo supercharger has the advantage of overboosting. It also has another advantage which is that running the supercharger at the pressure ratio only that is required for the altitude at which you are operating, you of course would not get the temperature rise that goes with operating at the higher pressure ratio that is required with straight mechanical. I do not intend to get into an argument here as to whether turbo or mechanical drive is better. That would be a subject in itself and could take up a couple of hours of discussion. It does not have that particular advantage and to obtain the same general filling out of the power curve with mechanical drive would call for three or four or five speed drives. Now we have actually developed such a drive but at the moment we feel that the weight is not worth the effect. Another method of obtaining it with what might be termed a mechanical drive is to use a hydraulic drive somewhat similar to that with which you are all familiar in the Chrysler automobile and also in the late Allison engine. The only catch there and one which must be fairly proven before we can go into the hydraulic drive is that the heat rejection to oil is of course very much increased due to the fact that when operating the supercharger at reduced pressure ratio it is necessary to increase the slip in the coupling. If you are operating at 50 percent slip you will put into the oil 50 percent of the supercharger power which is quite tough. That translated back again to what I started saying is that you do not necessarily want high indicated power that you can read in a torque meter or deduce from manifold revolutions per minute what you are really after is of course airplane performance and if increased heat rejection to the oil increases the oil cooler or the drag out of proportion you haven't gained anything and you might as well go back and take the simplest system.

"Starting at the beginning of this war the Navy really had no military engines. We had from time to time considered the possibility and had from time to time contracts for military engines. Fortunately or unfortunately, the Navy's engines at the beginning of the war were designed largely for commercial use. In other words, they were designed primarily for long life, for low specific fuel consumption that is in order to get good endurance, and good range. We didn't have what you could really call a military engine. We in the Navy still do not have what we could really call a military engine. Perhaps one of the reasons for that particular lack at the moment is the fact that we still feel that emphasis must be put on durability and reliability and upon the range of the airplane as well as the ability to deliver high powers at low weights. So far we have yet to build an airplane that I know anything about that has enough range. Always when the airplane gets out to the fleet it is very much short. Of course, I think it is obvious to most of you that if we built a 1,500 mile combat radius airplane when it still wouldn't have enough range as you are always trying to push it a little. However, notwithstanding the fact the increase in power in engines in this war has been not inconsiderable. Let me quote from specific examples which are not so startling in the air-cooled engines which have gone up on the order of 10 percent, and as high as 20 to 25 percent in others. That is safely usable horsepower which the Navy in guaranteeing at the moment. Some have been going as high as 50 percent. However as far as most are concerned they are still figures on paper. The liquid-cooled engines on the other hand have gone up considerably more than that. The Rolls-Royce was about a 950 horsepower engine at sea level. As you are aware it is almost double that now.

"In all consideration of fighter engines of course we come back to the one thing that is in all people's eyes at the moment and that is the jet propulsion engine. The number of projects carried on now in jet propulsion engines is probably 50 percent greater than that carried on in normal engines. By a jet propulsion engine of course I mean one that delivers all of its thrust in terms of jet. The gas turbine is usable as a normal engine with a propeller as well as making use of what jet thrust you have in the rear. In that case it then becomes not greatly different from the normal reciprocating engine. You take a certain amount of power out through the propellers and a certain amount out through the jet. In all cases, particularly where you don't have turbo superchargers of the GE type, you have the ability to recover thrust from the reciprocating engine. Incidentally, still going back to the airplane performance part of this don't let yourselves get fooled in the general case by looking at an engine power curve because an engine power curve only gives you brake horsepower. It doesn't give you what is usable in the airplane, which is usable thrust horsepower. They don't bear any resemblance in the general case as there are too many things thrown in. As you probably know the Navy undertook the development several years ago of a turbo supercharger which was designed to make use of what jet thrust you had left. That supercharger is now in production and will be put into a certain number of the F4U airplanes. To get back to the original subject the jet propulsion engines in general, as I said, would appear to have every advantage for certain types of use. Perhaps the field is narrow, or perhaps it is broad. We have not reached any conclusions yet as to what is possible or what can be done. In general I think we can say that the jet propulsion engine is applicable to interceptor fighter. Whether or not it will be acceptable to others is still in the laps of the gods. the jet engine at the moment seems to have very little limitation except in terms of materials and temperature which are always in our hair. There are possibilities of going further with weight and power for weight ratios but there seems little hope of getting better fuel consumptions at the moment, although we are trying hard enough. In considering the general application of jet propulsion engines to naval aircraft it was our conclusion a couple of years ago which still may have some promise that in order to obtain the things that we need in a Navy aircraft, that is good take off and plenty of power in the wave off condition and also good range we decided we would start on a program of a conventional engine in the nose and a jet engine in the tail. This may appear as if we were sticking our toes into water to see if the water was cold before we stepped in. But I assure you it was a well thought out move when we did it and I think the FR is a reasonable indication of the possibilities of that combination. We started with a more or less 50-50 combination and it was our hope to go to what we have calculated for our purposes as a better theoretical application which is about 30 percent power in the nose and 70 percent in the tail. Our first application of jet propulsion as a matter of fact was an attempt to be able to apply to service aircraft if and when we thought it necessary, the jet propulsion engine. As a consequence, the Navy developments in this line followed the theory that we would have small diameter which meant full compressors and would have a system whereby the jet propulsion engine would have enough gadgets to run itself but not run the airplane. However, it did indicate that if properly done we could actually send jet engines to the fleet and mount them on service aircraft to get improved performance.

"I think that is about all I had on my mind at this time. If there are any questions I would certainly be happy to answer them."

Air Commander BUCKLE: "May I ask Captain Spangler, in relation to this question of getting more power out of the engine, what the Navy's policy is regarding the use of special fuel, special anti-knock fuel for that purpose?"

Captain SPANGLER: "Our general conclusion on the use of special fuels for improving power operation in engines I believe is the same as anyone else's. We originally intend, when fuel availability rendered it such as to make it possible, to use a new fuel which would have 150 so-called rich mixture responsive rating. However, we felt that to take full advantage of that fuel we should have a lean mixture response rating of 120. We finally compromised with AAF on grade 115, and 145 fuel which will shortly be available within the limitations of production. Thank you, gentlemen."

Commander RAMSEY: "Thank you Captain Spangler. I would like to hear from some of the aircraft engine manufacturers – will one of the Pratt and Whitney representatives discuss his viewpoint on what Captain Spanger has said relative to the use of more power and different types of engines in fighter aircraft?"

Mr. Luke HOBBS (United Aircraft): "I don't quite agree with Captain Spangler on one of his statements that he had nothing but commercial engines when the war started. I think, however, that he did explain that when he said he knew no way of differentiating between reliability and durability, and we certainly do not either. That is how he happened to get the engines he did. The only other comment is that we have in the laboratory at last found out the limit of water injection. In other words, the internal cooling with water we have now gotten as high as it will go, and we are looking for something else to get more power. Better fuel will allow us to go higher and we have tried it and it will – (methanol or wood alcohol is much better than straight water) – allow you to go much higher. There are still two or three items to check and there is still a goal which we are after."

Commander RAMSEY: "I would like to ask Mr. Hobbs before I call on anyone else, if anything is in the offing at the present time to get greater power out of the present power plant."

Mr. HOBBS: "The answer to that is that outside of the water and alcohol and the various combinations there is nothing that we know of except more efficient supercharging. We are getting better and better on that, which means for the same compression ratio and supercharging, the same pressure in the manifold will lower manifold temperature. Actually this is effective and more so than any other one single factor, and with these small gains I think we still have quite a bit to go. We have the 2800 C. at 3,000 horsepower with a simulated turbo set up, that is with back pressure on the engine and fairly high carburetor air temperature around 130 degrees so we are not stopped yet, but outside of the small gains to be added to what we have, there is no one great big thing we are going to get that we know of to give us a big jump."

Commander RAMSEY: "Thank you Mr. Hobbs. I would like to call on a Curtiss-Wright representative to discuss Captain Spangler's remarks or Mr. Hobb's remarks."

Mr. Allen CHILTON (Curtiss-Wright): "We would like to ask Captain Spangler whether the Navy is going to consider the use of water injection to improve the economy for long range operation – if they have any thought of that in the near future."

Captain SPANGLER: "I think my answer must be that it appears that it is a desirable way of improving fuel economy. However, we have not been able to figure a way of improving the weight of the airplane. Our studies indicate a higher liquid weight usage by using water for cruising than by not using water. That of course means that the airplane has less range. However, we are not throwing away the possibilities of water recovery, although as Mr. Chilton is well aware the Navy has had the difficulties of water recovery for the last 25 years so far without any solution which could be applied to Naval aircraft."

Commander RAMSEY: "I would like to call on the Allison representative if one is present."

Mr. D. GERDAN (Allison): "There is another possibility that Captain Spangler did not touch on for obtaining rather greatly increased powers from present day engines and that is the possibility of gearing a turbine directly to the conventional engine and using the turbine to feed power back into the shaft. However, care does have to be taken in studying a combination of that type because of the fact that a certain amount of the jet thrust normally obtainable from the exhaust is lost and unless some means is used for obtaining some jet from the turbine itself, there may actually be net loss in airplane propulsive power."

Captain SPANGLER: "I would like to add just one thing to that analysis. Again getting back to the airplane performance idea. There is a very distinct possibility that gearing turbine power into the engine will greatly improve the fuel consumption and by greatly I mean exactly that. Improvements, specifically on the order of 15 to 20 percent, have been obtained, and even more than that appears logical. That in itself, even though of course you get no more power out of the engine, gives you better airplane performance because it means either a smaller airplane or less weight in the given airplane."

Mr. D. GERDAN (Allison): "With regard to obtaining improved economy through a means like that I think it is going to take some way of varying the turbine nozzle area; otherwise the maximum power will not be obtainable, if economy can be obtained. It will be necessary with very high power in that case to bypass the exhaust gases in order to avoid very high back pressure."

Commander RAMSEY: "Commodore Buckle, is there a representative here from the British Air Commission who would like to speak on the British endeavor to get more power out of their engines. Is there anything else the British are doing along the line of liquid oxygen?"

Commodore BUCKLE: "Well, I'm no engine expert but I can give you some general information on what our thoughts are. Although we rather felt that water injection was a clumsy way of getting more power out of an engine, we have concentrated on better supercharging and better after cooling to get the same results. The next move we made was toward higher grade anti-knock fuel. But we are now taking rather more interest in water injection and a certain amount of work on that is going on in England at the moment. I am not aware of any experimentation on the lines suggested by Mr. Gerdan at the present time. There may be something going on but I am not aware of it. So far as liquid oxygen is concerned or similar systems we have done some experimental work. The aircraft used was a Spitfire in the first instance. The results were not very encouraging. We were subsequently forced into the use of nitrous oxide on the night fighter, the Mosquito, in order to get the initial 410 nuisance raiders which came in at a considerable altitude and flew very fast and did not give as much time to intercept them. We managed to get an extra 24 miles an hour on the Mosquito by the use of nitrous oxide for short periods. That, of course, as Captain Spanger already said, is an emergency method. It is a messy thing and a nuisance and one does not want to go to it. We are conducting some more work on liquid oxygen. It is in the very early stages as yet, and there are no results to be had. In general terms, we do not like additional gadgets of this nature which only function for a few moments at a time and only add to the weight of the airplane. And of course the maintenance and servicing problems are made very much more acute. I think that's a fairly general picture of what is going on now. If you have any questions I will try to answer them, but I cannot guarantee to have the information."

Commander RAMSEY: "Commodore, what is the British point of view regarding the FR engine – the regular engine in the front and the jet in the rear? Have the British gone ahead on that?"

Commodore BUCKLE: "Well, we thought of that in the early days of the jet engine but we turned it down and one design which was started was stopped. We rather feel that it is a half-way sort of thing which isn't going to get you very far. If you are going to make a high performance airplane, which you've got to have if you are going to use jet propulsion (for normally when an airplane is traveling at a fast speed suitable for good jet operation, the reciprocating engine and propeller operation is apt to lose out on efficiency)-I think that if you are going to use jet engines, you've got to have a high performance airplane and it's got to fly fast. If you are going to need low cruising speed, a jet won't give you economy and you won't get the best performance out of the engine at all. We are working on the jet engine and hoping for some sort of an approach to reasonable economy."

Commander RAMSEY: "I would like to hear from the Army Air Force representative relative to the use of the air-cooled engine over the liquid-cooled engine."

Lieutenant Colonel TOUBMAN (Army Air Force): "All of this discussion going on so far is all well and good. I am sure as far as I am concerned, and most of the manufacturers are concerned, there are certain concrete facts I would like to have straightened out now. Since this conference is based on the war in the Pacific, we should be interested in critical altitudes. As far as I am concerned the B-29's are going to do the bulk of the destruction over Japan and China and we have only one fighter with large enough increase in speed over the B-29 to do that work in the Navy airplanes. In my estimation, if they had a large enough increase in speed they could very easily do some escort work in connection with the Army. However, right now they don't have anything and as far as I can see they won't have anything in the near future to do that. The war is going back up in my opinion – back up to 30,000 and higher. I don't think that this present sprint the Navy is putting on will continue very far. The Japs won't sit back and take it. We know that they have good fighters and they are not going to take it. A point I would like to get straightened out is about propeller-throttle combinations."

Captain SPANGLER: "Well, I might remark in reply to part of that comment that it is very true we have a habit of following the Army in some respects. However, it is not quite a fair statement to say that the Navy does not have a high powered, high altitude airplane. As mentioned before, the F4U certainly should be considered in that category because its engine will deliver rated power at 46,000 feet, which I consider to be well up in the air."

Mr. Fred BRYCHTA (Curtiss-Wright): "In the high speed field, we have tried to find out what the limit of propeller efficiency is. We haven't too much information yet, but perhaps it would be interesting to note that at least one airplane that we know of is flying with a propeller tip speed with a Mach number of 1.15 or 15 percent above the speed of sound. That was obtained at 35,000 feet and if it is safe to expedite that trend, we say that at the same altitude you can operate at the same type Mach numbers without running into excessive efficiency losses. That would be equivalent to a level flight speed at 10,000 feet about 560 mph. So then, we feel that the prop hasn't reached the limit of its possibilities as far as high speed is concerned. We don't know what is going to happen in the 600 mph bracket, but we certainly have something to do in that regard."

Mr. Carl BAKER (Hamilton Standard): "One of the trends that increases in power plant rating is developing into as far as the propeller is concerned, which some years ago, was considered quite doubtful as a satisfactory means of absorbing added power, is the trend to the very wide planform blades. This has developed as much possibly from necessity as it has from a research angle. Many of the airplanes in production have had increased power ratings applied to them which have in turn required increased ability on the part of the propeller to absorb that power. Because of structural limitations it has in many cases, in most cases, been impossible to increase diameter to get increased propeller area and therefore the means of widening out the propeller blade has been adopted. It has at the same time developed fairly generally in investigation work that, contrary to earlier thinking, propeller efficiency as such has not suffered severely or suffered materially from this trend in propeller blade design. For that reason the confirmation of the propeller performance is coming in many of these airplanes that were forced to adopt such a propeller. Some of the newer airplanes power plant or propeller requirements are being met by the employment of a much wider blade than was heretofore thought would give maximum efficiency. I think that it has been well established at the present time that this means is a satisfactory one of providing for increased power absorption while retaining certainly satisfactory and even excellent propeller efficiency."

Commander RAMSEY: "Thank you very much, Mr. Baker. I would like to hear from an Aero Products representative."

Mr. NEWDECK (Aero Products): "I would like to raise the question here of the opinion of the propeller manufacturers, Curtiss and Ham Standard, on the effect of the counter-rotating propeller to absorb this higher power output in modern engines and would also like to hear from the British. I understand that they were the first to try it."

Commodore BUCKLE: "We felt, or we know as a matter of fact, that as the engine powers have gone up and we have tried, as in the case of our Spitfire, to keep the size of the airplane down to what it was before, we have been faced with a considerable torque problem, a gyroscopic problem, diameter of props, and tip speeds going up. All these things have rather led us to feel that if we could reduce the diameters, cut down torque, as you can do with counter-rotating props, maybe additional weight and complication would pay off. So that is the direction in which most of our energies have been directed and we're achieving some success with it. The Spitfire 21 is going to have a counter-rotating propeller as standard. The take off qualities of that airplane have been considerably improved by the use of that prop. I don't think we have gone far enough to say we are absorbing all the power we want to absorb and getting the performance that we wish to get. I don't think tests to date are conclusive, but they do show promise. On the other side of the story, we are taking more interest in the wide blade, which until 9 or 10 months ago we had regarded as not much of a proposition. I think that over here in the States you got right away from us on that and have achieved some remarkable result, in particular, the results which recently have been obtained on the rather more square type of tip which has been experimented with lately. By and large our activities are in the direction of counter-rotation for single engine fighters."

Commander RAMSEY: "Thank you Air Commodore Buckle. Now I would like to hear from Mr. Brady on the question raised by the Aero Products Representative."

Mr. BRADY: "Our experimentation has extended to four different types of fighters and one bomber. The theoretical improvement in performance with dual rotation should be 5 or 6 percent, with dual rotation definitely being indicated, and a somewhat lesser amount in high speed. So far there is no direct comparison fully valid between a single rotation propeller and a dual rotation propeller, but indications that we have gotten are that neither of those increases in efficiency is being realized.

"We're rather puzzled by it, but it looks, as based on some tests with single rotation props, that the more limited efficiency at high speed operation is due to the fact that you have the drag on six blade shanks rather than three or four shanks and it seems that the use of a larger spinner to fair in the blade shanks would be desirable, or else the development of the blades having much thinner shanks. As far as the control is concerned, we understand the reaction of flying personnel has been very good so far as the absence of torque is concerned. There is a tendency toward negative stability which would require some redesign of the tail surfaces, but that is something you get into with higher powers and I think a single rotating propeller using the same blade area as the six blade dual would require changes. We don't know how fast the dual is coming in, but a year ago it looked as if the dual was coming in pretty fast. Recently it looks as if the development of the single rotation propellers was enough. I feel that for conventional types, it will be single rotation and for the unconventional types it will be the dual that will come in and pay for itself."

Mr. Carl BAKER (Hamilton Standard): "Theoretically, the adding of blades to a propeller, such as the dual rotation is, I think, the ultimate answer to absorption of power. Increasing blade width can obviously go just so far, and beyond that power absorption requirements in a given diameter must be met by the addition of blade area through additional blades, rather than increased width. The dual rotation prop has advantages, not only from the prop efficiency standpoint, but apparently from the airplane standpoint as well, such as the neutralization of torque reaction. However, there appear to be some disadvantages associated with it which must be further explored. Mr. Brady mentioned one – the destabilizing effect. There, of course, is the other angle of a dual rotation propeller being basically a somewhat more complicated piece of machinery and that may hold back its adoption in any general sense until absolutely forced by necessity. However, it does appear that when the limit of power absorption by increasing blade width is reached, the dual rotation propeller will come into its own with, hopefully, proper solutions to the destabilizing effect and acceptance of the more complicated machinery to deal with."

Major SCHOENFELDT: "I am wondering if either Commodore Buckle or the propeller people are working on the blinding of pilots in haze with the sun shining hard and flying in low formation. We had a lot of trouble at low rpms and long range flights with the pilot looking through the propeller. They seemed to feel that they were hypnotized by it."

Commodore BUCKLE: "I have no information on that. It is an interesting point and I'll make inquiries. We don't get very much sunshine in my country, whereas Eglin Field has rather more than its fair share perhaps. But it is quite a point and I will see if I can get any information on it. One point was raised by the third speaker back on blade shanks. Some calculations were done by one of our aircraft constructors three years ago in which he proved that the loss in efficiency due to drag of the blade shanks of the normal propeller was out of all proportion to the apparent area. Whether his calculations are correct, I am not able to say, but we intend to have large spinners to cover up as much of the shank area as possible to cut that drag down and all the new installations will have large spinners for that purpose."

Lt. J.G. GAVIN, USN (VF Design, Bu. Aer.): "In connection with the development of the counter-rotation prop, I think perhaps one of the black eyes it has received to date has been the fact that it has not always been used on an airplane specifically designed for it. That is particularly true of this decrease in longitudinal stability. I think, however, that there is one application for the dual rotation prop which will loom larger as time goes on and that is the use of the gas turbine as a power plant."

Commander RAMSEY: "Getting back to the question raised by the major from Eglin Field, is Lloyd Child here or anyone else who has flown the P-60C?"

Mr. H.W. THOMAS (Curtiss Wright): "I've flown the P-60C. What was there that you wanted to know about it?"

Commander RAMSEY: "The major raised the question of the blinding effect of counter-rotation propellers and the P-60C has a counter-rotating propeller. I wondered if that mesmerizing effect had been noted by any of the pilots who flew that plane."

Mr. H.W. THOMAS: "We have flown this dual rotation propeller for at least 50 hours around Buffalo and while we're not troubled with too much sunshine there, we have flown it when it has been very bright and have never been blinded. Looking at the airplane from the rear rather than in the cockpit, we can see the nodes – six of them I believe – where the props cross, but that is invisible when in the cockpit itself. None of the other pilots – several of us in Buffalo, at least half a dozen, have flown this plane – have reported being blinded."

Commander RAMSEY: "Col. Coats, did you have something to add to that discussion?"

Colonel COATS: "Only to say that in every case of the pilots down at Eglin Field – I have flown the airplane myself and the P-75 – and in every case the pilots all noticed the stroboscopic effect of the propellers and especially at low rpm and it is rather a psychological effect, perhaps hypnotizing and damned annoying for one-half hour at low powers. It may be that the props were not quite in synch. In the case of the P-75, it was never in synch and it just moved in front of your eyes like a stroboscope the whole time. It may have been the movement of the individual propeller blades in the constant speed position. I don't know the answer, but it was very annoying."

Mr. THOMAS: "We also had another fighter at Buffalo – the XP-62 – with dual props. The same effect was noticeable from the rear standing back 100 feet from it, you could see nodes all right, but there was no mesmerizing effect on the pilot in this installation either."

Mr. NEWDECK: "It might be pointed out here that in the case of the P-75, it was an in-line engine I believe, the cruising rpm was much lower than that utilized by the radial installation in the 62 and 60. The fact that the propeller rpm was much slower at cruising may have introduced that shaft hypnotizing effect of the opposite rotation of the props."

Mr. GERDAN: "I think in the case of the P-75 that it was the lack of familiarity of the pilots. That complaint never did arise at Fisher – some 700 or 800 hours of flight – and lack of familiarity might have had quite a bit to do with it. Actually what happened in the 75 was that you would not see the blade, but there was a shadow band of 1 inch or 2 inch width that was annoying, and I tried running the airplane up at our own plant with 14 to 1,700 rpm and that shadow form is quite hypnotizing. Whether or not you would become accustomed to it is hard to say."

Commander RAMSEY: "Any other aspects of that?"

Lieutenant Colonel TOUBMAN: "On a stability problem requiring dual rotation propellers, I would like to know if there are any manufacturers who have gone into the use of the gyroscopic trim to help a stability problem instead of going into dual rotation. Has anyone tried these gyroscopic trims? I just wondered if that would solve the problem or not. I know some manufacturers talked about not going into it for conventional airplanes; and if it is serious enough for the high-powered engines, I would think we could get away with it without going to dual rotation. I am all for it."

Commander RAMSEY: "Does any Curtiss representative know if there have been some such tests on the XF14C or any other planes with the counter-rotating prop?"

Mr. THOMAS: "No, we haven't run tests like that as such. I know that as far as the destabilizing influence goes, we have not conducted any tests to measure that either, but it was not open to any dangerous extent in any of the dual rotation installations. We had the XF14C, 60, or the P-62. I am personally interested in what this destabilizing effect is."

Lieutenant Colonel TOUBMAN: "What I had in mind was the sort of rudder trim that would take a pick-up off the needle and ball.

"We are experimenting with the reversible pitch motor which can be dampened to help the pilot solve the rudder stability problems. I don't know how successful it will be. Some manufacturers mentioned it some time ago, but I haven't seen it come out."

Mr. VIRGIN: "We have one interesting side light which might answer in part Thomas' question. In flying the XP-51F during stability checks, the directional stability was marginal. We are making some runs at the present time with a five-blade Rotal. The installation of this five-blade Rotal definitely made that plane unflyable, as a result of which it was necessary to bring the airplane in and add about 3 square feet of the vertical surface on the tail."

Commander RAMSEY: "Shall we continue this discussion on Sunday and Monday on that particular phase of power plants? I would like to bring to light some questions to ask the pilots: What type of supercharger do you want – two speed single stage, two stage? Do they want supercharger, jet stacks, etc.? What critical altitude will they want if they fight in the European Theater or Pacific Theater? Do they want water injection? Do they want electric starters? The answer is probably already known, but might be worthy of discussion. What type of warning do they want – lights, etc.? I would like to ask the questions again and Lieutenant Commander Owen, would you answer? What type of supercharger, what operating limits, what critical altitude, do you want water injection, what starter, turbo warning lights, etc.?"

Lieutenant Commander OWEN: "What I've seen of two-speed superchargers, particularly the 2800-C, I believe I would like. Regarding starters, I think it is fairly unanimous that we want electric starter. The direct cranking inertia starter seems to be satisfactory. The critical altitude is a big question and awhile ago the Army mentioned they wanted the altitude to escort the heavy bombers. You still need get-away speed at sea level to cover landing operations from carrier based aircraft. Having operated engines with critical altitude in the neighborhood of 26,000 feet some months ago in the types of operations that were taking place then, that was adequate. I have yet to see the Japs want to fight much above 20,000 feet. Of course, I haven't seen them all. I would say 30,000 for the immediate future. That's subject to adverse opinion, of course. "I have never had any experience at all in combat airplanes with the turbo supercharging. However, it certainly has its attractions and probably won't answer the altitude question. I'm rather inclined to agree with Commodore Buckle on the troubles of water injection. Whether or not it is actually worth the weight, installation and maintenance, I don't think has yet been completely proved. You might let someone else answer these same questions – it should be interesting to hear from someone who has seen a little different operation than I have."

Colonel COATS: "On the question of supercharging, I believe that it depends entirely on the type of operations. I don't believe there is such a thing as a fighter airplane that has everything from the ground on up. I believe you have to take them in steps. Both two-stage and turbo-supercharged airplanes have their advantages. On the question of starter, I believe we in the Army will stand by electric and hand-driven inertia starters for several reasons. In most of our operations, or at least a good bit of them, we are away from any bases where we have a source of electric supply or any means of charging batteries other than the ship's source. I believe, as far as the Army is concerned, we'll stick by the electric inertia starter.

"As far as warning lights in the cockpit, most of us feel that they have a definite advantage, especially for night operations. However, I'm not so sure we'd go fully with all lights, such as oil pressure lights, because in the event the light came on, you wouldn't know whether to jump out or try to come home. You would have no indication of whether you have no pressure at all or whether you are right on the ragged edge of the limiting pressure which puts on the light."

Commodore BUCKLE: "As to what form of supercharging, I think we are discussing it in terms of the Pacific war and I am very much in favor of the remarks made by the officer earlier in the meeting. If we are taking the offensive and proposing to fly around 30,000 feet, then the initiative is on our side and the defense has to follow as to what is done. They may not have aircraft capable of out-performing the B-29 at 30,000 feet. But they are going to try. As one never knows, perhaps we should go back here to the officer who said he must know the supercharging requirements for next year right now. I think the answer is that we never know. If we can get the answer here as we go to turbo blowers, or to two-speed two-stage systems as we have in the Rolls Royce engine it would greatly help the manufacturers. The Rolls Royce gives us two good altitudes – the medium high and the very high and we can use it as we like in the same airplane. Admittedly, we hall not get the same high performance at sea level, but we shall have sufficient performance at sea level and the primary object is to destroy enemy aircraft. So that we feel the answer on the supercharging is: Give us a supercharger set up to enable us to have good performance at medium altitudes, say from around 18,000 to 20,000 feet and again from 30,000 to 35,000 feet, and I believe it can be done on the same installation. Enlarging on that a trifle, we do like a warning light on our supercharger, but over and above everything else, we do like to have automatic change speed gears. I was flying an airplane here the other day which has three supercharging positions on the control. I am not clear, if you start a dog fight at 25,000 feet and the enemy moves away, whether you are to change gear each time as you go down lower, if you are supposed to stop what you're doing and change gear. If you are, then it's too much to expect a fighter to do and get on with his fighting. I think all that should be taken care of for you.

"As regards water injection, I think we have already made our situation fairly clear on that. It's a cumbersome method and puts up a complication as to weight and maintenance. If we can do it another way, we would rather do it.

"Regarding starting, we require a reliable self-contained method of starting the engine. The nearest approach to that is a small two-stroke gasoline engine which you start off and when it is running correctly, you turn it loose to turn over the main engine. We don't like electric starters for the same reason as given by Colonel Coats. You get out in the desert, your aircraft factories are not near and you don't have any facilities for charging the plane until you get flying, so we want a reliable, self-contained system which will operate anywhere and the two-stroke has promise."

Commander RAMSEY: "You have me on that two-stroke gasoline engine. How long would it take to start your engine with that?"

Commodore BUCKLE: "The gasoline engine I am talking about is the same as the one made by the same people who make the cartridge starters – Kaufmann, I think is the name of the firm. They can't get away from their cartridges. They start the small engine with a cartridge – it is perfectly true – and then the small engine starts the big one and you'll get a start all right, provided nothing goes wrong with it. There's no reason why it should. You get a start on that just about as quickly as you would with an inertia electric starter."

Mr. NEWDECK: "I think the Russians have an interesting set-up on this starting business. They attached a jaw clutch to the hub of the propeller and then they had a special truck with a gasoline engine which had to be jacked up to the airplane and engaged the clutch on the prop with this gasoline engine and when the airplane engine fired, the clutch is disengaged and they drive the truck on to the next airplane. If you are going to have a gasoline engine to start it, you might have to carry it around with the airplane."

Commodore BUCKLE: "I think we should take them up on infringement of patent rights. We were using that system with the Model T Ford in 1916 and when I learned to fly in 1926, the thing was still operating and the last one I saw was one which was destroyed by a Bristol Fighter which jumped the chocks and chewed it up."

Lieutenant Commander OWENS: "A situation like that might be all right on shore bases, but a truck would have difficulty getting around on an aircraft carrier deck. For that reason, I think you would have to have a starter right in the airplane and considering the deck-handling crew changing cartridges and everything about the inertia starter is more attractive."

Commander RAMSEY: "Dropping that particular subject for a second, I would like to call on aircraft manufacturers with a broad question, such as suppose you were given free reign to build fighters. Take first the Navy requirements of a small carrier fighter, to get a lot on board, folding wings, high rate of climb, high speed, good performance, good power and armor, without actually being tied down to using a certain power plant or being limited only by the space of the folded span and the length of the plane, what probable power plant that is out now and proven, would you use? That may be stepping on somebody's toes, so if it does, don't answer it. I would like to call on Vought first."

Mr. SCHLIEMANN (Chance Vought): "We're definitely on the spot with that question. We are presently building a ship-board fighter around a 2800 engine. Are you asking what power plant we would use if we started with new design at this time?"

Commander RAMSEY: "Yes, given a free reign, the Navy just comes along and says, 'Here's $50,000,000. Build us a damned good fighter.'"

Mr. SCHLIEMANN: "I'm afraid I can't answer that. I have been too close to the F4U-1. We are experimenting with jet-propulsion designs on the board and I think maybe in a few years we could give you a better answer. Right now the F4U-1 is our answer."

Commander RAMSEY: "Is there any reason why you don't mention the liquid-cooled engine?"

Mr. SCHLIEMANN: "The only reason I haven't mentioned liquid-cooled engines is perhaps Chance Vought's background of being a division of United Aircraft. However, that is not the entire case. We have experimented with air-cooled in-line engines and did not have very much success with it. We understand the radial engine and it is just our belief that that engine is more satisfactory for combat airplanes. We feel that we do not run into the problem of armor protection for the liquid-cooled, and we feel that that power plant is in the highest degree of development at the present time."

Commander RAMSEY: "I would like to hear from Curtiss Wright on that same subject."

Mr. G.B. CLARK (Curtiss Wright): "We are at the present time building a fighter around a 2800 engine. We are probably one of Pratt & Whitney's best customers and in addition to that main engine, we put a kicker in it and that again leaves us in the same position as Vought in answering the question. Our airplane is pretty big and we are designing it for high performance and we're pretty stuck on the type of power plant we can put in. I don't think we could say without knowing exactly what the purpose of the airplane was, exactly what power plant we would build the airplane around."

Commander RAMSEY: "I would like to call on North American, taking an average Army pursuit plane not particularly limited but for general Army use."

Mr. STEPPE: "Of course, all our fighter experience has been with liquid-cooled engines, first with Allison and later with the Merlin, and naturally we're much prejudiced toward it as Chance Vought is for air-cooled. I think we are firmly convinced we can build a smaller fighter with high performance and with longer range at a lot less weight, than you can with a radial engine. The 2800 is a good example of that. The aircraft gets very heavy – the amount of fuel required for long range operation becomes quite large. I think the P-51 has demonstrated that pretty well in the European theater. We have probably the longest range at the present time in the operational outfit and I think our performance will match anything else on operational duty right now. I think for any immediate future fighter design, we'll stick with the Merlin and later versions with a little higher powers."

Lieutenant GAVIN: "In connection with this discussion by the aircraft manufacturers, on the preference of an engine, I think it would be interesting if we could have a comment from some of the Army people here as to relative vulnerability of the liquid-cooled and air-cooled installations based on their experience in Europe."

Colonel GARMAN: "I can speak only for the African theater and for only a particular type of operation. The P-38 was used at low level on many occasions and we found that it was quite vulnerable to ground fire – any type of ground fire, even small-arms fire. But other airplanes also experienced that same ground fore and the radial engines brought the planes home. You can't lay down any hard and fast rule and say the in-line engine is no good at low altitude as far as ground fire is concerned. It all depends on the operation entirely."

Lieutenant Colonel TYLER: "We have data which shows that in the entire European theater the P-47 is much better able to take punishment and return after any sort of mission – either ground attack or any mission which incurs damage. That may be due to the P-47 airplane or due to the air-cooled feature. We don't know which, but it certainly can take it better than the other types."

Lieutenant Commander SUTHERLAND: "It's interesting to notice that the airplane that was designed to take the most punishment, that is to butt heads with ground fire, the Russian Stormovik has the liquid-cooled engine. It is well known that the thing is heavily armored – the whole bottom is covered with armor. I wonder just why the Russians chose liquid-cooled engines. I have a strong suspicion it was the only one they had. If they had a radial engine, and they have them in some fighters, you would have thought that they would have chosen that because the airplane was designed for nothing but scooting low on the ground. It doesn't even have high performance which the 38 and 51 do."

Commander RAMSEY: "I think it would be very interesting to hear from Republic, if they are not bound by customer relations regulation, as to why they chose the air-cooled engine. Why didn't they give some thought to liquid cooled?"
(There was no Republic representative available.)

Lieutenant Colonel TOUBMAN: "I think this vulnerability question, although I won't refute the evidence Colonel Tyler brought up, is entirely over-rated. I flew the liquid-cooled engine with a hole in the engine for about 10 minutes and I know of cases where they have flown for 30 minutes and they never stopped with a hole in the engine. If you get hit in the radial, you will go down. Similarly, if you get another hit in a less fortunate place, you can travel for quite some time. As regards radial versus in-line the only way I feel about it is that you can build a much smaller fighter and you can get the much needed view over the nose. I haven't seen a radial here where you get this view over the nose without paying for it in speed."

Mr. GUYTON: "I would like to ask Colonel Toubman if he gets more view over the nose of the P-51 than other planes in the Navy?"

Colonel TOUBMAN: "I guess I stepped into that one, but I think Chance Vought is paying for that raised cabin."

Commander RAMSEY: "The time is drawing nigh. Before we get into a discussion on performance, which we will take up Sunday, we must close this session. The weather man says we are in condition 1, which means that the hurricane is on its way toward us. We will put the planes in the hangars and we are incorporating a foul-weather schedule for tomorrow. There will be no flying. Tomorrow morning there will be combat films in the movie hall. The daily meeting here will be at 1030."

SIXTH MEETING
20 October 1944
1030-1215

Commander RAMSEY: "Gentlemen: We will call the meeting to order now even though the bus-borne people from Washington have not arrived. Due to the fact that the aerologists tell us we will have bad weather, we will try to get as much of the discussion over with today and tomorrow and use Sunday and Monday morning as clean-up periods for flying. Then Monday afternoon we will have the final session of the conference, to take up those points which we have not covered satisfactorily. "This morning we plan on speaking about bubble canopies, pilot's comfort, safety equipment, and fuel systems. This afternoon as you all know from the program we have a lecture on structural strength by Lieutenant Commander Conlon from BuAer.

"Saturday we plan on having in the morning discussions on compressibility, power plants and accessories, brakes, landing gear and in the afternoon Commander Monroe will speak on armament and armor. At any time please speak up if anybody has constructive criticism on topics of discussion or additions of topics we may have forgotten. On the subject of the aerodynamic qualities of the bubble canopy and wind screen I think I would first like to hear from pilots on what they would like in a bubble canopy and then go into the contractor's idea on how he goes about the tests he makes to determine whether it does ruin performance and stability. First I would like to call on the British delegation and ask them to discuss bubble canopies and the method by which they get better visibility out of the cockpits."

Lieutenant TWISS: "Our bubble canopies are very much the same as the ones you see down here at the conference, the exception being that the front plate windshield is usually three armored glass plates which are arranged in such a way you don't see any structure surrounding them, giving you all-around vision. The armor behind the pilot seems to differ inasmuch as there isn't too much of it and you get a better view. Most of the bubble canopies are hand-operated, not as in the P-47 with the electric motor, and are operated up to about 250 miles per hour but are not operated up to 300 or 350; they have to be jettisoned if you want it higher than that. The rear view is still helped out by having mirrors about the windshield on the outside but whether that is necessary or not is debatable. I don't think we have any more to say."

Commander RAMSEY: "Thank you Lieutenant Twiss. I would like to call on Colonel Coats to say a few words relative to the bubble canopy. Or Colonel Toubman."

Lieutenant Colonel TOUBMAN: "As far as the bubble canopy is concerned the Army is going for it 100 percent on Army planes. However, there are a few points I would like to bring up. A few of the bubbles have distortion and we would like to get that out as much as practical. As far as the bubble is concerned, they defeated the whole purpose by bringing the armor plate so close to the side of the bubble so that you still have very poor vision to the rear. Another main point concerning the bubble canopy is the matter of heat rejection: it is like a greenhouse sitting out there practically in the open. It is quite a serious point and therefore you have to go to the tinted canopy or a new type of plastic in the bubble that will reject the large amount of heat that comes in."

Commander RAMSEY: "I just wanted to ask the colonel if he had objected to the vision astern of the F8F. Was that his remark?"

Lieutenant Colonel TOUBMAN: "That is correct. Just looking around from the position on the ground, I didn't fly that airplane. I had to stretch myself and my neck quite a bit to see to the rear as much as I would like to."

Commander BOOTH: "I think that as far as I am concerned there's plenty of vision astern in that machine. I would say that it probably would be satisfactory. How about Colonel Meng, he flew the machine?"

Lieutenant Commander OWEN: "Of course all the bubble canopies we see here give the Navy a lot of ideas because so far we have no bubble canopy airplanes, and flying the P-51, P-47, F8F, and FG, our eyes are opened up considerably to the possibility. I believe all the Navy pilots will agree they want the bubble canopy. Considering carrier operation visibility astern even on a deck is highly important. You take off from the aircraft carrier with the canopy opened because of the possibility of hitting the water and necessity of getting out of the airplane. It is important that you see astern. Now that can be done. Some planes incorporate the rear vision mirror in the windshield in order to see the armor, and that means it can be adjusted to give you enough down vision directly aft. That is just a consideration. The bubble we have seen down here mentioned before favorably impressed us. I particularly like the P-47. The F8F I agree with Commander Booth gives sufficient visibility astern. I think the F7F looks particularly good. We tried to make tests on the bubbles down here and evaluate them at night and there the importance of good optical qualities comes up, but in some of the first ones we have seen the optical qualities at night have been pretty poor causing lights to be reflected into the cockpit in a sunburst or star effect. I would like to hear some discussion on the ease with which a bubble type canopy can be pressurized. I am not sure whether the P-47 has a pressurized cabin or not. I imagine that would make some difference in attempting to have a free blown bubble and would have to air pressurize the cabin. That is about all I have to say right now."

Commander RAMSEY: "I would like to call on the manufacturers present. We have here as you know, the XF8F with the bubble canopy and the F7F without the bubble canopy, the F6F with the flat windshield, F4U with flat windshield, FG with bubble canopy, P-51 and P-47 with bubble canopy and P-38 with regular enclosure. I would like to call on pilots first. Jack Woolams."

Mr. WOOLAMS: "We don't think we will run into any trouble pressurizing the canopy. We are putting a bubble canopy on our next fighter and it will be pressurized. However, there is a problem of explosive decompression if the canopy is shot and the pressurization we are putting on will amount to 2.75 pounds. However, the pilot cannot stand that very well if the canopy is shattered suddenly at high altitude and low temperatures. In order to take care of this situation we have put a valve in which when we go into combat relieves the pressure which of course makes the canopy less effective in regard to pressurization, during combat but relieves the pilot in trouble, in case of any sudden decompression. I believe that the bubble canopy is practically indispensable in regard to visibility. I think most all the manufacturers of Army airplanes are putting them on and our company for one considers it indispensable."

Commander RAMSEY: "Thank you, Jack."

Mr. HALL: "As far as the bubble canopy on the F7F goes, it was considered originally until it was decided to make all F7F's two seaters and after that we gave up the idea of trying to make a two-seater with bubble canopy. It looked like just too big a job for us to tackle now. We probably will have a bulged canopy on the F7F which is at least a half-way measure. I am sure all our single seaters will have a bubble canopy and we will not probably try to go back to put one on the F6F at this time."

Mr. GUYTON: "We have not done anything on the bubble canopy as yet, but we do have a bulge canopy which we think gives adequate vision to the rear. The thing we have not been able to find out is how much vision the pilot wants and how much protection he wants. The way we have the bulge arranged, with the flat front windshield, it hasn't cost us anything in speed that we can find by flight test. We did have one that is too much bulged and although we cannot find it by flight test, we calculated it would cost about 4 miles an hour. However, with the bulge that we have you can see across the vertical fin and completely astern on either side. I just want to know if someone would want more vision than that or if they want clear vision all around."

Commander RAMSEY: "I would like to call on Lieutenant Davenport to answer that question. He has had quite a bit of experience."

Lieutenant DAVENPORT: "In the case of the F4U, the bulged canopy is very much more attractive than the bubble canopy. With the radio equipment in the back the way it is, it is one of the best bullet tumblers we ever met up with. Even though the armor plate behind the pilot has never been built thick enough to stand a heavy 20-mm. shell, all the radio equipment back there would always start bullets tumbling and was more than adequate. And in the case of the bubble canopy you have nothing but glass and you would have to put in splash plates of varying angles not only to deflect the bullets but start them tumbling. I flew the first bulge canopy by Vought and as far as I can see it certainly is in the right direction, and gives all the rear vision that the normal bubble canopy gives."

Commander RAMSEY: "Thank you Lieutenant Davenport. Lloyd Child please discuss what your company is doing along that line."

Mr. CHILD: "We are using the bubble canopy on our new Navy fighter and I think it is important to have the bubble bulged out in circular sections so that the oxygen mask has enough room. I know Air Commodore Buckle was very prone on that point. He said it was too small so that the pilot could not turn his head without hurting it or hitting the oxygen mask on the side. I certainly like the bubble canopy flown here. The electric device on the Thunderbolt is especially good so you can open it with a switch. I would be interested to know how much weight that costs, if any."

Mr. WOOLAMS: "I wish to disagree with Lloyd on this business of bulging the canopy out. At speeds now obtained we are running into pretty severe drag losses due to the bulging. I think that Joe Parker will bear me out on this due to the bulging of things such as the canopy. We are liable to run into very severe speed losses. I would much rather have an airplane that went 25 miles faster in the vicinity of 500 miles per hour and hit my oxygen mask when I turned around."

Mr. VIRGIN: "We had read of quite a few flight tests in England on the P-51 with a molded canopy more or less on the same plan form of the P-51 versus a free blown one. Actual flight test showed that there was no difference in speed. We could not find it at all."

Mr. FOOTE (Eastern Aircraft): "We are doing more listening than talking on this item so far, because our new fighter will be our first experience with bubble canopies. We had experimented somewhat and the thing is still being glued together up there. We are at present experimenting with the bulge on the FM-2 canopy. We had comments from pilots at the fighter conference on the FM-2 and they weren't any too good. It seems that it was too flimsy so I don't know how this bulge will work out on it. I would like to get some more comments on this canopy. We found that it would add weight and we do not know whether it would be quite advisable as our plane is primarily a fast climbing ship and we would have to give up the idea of top speed on it. Could I get a few comments from people who have had experience with our canopy to see whether they favor bubbles or a bulge with the idea of saving weight?"

Commander RAMSEY: "I would like to call on Commander Booth, who has had experience in the African theater with the F4F which has practically the same type of canopy."

Commander BOOTH: "I should think that if you are going to make changes in the FM2, a bulge would be the sensible thing. It would require a lot of redesigning with all that turtle back structure in the present FM2 and it is quite clear that there is not enough visibility aft. A bulge would give you a lot better out of the airplane. I think it would be unsensible to tack a bubble canopy on that airplane."

Commander RAMSEY: "I am inclined to agree with Tommy. I had a little experience with the F4F type in the Pacific and I believe it would be better not to get the bubble canopy in the FM2. However the F2M with the bubble canopy would be something to shoot for. I would be very happy to have that plane down here soon. Is Connie Converse here for Grumman? I would like to have some Grumman pilot volunteer his remarks. Are there any Grumman pilots present?"

Mr. ROWLEY: "We have had some experiments with these bulge canopies on the F6F. We had a little trouble at the higher speed, but I think it improved the visibility quite a bit. We have one now that we had hoped to send down here very shortly. However, I don't think they are quite the answer. I think the bubble canopy is the final answer as far as visibility is concerned."

Commander RAMSEY: "I would like to call on Bud Martin from Lockheed. I know you are not particularly interested in the bubbles in fighters, but can you give us some comments on them?"

Mr. MARTIN: "We have heard about the weight penalty that you might take from putting on a bubble canopy and the possible increase or decrease in speed, but in some planes here there is one noticeable feature I have not heard comment on. When the bubble is on a plane it appears in general that the vertical area behind the CG is reduced and I wonder if any of the people who have put the canopy on the planes have any good flight test information as to what happens to their directional characteristics."

Commander RAMSEY: "That is a very good point. It will come out very soon. I would like to call on Mr. Burke from McDonnell."

Mr. BURKE: "We have more or less adopted a standard policy that any design for fighters must have the bubble canopy. However, we have a theory which may or may not prove correct on the XFD that the type of bubble used on present airplanes is not the best possible solution. By that I mean we may be able to work with variations in plan form, and achieve a canopy which will have less drag than the present bubble while sticking approximately to the circular cross section of the bubble canopy and a built-up canopy will come out equal because a built up canopy for pressurizing becomes a quite heavy structure due to the necessity of holding in the edges of all the plates and I think as we go into pressure cabins we will probably come out more nearly equal on those."

Commander RAMSEY: "I would like to call on Mr. Virgin again for North American. But before he speaks to those pilots who are flying the P-51 I wish to call attention again to the fact that you must throw the stick pretty far forward to release the tail wheel locks. Yesterday I noticed some people (I won't mention any names) who evidently were not checked out thoroughly enough, even though the stick was about neutral it was evidently not forward enough and dragged the tail wheel around and almost broke the post. Please throw the stick way out. We only have two P-51's here for everyone's use."

Mr. VIRGIN: "Commander, we never did run a comprehensive test on the P-51B. I can't tell you exactly the difference that was caused by that canopy. The time the B came out, instrumentation was not such that we could get what I would say is a comprehensive flight test. Since the advert of the D with the canopy we, in conjunction with the Ames Laboratory on the West coast, got the directional characteristics of the airplane pinned down. So I can't speak on the actual difference as caused by the bubble canopy."

Commander RAMSEY: "Thank you very much Mr. Virgin. Mr. Lindbergh, from United, are you doing anything on that bubble canopy?"

Mr. LINDBERGH: "I'll have to refer you to Mr. Guyton who has been working with that."

Lieutenant HALABY: "It seems to me that this discussion on directional characteristics with bubble canopy should also consider the fact that for example on the P-51 they also put on a four-bladed propeller and they also move the CG aft and it is quite possible that those additions may have caused the deterioration in directional stability to a greater extent than the bubble canopy. But we certainly think at Flight Test that those changes should be evaluated before much improved visibility had been bought with directional stability. It seems to me there might be a distinction drawn between ships with a short configuration coupled like the P-47 and P-51 and an airplane with a longer fuselage like the F4U, more particularly the F7F or Mosquito. For a single place, the F7F bubble canopy can be employed with much less deterioration in directional control and a lot more visibility and pilot comfort. As the discussion goes on I would like to hear some thought on two other items. One is what protection is there offered in the bubble canopy in the case of an overturn? For example, the armor plate in the P-51 or 47 is stressed to take the load on the pilot's head in an overturn.

Now, not only is it important to the Army but to the Navy as well because of the barrier crash condition. Although some people say it is not important on a tricycle gear I believe that the carrier pilots think that with a tricycle gear on a carrier the nose may be wiped out by the barrier because of the CG position and be even more likely to overturn and require overturn protection. One other question I would like to hear some thought on is, at what speed should the pilot be able to operate one of these bubble canopies. On almost all of them the air pressure appears to want to close the canopy and as the speed gets higher, the pressure seems to want to increase and without an electrical device or without a big handle with a big mechanical advantage, I wonder if you are going to be able to bail out of such an airplane. Merely pulling the escape mechanism without some spoiler arrangement or something, may not pull the canopy off because of the air load tending to close it. Could we have some thought on these items when it is appropriate?"

Commander RAMSEY: "I would like to call on Joe Parker for Republic."

Mr. PARKER: "As the pilots come back that have been shot down, they say they never saw what hit them. Now if you can see an airplane making a run on you, surely you can take some evasive action especially in the P-47 with its dive qualities, but first there are some things the present bubble does not incorporate. Talking about stability, it has ruined our stability to the point where erratic or on coordinated flight will get you in trouble and I want to warn everybody that flies the P-47 not to yaw the airplane over 10 degrees especially at high power and high speed. Now we have gone to a dorsal fin which is very satisfactory at high altitude and high power which are the most critical conditions on dorsal fins. You will find help in high angles of yaw. My third note that I put down was speed. The first bubble canopy we put on was a molded canopy and caused us no loss in speed; however, we have gone into a blown canopy which has cost us speed. By blown canopy I mean cockpit enclosure, and not to include the windshield. However, I am a firm believer that if you start to bulge one thing, you should bulge everything in conjunction with it so that you have a nice smooth curve. That part of the windshield should be curved to come in line with the canopy enclosure. I have also had quite an experience down in New Guinea with a bubble canopy and I believe they should be tinted to stop some of the reflection. It gets quite uncomfortable down there at low altitudes. You also have a very definite change in temperature in high altitude dives. The only thing I can see about the bulge we have on the present canopy is that the bulge is put in there to give you more forward visibility, when you turn your head over to the side. I am talking about the bulge in overhead of the canopy and not at the side. And regardless of how clean the mechanics in the field try to keep that canopy by wiping it off with wet rags, there is still a light film of grease or some matter on the windshield and as soon as you taxi out in the runway, some other airplane blows dust which, instead of only 1 1/2 inch framework up there you can't see through. Another thing I noticed on a Navy fighter, I forgot which one, was that it had a rear view mirror which could only be used with the canopy closed. I think a rear view mirror is absolutely necessary especially for taxiing on the ground. We have a rather nice arrangement of the rear view mirror in the front part of the windshield and you have use of it with the canopy opened or closed. Our armor plate will not provide adequate protection for turnover; however, it is so designed that if you do turn over, the armor plate will not collapse immediately, and it will bend the pilot's head forward enough to save him from being hit by the ground. That is all I can offer. However we have not had very many causes of absolute turn-over on the P-47 and the pilot's psychology of that is that he asks about turn-over, and makes a mental note that he is going to be damned careful from then on. Again I want to warn you about all of our P-47's. Don't use full rudder force. Plain coordinated flying is all that is required and you won't get into any trouble with a 47."

Mr. VIRGIN: "I would like to answer Lieutenant Halaby on the stability of the P-51B. We had experienced deterioration in stability before we arrived at the P-51D and it first became evident on the P-51B which did not have a free blown bubble canopy. On the P-51D we do have a complete picture of the stability of that airplane and every one that left the factory with the exception of a very few were equipped with dorsal and reversed boost rudder. We do have a complete picture of that airplane with stability curves. I might have created an impression that there is no difference between the P-51D and B in speed. There is. We figured that the bubble canopy costs us 5 miles an hour. The point I made before was that there was not difference between the molded canopy bubble and the free blown canopy."

Mr. STEPPE: "There are a number of questions with reference to turnover characteristics that come up on this canopy and I think we can elucidate a little bit. One of them is on the business of tinting canopies. We have done quite a bit of experimentation on that. We equipped one airplane with tinted canopy and made a bunch of samples. The pilot's comments were quite good and they felt that the reduction in light transmission was just about right. But we furnished the materials to the Aeromedical Lab at Wright Field and they published the report a couple of months ago. They felt the thing was completely unsatisfactory. I don't know a lot of the details about it, except it shut out quite a bit of white light, but the rays which were harmful to the eye, particularly at altitude came through the thing with 100 percent transmission. Aeromedical felt that it would actually give the pilot a false sense of security and the harmful rays at altitude would still actually come through and we were getting no value out of the tinting business. It is my understanding at the present time there is no suitable dye for this or flexible glass that will give the reduction of and light transmission that glasses would. And it is their recommendation to lay off the thing. Another question brought up by Lieutenant Halaby on the opening of the canopies in flight and the closing moment that is apparent on them. This closing moment is inherent on the P-51 canopy. In other words, if you let it float out of lock, it will suck closed to a position 1 1/2 inches in back of the windshield bulb. As far as jettisoning goes we actually had to jettison in flight and the thing came off very cleanly. It has a very high pressure peak. The thing came off and missed the tail by 15 to 20 feet but it jumped straight up. I think you can check me. I think it was a release at 200 indicated or 150. But I think at that rate of speed if the thing is released it will jump back enough to clear the rear of the windshield. We have also made some checks flying the thing with the canopy open and we have taken the thing up at 350 indicated or at least cranked the thing open at that speed and it took about a 25# force on the handle. The limiting factor seems to be that it sucks your head phones and everything else at that speed. The other point in overturn protection we have known in the P-51 is that the armor plate might give you a little bit, but I think it is very doubtful for we used to have an overturn truss in the P-51B and we took it out on the Air Forces' request. There is one point that Joe Parker brought up on this apparent bulge or discontinuity in the free-blown canopies from the windshield. That is caused by purely mechanical reasons. We experimented with a number of blowing processes and the first P-51B we delivered with molded canopies. The optical qualities of it were not quite as good as the free blown ones and the Air Forces wouldn't accept it, so we were forced to go to the free blown ones. However, I disagree with Joe that the windshield should be bulged out to give you a smooth line on that, and actually due to the shape of the windshield there is very little speed lost on it and our comments from the European theater on that bulge is that the few early airplanes that got over there with the flushed windshield the pilots discarded with the least excuse. In fact our field service representatives reported that if fliers were scheduled for a mission with a P-51D with a molded canopy and there was an airplane sitting around with a free blown canopy, he swapped enclosures. They like that extra vision because of the fact that you can get your head out and see straight down alongside the nose of the engine particularly on ground strafing missions.

Commander RAMSEY: "I would like to take up each plane that has a bubble canopy or just enclosure and get comments from the pilots who have flown them, so we can arrive at a decision on which one has the best bubble canopy without hurting anyone's feelings. We would like to take up the F8F."

Commander OWENS: "There is one feature on the F8F which I liked very much. That was the locking device and the opening and closing device. It is a regular hand crank that locks the enclosure in any position. That is without a pin affair, you merely stop cranking and the canopy stays in any position. It seemed to open and close readily at any speeds up to which I tried it – around 200 miles per hour. I can't speak too intelligently on bubble canopies because I haven't had too much experience. I think the F8F is very nice and I think it provides adequate vision aft. One feature which is attractive in any bubble canopy is the electric opening and closing device such as the P-47. The F8F being a smaller airplane, you do feel that you have too much room to turn your head. I haven't worn an oxygen mask yet, and I don't know whether that would interfere, but it appears however that it would not. I think Commander Booth could continue with some remarks there."

Commander BOOTH: "I don't think I have very much to add to that. I was arguing with Toubman about the visibility aft. I think that cockpit is considerably narrower than the 51 or 47. You probably have to be a little shifty to turn your oxygen mask around in there. Actually it is very nice, good visibility, good vision downward and to the side and the closing device seem quite simple and very satisfactory."

Commander. RAMSEY: "I agree with Ed on that closing device. The P-47 closing device appealed to me very much. I would like to know from Bob Hall if any thought was given to an electric-closing device on the F8F. I don't mean that I did not like the closing device on the F8F. I think it is excellent and I like the feature of not inserting a prong in a hole to lock it in any position. Bob, will you carry on?"

Mr. HALL: "Before we had finished the design of the 8F, we had given serious consideration to electric operation of that canopy, but we didn't adopt it because we were fighting weight on this airplane. We tried to make the smallest and lightest fighter we could and we didn't add any weight which was not absolutely necessary. Our F8F canopy is a little smaller than the P-47 and therefore we were confident that if we put a sufficient hand crank in the airplane, we could operate it satisfactorily. I don't believe it is possible to open the canopy at speeds much higher than 200 miles per hour with one hand because the closing forces do build up with speed very definitely. However, we have felt that 200 miles per hour was a reasonable speed to operate the hand crank. There is a jettison device not yet tried, but it releases the canopy from all those supports and we do feel confident that it will come off satisfactorily in an emergency. I would like to mention a few things about it that have been discussed on other airplanes. In regard to overturn structure, the airplane that is down here now has none. But we have provided a removable overturn structure, again with the idea of saving weight. For training purposes, it can be supplied and removed easily which will take specification overturn loads. The armor in there certainly will not. We rather hope that a four-blade propeller will help in that respect and it has in some instances occasionally prevented the full weight from landing on the bubble alone. I think if you will look at the canopy on the F8F you will find there is almost no discontinuity between the windshield and the canopy – we played around a long time to get what I would say is halfway between a full blown and a molded. Actually it is full blown and it is restrained and cut off at the right place so it does blend with the windshield fairly well. We were fortunate in having no deterioration in control with it open. We really expected it but it didn't happen. I am sure, however, that it did cause some deterioration in directional control and according to the percentages of fin area and the size of the airplane, it should be exceedingly stable directionally. But as Joe Parker told us in the case of the P-47, at high angles of yaw it was evident that it was not the same as other airplanes we built, and again that may be due to the four-bladed propeller, but I personally think the bubble canopy had something to do with it. You will notice that we added dorsal and ventral angles to alleviate that. We also reduced to some extent the throw of the rudder and we believe it is satisfactory, but I am sure that it did have some effect on directional stability at high angles of yaw."

Mr. GUYTON: "We were worried about the same thing that Mr. Hall was in the bubble canopy and it looks like we have to take a loss in overturn structure in order to get some. As

Commander Owen said he was worried about barrier crashes, etc., and it was one of our foremost thoughts. At the same time, I would like to know from someone who has been shot at from behind whether they would like to have the head cut out or full armor protection aft.

"On the F4U4 a proposed armor protection behind the pilot that we have is 50-caliber protection and retains the same that we had before; it will have the bulge canopy, so you can see to the vertical fin. We would like to know just how much vision you would like aft, what percentage of overturn structure you want which we incorporated in this airplane that is coming out now."

Commander RAMSEY: "I will speak for myself. I was unfortunate to have a few Japs get on my tail and the protection was sufficient – I have no scars. The protection was from 20-mm. shells which evidently came in from quite a bit of an angle and I think that no less than the protection in the F4F should be in any plane. I have forgotten the exact angle; we can get that out of the blueprints, but that protection I feel is just about sufficient because some of the bullets whizzing by me got the windshield from each side. But it was sufficient to give me protection. As far as visibility astern I would like enough to look around my head rest and see my vertical fin and nothing less than that. It may be quite a problem to get both the protection you want and also the visibility astern that you want. I would like to call on Colonel Toubman."

Lieutenant Colonel TOUBMAN: "Naturally we all can't have our cake and eat it too. I think the first requirement of all fighter pilots is that he wants visibility and I think fewer of them would be shot at, at the rear, if he had rear visibility. Going back to the F8F, I just checked out in that and that armor plate is carried up 6 inches above the base line of the canopy and being a rather small fighter pilot, I had to stretch my neck. When I spoke to Colonel Meng after his flight, he felt he had to stretch quite far to see to the rear. Now it is just a question of what we want. We either want protection or visibility. The P-51 has just the bare minimum I feel but I can turn my head to the rear from just the normal sitting position in the cockpit. I could look over my left shoulder and see my right stabilizer. However, I agree we don't have the protection and it is just a question of whether you want protection or you want to see the air, or whether you want to see something before it gets on you. As far as overturn structure, that also holds true. You can't have your cake and eat it. You will have a few losses, but it is something you must contend with."

Commander RAMSEY: "I agree with Colonel Toubman that the paramount interest should be shown in protection first then the visibility. As far as overturn structure goes, I agree with Bob Hall that there should be overturn structure especially during the operational training period, but it should be removable."

Mr. BURKE: "I wanted to make a comment on Mr. Hall's statement, that in regard to the F8F, although the percentage of fin area was consistent with other airplanes, they had a loss of directional stability. The loss in going from the conventional canopy to the bubble canopy may be far in excess of the apparent decrease in fin area because you go from a deep flat sided fuselage to a fairly circular cross section which when you yaw it off, gives you practically no restoring effort. Now that might have been particularly marked in the P-51 which was quite flat sided before and now is approaching a more streamlined shape in yaw.

"Also in the F6F as against the F8F I think that the fact that on the F8F the spoiler angles which add practically no fin area, but spoil the fuselage when you yaw it, the fact that they were the correct measure may indicate that was the main loss that they were running into."

Commander RAMSEY: "Bob Hall, do you have any thing to add or subtract? Mr. Hall says he agrees with that statement.

"I would like to call on Owen to answer some of the queries which were brought up by Booth, Guyton, and Colonel Toubman."

Commander OWENS: "In connection with armor versus visibility astern I think it is pretty marginal there because I think you can have sufficient visibility astern and still adequate armor. If you are looking around over your shoulder, you are only looking in one direction and you have to remember that's just when somebody starts shooting at you from the other side. I would never say sacrifice as much armor as we have in our present Navy fighters today. Don't cut down on that to increase the visibility. I don't want to sacrifice the armor because you lose your earphones every now and then."

Mr. STEPPE: "In connection with this protection versus visibility we have been making some studies on it and we have an idea now that might be of interest to you. We are proposing to hinge a headplate so that when the canopy is closed the armor swings down very close to your head. As you probably noticed in most of these airplanes due to the limiting space for exit, the armor is quite far back from your head in normal flying position. Of course that has to be maintained, but the thing we are working on is to hook the thing up to the canopy so that when the canopy is closed, the armor-plate hinges forward and comes very close to your head with the resultant protective angles but still maintains the basic head and shoulder silhouette that we have in the P-51 now. If the canopy is jettisoned the armor-plate snaps back out of the way and in fact improves the emergency egress."

Commander RAMSEY: "That seems like a very good idea."

Mr. FOOTE: "It has been stated that we will have to choose between protection and visibility astern. We have something in mind, that is in the specifications that have been given by the Navy for protective angles, that is for normal flight.

"Say that you are shooting at someone else, people are pretty much in the same position in the cockpit-as close to the bulletproof protection as they can get. Now I think we have a particular advantage in this bubble canopy in all positions. We can provide for protection when the pilot is up against his armor, and we can have vision astern when you want to look around. Now as we have suggested the chances of getting shot just the moment that you are looking around is nil. So I think we can reach a compromise there by getting full protection when we are on the offensive and when we are looking around we can disregard it."

Commander BOOTH: "Well, I was going to remark that on this question we have not shown much imagination, but I am glad to see that Steppe's remarks were excellently taken. I think that if we could use the armor plate, which has always been very difficult, it is a very good step in the right direction. It seems to me that a study of the cut-outs could be made, to give you some vision at your eye level and still give you protection where you want it – mainly in back of the head and shoulders."

Mr. BECK: "I would like to ask Mr. Hall what the direction of motion on the jettison is, does the bubble lift up, or move aft? If it does move up – how do they take care of the small lip that they have which the canopy fits in under?"

Mr. HALL: "We are not at all sure of the direction of motion, as I have said before. We have not yet jettisoned this canopy but it has been our hope that it will lift up from the rear, thereby allowing air to get in under and then blow up and aft with the canopy completely released all around. We are not at all positive that will be the case and if it is not, we may have to relieve the groove into which the forward edge of the bubble fits at the windshield frame. That is not far enough along in our experience to answer completely."

Mr. BECK: "We ran into the same trouble with the FG1. Our directional motion on jettison is vertically upward and we also had a slight lip on the windshield frame. When we tried to jettison the thing it would not release. We tried it on the airplane with some power on, we tied the canopy down, so in case it did try to go away, we'd save it and it did under full power. On the F2G it did release and did move aft into a position that would have carried away satisfactorily."

Mr. BURROUGHS: "It seems that one point in connection with bubble canopies and armor protection has been missed. We have a little piece of head armor directly above the pilot's head in the Corsair and in the South Pacific it is very popular and nobody wants to dispense with it. Of course that means you can't have a pure bubble type canopy. We have it on a bulged canopy and it is impossible to support it on the bubble type canopy. It is the same idea as the hingable armor plate because it moves aft when the cabin is opened."

Mr. VIRGIN: "I might add a few words regarding Guyton's discussion. I happened to experience the jettisoning of one P-51 enclosure. It was jettisoned from a full closed position by just using the emergency handle. Mr. Steppe made the statement previously that it raised straight up and just missed the tail of the airplane by about 25 feet at 150-170 indicated. We have a P-51F at Englewood which has the bubble canopy support in three places. They have a forward support on each track. They have a center rear support. When you jettison the canopy you rip the rear lock which allows the canopy to raise from the rear and becomes unlocked from the front track. We became skeptical about whether or not this would jettison properly so we ran a pressure distribution check on the canopy. We found that that particular idea was very bad. It will not raise from the rear according to our pressure distribution."

Mr. STEPPE: "Just to elaborate on that pressure survey a bit, we found that we actually had a down load on the back end of the canopy, a little more than we would like to have for aerodynamic reasons. In particular all these fighters now get a negative pressure peak just aft of the windshield bulb which tends to lift the front end of the canopy very quickly and there is a down load on it. I think just from looking at this F8F, you will find the same thing."

Commander RAMSEY: "Colonel Meng, we were talking about the different types of canopies and we started out with the F8F first. Will you discuss what you thought of that particular canopy."

Lieutenant Colonel MENG: "I'll just probably repeat what Colonel Toubman said. I talked to him about this rear vision. The canopy is nice and it has one thing that several of these bubble canopies do need, especially in the Army airplane. It has very nice handles for locking. I believe you can lock stop it in any position. However, visibility to the rear brings up a question as to what you have a bubble canopy for, if you are going to block it off from the rear. That is a question. You put the canopy in and you go ahead and run into stability trouble and loss of air speed. I heard some people talking about putting one block right across the rear of it. Now on the F8F I might have been sitting a little bit low, but to look behind, I would have to stretch up and make some effort to look back. The canopy was very nice with large head room. In fact you have more head room than shoulder space. I was wondering why some people put the bubble canopy in and close it from the rear. It looks better and the airplane is prettier."

Commander RAMSEY: "Bob Hall, do you wish to refute anything Colonel Meng said? I might add that I don't totally agree with the colonel. The few times I flew it, I tried to keep in such a position that my eyes would be looking directly into the sight. I had very little difficulty seeing to the rear; as a matter of fact, I could see just the rudder."

Mr. HALL: "As probably has happened in many other airplanes, the shape of that silhouette armor in the back has changed at least four times since the airplane has been under design and it was not possible to change the width of the bubble canopy quite that many times. As we originally proposed it, it was a pure silhouette and there was considered to be enough room to turn and look aft and almost see the tip of opposite stabilizer, but with the present configuration of that armor plate it is impossible. It is possible to see the vertical fin and perhaps 4-5 inches behind it on the other side. Many people who have sat in the airplane at proper height have told us that they considered that sufficient and if they think more is desired, the canopy will simply have to be widened. We don't like that idea very much because it was a hard job to develop that shape. But it is a compromise as far as armor plate goes. It is not a pure silhouette and it is not complete blockage."

Mr. WOOLAMS: "In the discussion of the bubble canopies, I think it should be emphasized that the characteristics of each individual design of the airplane should be taken into account. In the case of turn-over structure, we have given a great deal of thought to putting a bubble canopy on the P63 and after checking into accidents on the P39 we discovered we only had 3 turn-overs out of 10,000 which were out in the field. This to us did not indicate that it was necessary and nobody asked for it when the records were shown. On the other hand, if an airplane is very apt to go over on its back and if the C.G. will allow it, it might be possible to put adequate turn-over protection in the vertical fin, provided the C. G. will allow you to put structure back that far. In regard to the bulge in the canopy, if you have an airplane which has relatively poor visibility, and you need to stick your head out for ground strafing, it may be worth your while to put the bulge in the canopy. If you have the bubble canopy with a very narrow cockpit, with an oxygen mask, it may be necessary to put a bulge in the canopy. If you have good forward visibility, wide enough cockpit and plenty of room so you can turn around, I think that in future designs it would be a great mistake to put a bulge in a canopy. The airplanes we have now are not flying at a high enough Mach number to really run into too much trouble, but I believe in the future when we start flying at Mach numbers in excess of .8 in level flying, slight bulges on places such as the bubble canopy and other parts of the airplane will be a very great factor determining the top speed of the airplane and I think they should be avoided if possible. There is one other point on the jettisoning of the canopy; due to the high negative load on the forward part of the canopy, several people who have been working on it have found a great deal of difficulty with high loads on the jettisoning lever. We have run into that also. One possible way of doing away with that particular difficulty is to have a spring loaded pin in which a very strong spring does the actual pulling of the levers. The pilot does nothing but pull the pin out, which releases the spring. Some of the jettison loads run so high at such high speed that there should be a very long lever in the cockpit to give the pilot a mechanical advantage."

Commander RAMSEY: "Jack, has your company ever given thought to the German method of using shells?"

Mr. WOOLAMS: "I would like to ask the fighter pilots what they consider the best use is of this bulge on the side. They have been talking about vision aft but how about vision down or to the side? The F8 has comparatively small bulges in comparison to some bulges and that will restrict your down vision to the side. I wonder if that is a desirable feature. Also, is it desirable to jettison the canopy at very high speeds at which loads may build up where some device for jettisoning is required."

Lieutenant Commander CALLINGHAM: "On the question of the compromise between visibility and protection, I would like to know if any of the manufacturers have given any practical consideration to the combination of armor and bulletproof glass. It is provided on the Mosquito for the observer, not in the one we have here, but operationally. He has a folding panel behind him of armor plate which incorporates a circular section of bulletproof glass so he has full rear vision."

Commander RAMSEY: "That is a good point. Before we go into the answer, from Mr. Beck of Goodyear, I would like to call on Mr. Steppe."

Mr. STEPPE: "I was going to answer Mr. Woolams on this emergency release. I don't know if you are familiar with the mechanism of the P-51. Our release is very easy even though the actual loads on the tracks are very high. For example, on our rear track the design load is about 2,300 pounds. We use what actually amounts to a bomb shackle release on it with a sliding bar operated by the pilot."

Commander RAMSEY: "Have you ever gone into the thought of using cartridges?"

Mr. WOOLAMS: "Yes, we have. We checked into this Focke-Wulf rather thoroughly and what scared us was the maintenance. It just didn't seem to be worth while. It made another item that had to be maintained and watched very thoroughly. We finally hit on the mechanical means of doing it. It seemed to us to be a better idea."

Commander RAMSEY: "Relative to Mr. Beck's question on the high-speed release down alongside, I for one would like a bit more visibility at the side than the F8 has. However, if it means a decrease in speed or decrease in performance or lack of directional stability, I would say let it go as is. The high-speed release I think definitely we want."

Mr. BURROUGHS: "I would like to suggest that all manufacturers be required to jettison cabins in flight at three different speeds and photograph the track of the canopy as it leaves the airplane. I jettisoned a canopy from a P40; it didn't go upwards and it didn't go backward, but it went sideways and it caused considerable damage to my head. And I thought that if I jettison another canopy I will disappear in the bottom of the cockpit for a while. And I think it is very possible that many of our canopies will release unsymmetrically. It is very possible that it would go sideways and take the pilot's head off if the airplane is yawed slightly, or if one side jumps the track."

Commander RAMSEY: "You had better not let the Bureau of Aeronautics and the Army Air Force hear about that, Dick. They will probably put it in the demonstration requirements."

Mr. PARKER: "We have had several pilots in the Army jump out of the P-47's at an indicated speed of 500 miles per hour and had no trouble at all. They couldn't release it with the electrical gadget or hand lever, so the jerked the jettison and it came right off."

Mr. WOOLAMS: "I think Dick's point is very well taken on a canopy swinging sideways through a cockpit. We are fairly certain that one of our pilots at the factory had trouble due to just such a thing. He had some engine trouble, and never got out of the canopy and drove straight into the ground. We feel that the canopy swiped sideways through the cockpit and cracked him in the head. We think that is a very important factor, something which should be tested or pilots should be instructed that if they are going to release this thing, to get their heads way down and pull the release lever."

Mr. FOOTE: "I would just like to put a further endorsement on this trouble that has been had in jettisoning a canopy. It would seem very difficult to instrument this business and we have had the same trouble. The difficulty there is that the pilot is generally so excited he cannot give an explanation of just what happened. We had cases of canopies going off like that, very few cases; and we didn't have any story of what happened. We found quite a few scratches on the left-hand side of the cockpit and nothing on the right to indicate that it went off on the left-hand side in a twist. Another thing is that if the release is on the top of the canopy it would be very difficult to reach from that position. Perhaps we will have to go to a lower control."

Mr. VIRGIN: "We have considered demonstrating the jettison of the enclosure in the air. But I disagree with Burroughs because there are an infinite number of positions the plane can be in. If you wanted to demonstrate you couldn't do it in all. You would have to do it on your back and yaw it at different speeds. It is something we would be glad to do if the Government is willing to take the responsibility of writing off a few airplanes. But jettisoning two or three enclosures would not prove anything. I feel that Woolams has the right idea in trying to educate the pilots in the way we know it will jettison and to instruct them to put the plane in the right attitude, for instance the zero yaw."

Commander RAMSEY: "I am inclined to agree with you, Mr. Virgin. Also I think it is a good point about having the handle for the jettison in the proper position; I don't think it has ever been set in the Army or the Navy. It is in one location or another."

Mr. STEPPE: "I would like to suggest that possibly some of these screwy releases that have been mentioned here in rolling off to one side are caused by mechanical reasons rather than aerodynamical ones due to the fact that the release does not disengage at all points at the same time."

Commander RAMSEY: "The questions raised by Mr. Beck are still open: the high-speed release versus slow speed, the visibility down and alongside. What do you want in a bubble canopy? And also the jettisoning as part of the demonstration. Joe Parker, have you anything to say about spinning characteristics with the bubble canopy?"

Mr. PARKER: "Of course we did lose some of our directional stability. We tried spins and that gets back to a long story on spinning a P-47. If you don't try to do a proper spin, you have high angles of yaw and they are exceptionally noticeable with the bubble canopy. I have spun it without the dorsal fin and with the dorsal fin and the dorsal fin allows you to make more mistakes in the correct method of spinning a P-47."

Lieutenant Colonel MENG: "The pilot who jumps out very often does not have control and he wants to get out immediately. A large number of airplanes have been bailed out of in combat where there is no control. On this other point, about whether you jettison or not, I think that it should be dropped, if only once to see that it will release. That should be a requirement and also brought up to instruct the pilots how to do it."

Commander RAMSEY: "Colonel Meng has a very good point. Often the plane may not be properly trimmed for straight level flight when you do want to release the canopy. The plane probably is in yaw or a high-speed dive or some other attitude. Any other question opened on that? If not, we will go to the next item which is pilot comfort."

Captain TURNER: "I would just like to say that the Army is going all out for long-range missions, particularly in the Eighth Air Force. As a consequence of that in the South Pacific it has come to a point where some of our missions have a chance for running into in excess of 11 hours. Just recently we have been running tests on the P-47 aircraft and there are several things that have come up which have always been in mind but have not been of any importance on normal missions. The P-47 I would like to take up because that is what I know the most about. First, it seems as if the manufacturers have ignored the seat arrangement, how to make a proper fitting seat that will fit your dinghy and parachute. Always after about 4 hours little things become noticeable and after 6 hours everything becomes noticeable. In the airplanes that I have flown most prevalent was the cockpit ventilator. In most Army airplanes the ventilator of the cockpit comes from one central place. In the P-47 they have a single blast of air that goes up between your legs and in warm climates that usually is a hot blast of air. We have been running extensive tests on the P-47 where the temperature was between 80 degrees to 85 degrees at ground level. There should be a way of distributing cool air around the cockpit so that the pilot has uniform cooling all over his body, rather than just a single blast. After a certain time in the cockpit you get so you want to stretch your legs every now and then; but in the Army, due to the small pilots, usually in fighter airplanes they have the rudder so you can't ever adjust your legs to the full length. Most of the pilots that have flown long range missions say if the rudder pedals could be more adjustable to fit any pilot to the full length, so that at times in flight you could still get your legs stretched out, it would mean a lot.

"The day before yesterday we had a conference in Washington with the Air Surgeon and the Air Surgeon from Randolph Field who has been doing work along this line. The majority of the discussion dealt with pilot comfort and what they were going to do to make him more comfortable. It was decided that mental strain had more to do with the pilot's fatigue than anything else. So the suggestion of an automatic pilot on missions in excess of

6 hours was brought up and O. C. & R. in Washington said they were working on that; at the present time we have not got an automatic pilot that would give us the proper amount of performance without loss of weight. The automatic pilot we have been working with weighs 90 pounds; most pilots don't want to sacrifice that much weight for an automatic pilot. Also in Washington they brought up the fact that a different design of the pilot's seat should be made and the possibility of hanging the dinghy or parachute on the frame of the seat to take the weight of the cylinder and with a canvas type arrangement for the pilot to sit on which would take his weight off the dinghy. Right now Wright Field is going all out to get that made. One of the other things that causes a lot of mental strain is the thought of navigation. Now the Army does not navigate over water as much as the Navy and that was considered pretty much of a problem. If these boys are going to run missions between islands there must be some homing device. The bombers that they will be escorting have sufficient speed at altitude, and it will be necessary to have some means to home on the bombers. There should also be some homing device they can use within 100 miles of their home base. A suggestion of a radio compass was made. The other thing we have just been discussing is the canopy. Those canopies on our normal missions you don't notice very much, but down where the sun gets pretty hot after a certain length of time, the sun beating down on that canopy gives you headaches and causes your eyes to hurt, even with sun glasses. The O. C. & R. in Washington said they are starting to work on a canopy that is still clear, since Wright Field said that the tinted canopy was unsatisfactory, and would eliminate about 60 percent of the ultra-violet rays. The other complaint was the fact that on your silver airplanes with the bubble type canopy, after flying out at a certain distance, the sun is going to give you a definite reflection off your wing, and no matter which way you turn your head you will get that effect. They were thinking that possibly some dull paint should be put on top of the wing to cut down the reflection of the sun. That is about all I have."

Lieutenant Colonel TOUBMAN: "I could extent it just a little bit further. Going into this long-range operation with the P-47, 51 and 38, there are usual problems that will interest the engine manufacturers. Up to this time most of the pilots have brought their RPM no further back than the green line and it is going to take quite an extensive training program to get them to use these low RPM's in the 15, 16 and 17 hundred RPM range. We have been running into plug troubles. Furthermore, in all these propellers, especially paddle blade propellers, you will find rough spots in one range or another and you find in certain airplanes definite rough spots at the RPM range we like to run at. So much for that. I would like to find out if any of the manufacturers have thought about air-conditioning the cockpit? I think you find a good bit of heat is not just from the engine but due to the high speeds, especially these high speed dives. Also, there is the matter of providing food for the pilot. He is going to have some sort of nourishment; we don't have to give him a whole 7 course dinner, but he has to have something for 12 hours."

Mr. WOOLAMS: "There is a problem in regard to proper ventilation when you get into pressure cabins. If your cabin is sealed off there is very little circulation of air. The air has a tendency to rise in your cabin and your head and shoulders get as hot as the devil while your feet are not. It has been quite a problem on our P59. We have put a couple of electric fans in the cockpit alongside and underneath the seat to stir the air up, and find these fans don't do the trick."

Commander RAMSEY: "I would like to hear a little more comment from Captain Turner on pilot comfort: 1. Has he experienced any detrimental effects due to reflection of propellers? 2. The Navy in some cases have selected power and engine combinations with fairly high gear ratios and taking somewhat of a drop in performance. This is due to the fact that they climb to tip speeds which are extensive and disturb pilots from the noise angle. In that case a lower gear ratio in the engine and perhaps a slightly larger propeller diameter would change those tip speeds. I would like to hear some comments from Navy pilots on that point."

Captain TURNER: "I would just like to say that all of our missions were run with propellers that were painted and the RPM's were running down as low as 1,000 and 1,550. At this speed I did not experience any difficulty with the propeller, no annoyance of any type."

Commander RAMSEY: "Does any Navy pilot have anything to say on range hops? As a matter of fact, Navy hops are limited to 4-5 hours as the maximum."

Lieutenant Commander OWEN: "Our cockpits on present fighters are noisy and after 2 hours you are very much aware of that. You pad your ears with cotton, but you still hear the engine running 3 or 4 hours after you get back on the deck. I wish we had a Navy fighter today as quiet as the P-47. We don't have a quiet Navy fighter that flies today, that is all there is to it."

Mr. STEPPE: "There is one point of interest in regard to cockpit ventilation. As you all probably know, the P-51 has never had a ventilation system to brag about. One of our pilots hooked up a flexible hose we were experimenting with to find out how the cockpit would feel with cooling air. He didn't have much luck so he stuck it in his flying suit. The suit was all closed up and puffed him up, like a diving suit with not much pressure. Of course, while it did not increase the pressure very much it increased the vaporization. This made it very uncomfortable."

Lieutenant Colonel CANAVAN: "I think this discussion on excessive cockpit temperatures is very interesting. I think the tinting of canopies is rather begging the question. I think you will find about 25 percent of cockpit temperature could be absorbed from the canopy, about 25 percent from engine heat, and 50 percent is caused by air friction heat when you go over 350 miles per hour. I don't think you can induce flow through the cockpit properly to cool the pilot, and in some of the present airplanes (the P-51 for example) if operated in tropical areas you will find that you actually get a temperature of the body over 110 degrees in certain operations. The pilot cannot survive under these conditions and I think the contractors are going to give more serious consideration to mechanical means of lowering the temperature."

Lieutenant Colonel MENG: "Talking about propellers, we have at the present time in the Army a bomber with the engine located approximately the same distance forward as on the F7F. It is not the blade that annoys the eyes so much as it is the high spots on your cowls and other spots on the airplane which reflect the sun; also the blade keeps flickering. After flying the F7F for 15 minutes, it got so I didn't look over where the sun was shining. Maybe I was susceptible. I wonder if anyone else noticed that on the F7F?"

Commander RAMSEY: "Are there any comments on that? Then let's continue the discussion on seats."

Mr. A.B. HELLER: "Because of my height I have a great deal of difficulty in fighter aircraft to get suitable room for my legs. The adjustable pedal idea in allowing a full swing is very satisfactory. In landing and taking off adequate brake pressure would require having the pedals back toward the pilot; in any long dives, it is satisfactory to have the pedals well forward. I am particularly interested in seats and I think I would like to offer a suggestion of having the seats wider so that in any long missions you could turn in your seat. Usually you are fastened to a parachute harness but if the seats were wider, it would give you a chance to turn sideways and give you some relief. If the arm rests were adjustable up and down it would give a measure of relief in long distance flying. In my own particular case, it may be against regulations, but I have seen it happen, particularly during low altitude flying, you tend to have a cigarette and there seems to be no ash trays or facilities for that. Also regarding particular glares on component parts of the airplane, some small adjustable glare shields seem in order that could cover various sections. Ventilation in cockpits doesn't have adequate control and if more than one source of air intake and possibly some exits would be considered, the pilot could control to a more measurable degree the ventilation he wanted. I am inclined to think that this fan idea with some adjustable shutters would be advisable."

Mr. WOOLAMS: "I don't believe that such things as ash trays and arm rests and glare shields have any place in a fighter airplane. I think it is an unnecessary amount of weight. In regard to glare shields, the pilot can wear glasses and get the same effect and I think they are just unnecessary.

"I would also like to say that the adjustable rudder pedals are not the complete solution to the roominess in the P-38. It is not large enough for more than an average size man, if he is carrying a one-man life raft, plus the necessary jungle equipment. There should be some method of adjusting the seat back and forth, so if you are full in the legs and large all around you can make adjustments."

Mr. MARTIN: "I think it has become increasingly apparent to everybody in the aviation industry in the last year or two that airplanes performance-wise are rapidly surpassing the performance of the pilot. My slurring remark on aerodyamicists applied to people who did not apparently put the proper price on some of the necessary adjustments which do involve pilot comfort.

"We at Lockheed had some conferences with Dr. Ross MacFarland of Harvard University in the last year. Dr. MacFarland made a survey in the South Pacific, and his basic comment was that pilot fatigue and crew fatigue were responsible for more accidents out there than Jap bullets. It is very difficult for us at the moment to put any mile an hour price on an article such as ventilation. I would like to point out that in the case of ventilation for instance, some people have tried to kill two birds with one stone. Heat, for instance, may appear to be an easy solution, but I don't think so. I think the problem of ventilation requires that a blast of air touch the pilot, with one very definite exception of the eyes. You cannot have any air flow across the eyes. They dry out in a very short time and the pilot's vision will actually be impaired. The heat problem on the other hand is quite the opposite. Nobody minds a blast of hot air on his person. For the heat problem we must have the air entrance to the cockpit and the flow directed away from the pilot.

"The actual determination of fatigue has not progressed very rapidly. It is difficult for us to tell whether vibration or noise or heat or mental worry about weather, a rough engine or navigation are prime causes for this very serious fatigue condition. It may be that we could do a good job with soundproofing and vibration proofing and ventilation of an airplane, but on an uncomfortable seat in the airplane the pilot may still get very tired. The ability of a human being sitting in one position is not very great. We all may have our personal opinions on our personal dislikes about comfort in airplanes but I think it rests with the medical boys to tell us actually and to put the proper price on those articles which do beat down our ability to sit there and take it. My last chat with Dr. MacFarland indicates no serious concrete results of that nature and it remains a battle now for comfort between the pilots who are flying these airplanes and the designers who are looking for speed, rate of climb or great weight reduction, to battle it out between themselves. Perhaps it is up to the Army and Navy and their medical units to tell us what they want in an airplane. I don't think pilot fatigue can be taken lightly."

Commander RAMSEY: "Gentlemen, I would like to call the meeting adjourned until 1415 this afternoon."

SEVENTH MEETING
20 October 1944
1530-1730

Commander RAMSEY: "This morning we got pretty well involved in discussion on bubble canopies and pilot comfort in the cockpit. We just heard from Lieutenant Colonel Tyler and Bud Martin of Lockheed on pilot comfort points – the characteristics that manufacturers should remember in designing the cockpit. The fighter planes in different theaters and for different uses must have different types of cockpits. The short range fighter does not have the same comfort features as the long range fighter.

"This afternoon we have a speaker from the Bureau of Aeronautics structures desk and also four combat pilots – two Army and two Navy – who are fresh back from the theaters of war. We would like these fighters to talk first. I want to call on Second Lt. R. Eluhow, AAF, who is just back from a combat tour in Europe."

Lieutenant ELUHOW: "Thank you Commander. I don't know just what you want me to tell you. I'm a P-47 pilot. As far as I am concerned, that is about the best plane for the type of work we have been doing. There are a few things we would like to have corrected, but after speaking with different members of this conference, I find they are being corrected. The enormous and unnecessary size of the bomb releases could be narrowed down to reduce drag. We don't see why they are so big under the wings. The glass elbows cause quite a number of planes to miss a mission due to breakage by vibration, etc. Another thing, a bubble canopy gives very good visibility, but it also acts as a hothouse up there letting the sun in from all angles which makes it hot inside the canopy. I don't know if anything can be done about that. I can't think of many things right off the handle that are bad about the P-47; but on the whole, I enjoyed it very much. As for pilot comfort in the cockpit – it would be good if any fighter plane could have a little pocket on the side for cigarettes and candy or something, because on a mission it eases your nerves if you can eat something or smoke. Pilots usually have their crew chiefs put that on for them. We have been carrying bombs, dropping bombs, skipping them, etc., and the plane I flew did a good job. The D-25's, according to my viewpoint, were not much better than the old type. The D-25 mushed more in taking off and landing and was a bit uncomfortable to fly. It trimmed a little harder. Some of the pilots get used to that, but I never did and stuck to my D-15. The new one has a small dorsal fin, and it would probably help in keeping trimmed better. There are quite a number of additions we do not know about over there and they may make it better. On the whole, I found the 47 to be very good in everything."

Commander RAMSEY: "Thank you Lieutenant Eluhow. Joe Parker, would you care to make any comments on his constructive criticism of the 47?

"I think the pilot commenting on the P-47 bubble type over the D-15 has reason to make those comments because the wing loading went up and consequently you had a heavier airplane for the same amount of wing area. However, people in the South Pacific used their war emergency rating, which was 2535, to a better advantage for take-off's especially. I thoroughly agree with his other comments on the comparison between the two planes."

Commander RAMSEY: "I would like to call on Capt. John Kessler, AAF, who has just got back from the combat area."

Captain KESSLER: "I am going to have to disagree with my colleague here because I flew about half in 47's and half in 51's. We went all through last winter with 47's and we were the first group to convert to 51's. We were sort of put out about it. We felt if you got hit in an in-line engine, you were surely not going to get back, where you would with the radial. We got them there anyway and our victories soared. I will always say if you put a good pilot in a P-51 and an equally good pilot in a 47, the 51 will come out on top. Of course, when you get over 30,000 feet it is a different story. The P-47 is strictly for that altitude. I really didn't find much wrong with the 51. Not much criticism for it. I liked it. When the new D's came in, I had been flying B-2's, B-5's, B-7's and all along the line and I had converted my B-15. I had got hold of a British canopy – a Spitfire job and got very much visibility, as much as the boys with the new D canopy. Of course, we didn't get a good rear view mirror and that is one of the most essential things in a fighter aircraft. We depend on each other for cross cover. As long as you have two guys, you don't have to look behind you. The other guy covers your tail. Once in a while you are by yourself coming home. Once my rear-view mirror in the 47, which I think is one of the best, saved my life. If they could put a good mirror on that D canopy, I think they would really have an ideal situation. There is a little distortion. But I really can't say there was anything drastically wrong with the 51 as I found it."

Commander RAMSEY: "Thank you, Captain Kessler. Mr. Steppe, would you like to comment on that remark on the rear vision mirror on the D canopy?"

Mr. STEPPE: "We really haven't done much with the rear view mirror. It was in the early P-51A's and the A-36's and most of the comments were bad. We dropped it beginning with the B's and have not done anything with it since."

Captain KESSLER: "When we first got the old B's with the folding canopy with sliding windows, we got some Spit mirrors which we mounted right above, directly flush in the middle as we looked up, and that gave us very good cover under our tail as well as in back of our tail. You do not need cover up and behind as much as down and below. If they could design some kind of mirror like the Spit mirror, I think they would really have something."

Lieutenant ELUHOW: "I just noticed on a gun sight that the British don't use an extra glass for the reflector for the reticle. They use the window for a shield and as a reflector for a reticle set at a 45 degree angle. The D-25's have their glass set at that angle and they have another glass mounted on the gun sight which they use for the reticle. It gives double vision because the light shines right through to the other glass and you have two reticles. That glass could be used to give more vision in front. It obstructs vision a bit. I noticed in the British planes that they do not use that glass and you get a bit more vision. We could get the same idea and use it.

"The safety belts have straps across the pilot's shoulder and that wide strap across the pilot's lap. The British have the triple strap. You are held in by your shoulders. When I flew the Spitfire I noticed you can move the lower portion of your body around and move it from spot to spot, where with the strap across your lap you can't move around much, you get blisters."

Commander RAMSEY: "Thank you. Would you care to comment on the history behind the sight using the special glass for the reticle, Joe?"

Mr. PARKER: "The only experience that I have had with that gun sight is practically the same as the pilot who spoke. We found the same double reflection on the upper portion of the flat front windshield. Way back in the progress of moving gun sights up and down for a satisfactory operation, we suggested reflecting the gun sight on the flat front windshield and we had an investigation about grinding the flat front windshield for perfect vision. We found that the glass we were using was more adaptable for that use than the reflector. However, I don't know if we have had any changes in that. But the 348 Fighter Group in New Guinea lowered the gun sight exactly 180 degrees from what we had been doing, got down near the floorboard, reflected it onto the flat front windshield which gave more vision, and more pilot safety. I don't know what's going to be done about that."

Commander RAMSEY: "Is there any other contractor who knows about some thought being given to that particular feature?"

Mr. STEPPE: "Sometime ago at the request of the British we made an investigation on this and the glass manufacturers practically gave us the bum's rush. They said they couldn't hold the spec tolerance on the angle of refraction through the glass. In other words, they couldn't hold the spec tolerances within the limits of parallelism that we gave them. The attempts we made on it got terrific double images. We more or less dropped the thing and will go back into production on the lead computing sight. It will work reflected on the armored glass."

Lieutenant Colonel TOUBMAN: "I want to ask the Grumman representative what they are doing on the F8F. I got double images on the glass they are using in the windshield for a reflector. Is that optically ground glass or are they trying to get away with an old bulletproof windshield?"

Mr. HALL: "The F8F windshield is not specially ground bulletproof glass. All the 6F's are reflected on the bulletproof and most are pretty good. We are able to get glass ground to the specification limits. We have considerable trouble in getting it, but we are able to get it in production. We have found other things affect the double image besides the actual accuracy of the bulletproof glass and some of the sights themselves will produce the double image. In the case of an unacceptably bright second image we have changed sights and eliminated it without changing the glass at all. We believe we can reflect the sight image on the windshield successfully even with a lead computing sight and intend to do so on the 8F."

Lieutenant Colonel TERHUNE: "I might add that a while back we got the idea of putting this perfect glass in the front of the windshield from the British and other people interested in the idea and, as the North American representative stated, the glass manufacturers were positive they could give us a satisfactory glass. On the experimental planes we are now buying we are using that type of glass. The K-14 compensating sight changed the picture and we do not need it right now. But if that sight goes out and a new one comes in enabling us to use the windshield, we will probably use it because the planes are all made so we can. That does not include planes in service now, but it is a thought on the subject."

Mr. N. SIEGAL (Chance-Vought engineer): "We have had a lot of good luck using bulletproof glass as a reflector for the gun sight in the F4U1 airplanes and the practice is continuing with the F4U4's. I would like to ask Mr. Parker about the installation on the front of a P-47 that was attempted. These gun sights are definitely restricted in the distance from the pilot's eye to the gun sight itself. With the limitation on that distance, I don't see how you can get efficient operation of the sight by lowering it to the floor board."

Mr. PARKER: "I don't think that that was a perfect method of gun sight mounting. Several pilots took it out on ground strafing missions and said it was satisfactory. It was not on the floor. It was about where the parking brake handle comes out of the lower part of the instrument panel."

Commander RAMSEY: "Is there any other comment regarding using the front flat surface of the wind screen as the reflector for the reticle? How about the curved bulletproof glass the English have now? Is there any attempt to reflect off that?"

Lt. L.P. TWIST, RNVR (British Air Commission): "I am afraid we didn't know they had any curved bulletproof glass."

Commander RAMSEY: "Any other comments on that subject? If not, it might be of interest to the manufacturers to know that with the work on long-range fighters the Army has made great strides. I am afraid the Navy has been asleep on that, but our needs are not as great. However, Mr. Gates, the Assistant Secretary of the Navy for Air, has assembled a board of advisors on increasing the range of Navy aircraft. I am going to have Lieutenant Wilds take the floor for a few minutes and tell what is meant along that line."

Lieutenant WILDS: "This range panel which Mr. Gates has formed may include representatives of your companies. They are bombarding the question from a number of angles, not only the obvious ones respecting performance of the airplanes and engines and propellers, but from the aspects of pilot training, maintenance, traffic patterns, fuels, safety factors, etc. The problem centers chiefly around first what can be done to narrow the spread between the rated performance of carrier types and the actual combat performance of those types, and second what can be done to build in more range. This subject obviously is of increasing importance to the Navy and as it bears on the topics that are discussed here I think it would be valuable to that range program to have in this record any comments, experience, ideas which would benefit the subject."

Commander RAMSEY: "Does anyone want to comment along that line at this time? I take pleasure in presenting Lt.Comdr. E.W. Conlon of the Bureau of Aeronautics, who will give a talk on airplane structures."

Airplane Structural Design

Lieutenant Commander CONLON: "Commander Ramsey. I don't want to make a speech, I shall just give some comments intended to provoke discussion.

"Perhaps the best way of opening this session would be to review briefly some of the major changes affecting airplane structures which have occurred. Just prior to the national emergency which preceded the war, fighter airplanes were required to be demonstrated to terminal velocity in an 8g pull-out. One of the last airplanes to be demonstrated under this procedure was the F4F. About the beginning of the war something happened, and fighter airplanes refused to be demonstrated in these conditions. Although the stress analyses showed that 8 1/2 g was available at the original design gross weight and at speeds which approached 550 miles per hour, the test pilots found that it was unsafe to approach these conditions. Various airplanes encountered buffeting and loss of stability and it appeared that some phenomena not thoroughly understood by airplane engineers were being encountered. As a result of the urgent need for airplanes by the Navy, production was authorized without a formal demonstration of the airplane, and the airplanes were restricted to limits recommended by the contractor.

"Most of you will recall that the demonstration is conducted by the contractor for BuAer, to demonstrate that the airplane is ready for trials which are conducted by the Board of Inspection and Survey. During the first 2 years of the war, the Bureau was frequently reminded by BIS that several thousand airplanes of a particular model had been delivered to the combat forces and as yet the airplane had not been demonstrated. The Bureau and the contractor engaged in lengthy correspondence on the subject but it was not until the parties concerned initiated structural flight testing on well-instrumented airplanes, that it became apparent that there was a boundary for each model airplane which could be defined in terms of CL and Mach number, or speed or G.

"The first slide (copies of the slides follow the pages reporting this talk) shows a limiting curve for an SB2C airplane. You will note that at the boundary conditions the pilot experienced fin oscillations in the low-speed high G range, tail buffeting and stall at slightly higher speeds and that as the speeds were increased the severity of the buffeting increased. It will also be noted that at the highest speeds obtained and at a Mach number of about 0.69, the fin oscillation as well as tail buffeting again became apparent. Slide 2 depicts similar boundaries for the F4U, F6F-3, F6F-5 and F7F airplanes, and in addition it gives a few clues as to how the boundary may be moved outward, thereby increasing the combat effectiveness of the airplane. You will note that the buffet boundary of the F6F-5 is appreciably higher than that for the F6F-3. This improvement was accomplished by an extensive structural flight test program conducted by the contractor. Initially a stabilizer failure in the final demonstration in the low-speed high G range made such an investigation imperative. The first flight tests on an instrumented airplane indicated that a vibration at 3,600 cpm increased the stress in the stabilizer 30 to 50 percent over that due to the normal loading. A reinforced stabilizer was designed with approximately 40 percent increase in strength and with a natural frequency of over 4,000 cpm. (The original stabilizer had a natural frequency of approximately 3,600 cpm.) The test pilot duplicated the demonstration conditions which had previously produced failure and once again the stabilizer failed but this time adequate records were obtained. These records showed that the condition which produced failure was a dynamic loading at about 800 cpm which increased the stress over 100 percent. This information cracked the problem. By duplicating flight conditions on the ground with a dynamic loading device which looked like a Rube Goldberg set-up, the contractor was able to try various structural modifications to determine one which would minimize the response of the structure to the exciting forces. The stiffening of the fuselage was of primary importance rather than increasing the stabilizer strength. The outward movement of the buffet line of the F6F-5 represents the tangible results of the investigation. Particular attention should be given to three points:

"(a) The problem was cracked by flight tests in an instrumental airplane to the actual conditions which produced failure. Approaching these conditions did not yield the necessary data.

"(b) This flight was possible because the type of structural failure was not catastrophic. It is proposed to discuss this subject a little later.

"(c) The engineers solution was obtained by duplicating flight conditions on the ground where the engineer could actually see what was occurring.

"The F7F buffet lines at 10,000 and 20,000 feet illustrate a second point of interest in the general subject of buffet boundaries. These lines represent approximately the same intensity of buffet and the fact that the 20,000 foot line is appreciably above the 10,000 foot line indicates that the intensity of buffeting is a function of the dynamic pressure "q." At higher altitudes "q" is lower for the same CL and Mach number and the exciting force appears to be proportionally less and develops at a slower rate during a maneuver. Therefore an initial exploration of the limitations of any particular airplane can be made at high altitudes, thereby considerably increasing safety to the pilot.

"If we may conclude that the buffet line is not fixed but can be moved, although such movement may entail considerable research and modification of the airplane, the reason becomes apparent for the present clash between the Bureau and contractors as to what limitations the contractor should guarantee for new airplanes. Actually, the Bureau and contractor wish to achieve the same objective, that is to keep one jump ahead of the enemy and to maintain the 20 to 1 ratio of superiority which has been reported by the newspapers as the result of recent engagements in the Pacific area of operations.

"The data which engineers need to design airplanes with higher buffet boundary are not immediately available. Unfortunately, the correction factors for high-speed wind tunnel data have not been determined and therefore these data cannot be used for structural design of the airplane. It is anticipated that this condition will be corrected in the not too distant future, and in the meantime the Bureau is cooperating with the NACA and Army in a compressibility research program. The NACA is dropping models representing surfaces of revolution to determine drag at high Mach numbers and is conducting instrumented flight tests on a P-80 furnished by the Army. The Navy is constructing free-fall models of airplanes to be dropped from high altitude. In the latter tests the models will be radio controlled for the purpose of making high G pull-outs and the data will be transmitted to the ground by telemetering. Flight tests under radio control are also being conducted on complete airplanes but to date the most efficient process for solving specific problems has been made by flight tests of well instrumented airplanes under pilot control.

"The preceding discussion has dealt with the buffet boundary in general terms but has not dealt with the general cause of the phenomena which are commonly designated as compressibility effects. In a recent NACA Advanced Report by Rodey, entitled "Correlation of Flight Data on Limit Pressure Coefficients and their Relation to High Speed Burbling and Critical Tall Loads", this problem is discussed in more detail and the effect of skin bulging and wrinkling is considered. As wrinkling occurs at the high load factors, the lift coefficients at which burbling occurs are reduced and the maximum load factor which can be obtained is thereby lowered. The existence of bulging and wrinkling makes it practically impossible to predict the exact pressure distribution at higher loads and speeds. It would seem therefore that every effort should be made to design wings of high performance airplanes to have rigid coverings rather than to permit wrinkling and bulging of the skin. This is the reason for the Bureau's introduction of nonbuckling requirements for fighter airplanes. Considerable progress had been made in achieving the desired results without prohibitive increase in weight or cost of manufacture. The use of more spanwise stringers of smaller area and heavier gauge skin has been of material aid. The Vought Company has attacked the problem from a different angle and has produced a sandwich material which they have designated as Metalite.

"The Bureau is procuring a number of experimental F4U stabilizers of aluminum-alloy balsa sandwich-type construction from the Chance-Vought Company. It is believed that a brief description of the sandwich material developed by Vought, and some remarks on the stabilizers and their advantages over the prototype metal stabilizers would be of interest to this gathering.

"The sandwich material is essentially as shown in Slide 3. It consists of thin sheets of aluminum alloy separated by a relatively thick, low density core of balsa wood whose grain is perpendicular to the metal faces. Moderate heat and pressure are used to bind the balsa to the face sheet. The bonding operation is accomplished with parts or assemblies in a mold of the desired shape. For flat work, the parts are normally put together on a bench and the whole assembly placed in a mold afterwards. For curved work, both single and double, several different methods can be used to form the work. Where only gentle curves are required, the work can be assembled flat on the bench and then the entire assembly placed in the mold and forced into the desired shape by the application of pressure. Where sharp curves such as are encountered in the leading edge of a tail surface or wing are to be formed, it has been found expedient to bond the metal sheet which is to be on the outside of the curve to the core while flat, and then form to the desired shape. The inner sheet is then bonded in place in a second operation.

"Due to the thickness gained by the light core, the bending stiffness of a sandwich panel is many times greater than that of a simple sheet of metal of the same weight. Because of the high strength obtained from the metal faces and consequent reduction in supporting elements, a saving in weight over conventional structures is generally obtained. Moreover, the continuous support of the metal face sheet afforded by the balsa makes possible a surface which is markedly free from wrinkles when under load. In a good may instances, test specimens in compression and shear have been loaded beyond their yield stress before wrinkling occurred.

"The experimental F4U sandwich stabilizers when compared to the prototype metal stabilizers will have an increase in bending strength of 20 percent, an increase in torsion strength of 80 percent and an increase in torsion stiffness of 50 percent, all this with an estimated zero increase in weight. The increase in torsion stiffness is of particular significance since it will preclude undue twisting of the stabilizer tips under load and resulting outward shift of spanwise center of pressure. Of further significance is the fact that the stabilizer surface will be nonwrinkling which should result in improved performance at high Mach numbers.

"The stabilizer consists of an integral outer shell of sandwich material comprising 0.012-inch aluminum face sheets and 0.25-inch balsa core. The balsa is completely sealed by extending the edges of the metal face sheets beyond the balsa core, and bonding these extended edges together. Slides 4 and 5 show the degree of surface smoothness obtainable in a shell of this type. This shell is stiffened by three intermediate ribs and an outer-end bulkhead at which the stabilizer tip is attached. No longitudinal stiffeners are required.

"The three intermediate ribs are shown in Slides 6 and 7. The method of attaching the ribs to the shell is interesting. Two T-section extrusions are bonded to the upper and lower surfaces respectively. The rib is then riveted at top and bottom to these extrusions, thus assuring proper fit.

"The next slide, number 8, shows the stabilizer spar and three-point fuselage attachment fittings. The front attachment fitting is connected to the shell in a manner similar to the intermediate rib attachments, that is extruded T-sections are bonded to the top and bottom skins, and the web to which the front fitting is attached is riveted to these extrusions.

"The rear spar is riveted to the shell in a conventional manner.

"Various problems will undoubtedly arise in connection with the maintenance and repair of these sandwich stabilizers. It is expected however that the merits of the stabilizers will be sufficiently outstanding to warrant such extra maintenance and repair measures as may be required.

"In the front part of the discussion it was pointed out that the F6F stabilizer failure was not catastrophic. That brings up the question of controlled failures, and especially breakable wing tips.

"The breakable wing tips' are tips designed to break off when a specified load factor has been attained, leaving the airplane with adequate wing area and control to effect a safe return to its base of operations. The special significance of the breakable wings tips lies in the fact that the use of this device permits an appreciable saying in structural weight of the wing. This can be illustrated best by referring to the Model XF8F airplane which incorporates breakable wing tips. The XF8F wing is designed for an ultimate load factor of 12g with the wing tips designed to break off at load factor of 8.6g. Compare this wing with a hypothetical 12g wing of same dimensions but which does not incorporate breakable tips. As the air loads on XF8F wing and hypothetical wing are increased, the wing bending moments will increase in identical fashion up to a load factor of 8.6g. At this value of g, the XF8F wing tips will break off, with consequent inboard shift of the spanwise center of pressure and marked reduction in wing bending moments. As the wing airloads for XF8F and hypothetical wing are increased further to 12g, the wing bending moments for both wings will increase, but the bending moments for the hypothetical wing will continue to remain about 20 percent greater. It is this difference in bending moments which accounts for the saving in wing structural weight which can be realized by using breakable wing tips. In the case of the XF8F airplane, this saving will amount to 150-200 pounds.

"The use of breakable wing tips should not affect appreciably the combat effectiveness of the XF8F airplane. The breaking load for the wing tips of 8.6g is in excess of any load factor a pilot will normally attain, and it is anticipated, therefore, that very few wing tips will actually be broken off in service. In the event the tips do come off, the pilot nevertheless will be able to return to his base safely. Certainly any pilot prejudice against flying the airplane minus its wing tips would be overcome by the increased airplane efficiency resulting from the saving of 150-200 pounds in weight.

"At the time consideration was being given to the incorporation of breakable wing tips in the Model XF8F, it was decided to prove the practicability of the breakable wing tips concept by conducting preliminary tests on the Model F4F airplane. The model F4F-4 airplane was chosen for these preliminary tests because of its similarity to the Model XF8F-1 airplane in size, general arrangement and wing structure. (See slide 10.)

"The procedure followed and the test results obtained on the Model F4F investigation are described below. Calculations were made to determine the shears, bending moments and torsions at the breaking section. The results of these calculations were compared with flight test checks on shears, bending moments and torsions. On the basis of this information, the breakable wing tip was designed. (See slides 11 and 12.)

"The wing tip was designed to fail at station 186 at the first rib outboard of the center aileron hinge. (The wing tip area therefore constituted approximately 15 percent of the total wing area.) All spanwise structure was cut at station 186. The wing tip joint was designed exclusively of riveted splices with excessive strength on the inboard side so that failure would be produced by shearing of the rivets in the outboard side of the splice. To do this, a wide-flange pressed rib was used to back up the surface skin splice with a double row of rivets inboard and a single row connecting the outboard skin to the rib. A splice plate was used to tie the main beam together, with a similar splice plate used on the rear beam. The aileron was completely cut on the outboard side of the center hinge. A rib was inserted on each side of the cut so that the two portions of the aileron could be joined together by dowels through the web of the ribs. This dowel connection permitted the transmission of torsion but would fail on application of bending moments. The two portions of the aileron were covered separately, and the space between the ailerons covered up by doping on a strip of fabric.

"The wing tip joint was designed to fail initially on the under side of the joint near the leading edge, followed by failure of the front beam splice. The rear beam was designed to fail last with the wing tip portion of the aileron breaking off at the same time. This sequence of failures permitted the leading edge to twist up first upon application of the failing load, to make certain that the wing tips would clear the airplane proper upon breaking off.

"The wing tip then was static tested to failure successfully, the failure occurring at 4.4g. Slides 9 and 13 are photographs taken after the static test. Subsequently, the wing tips were installed on a flight test airplane and subjected to a symmetrical pull-out. Stills taken from a motion picture of the flight tests are shown in slides 14 and 15. Slide 16 shows an F4F on the ground without wing tips. Results of the flight tests were as follows:

"(1) The wing tips failed cleanly with a combined bending and twisting action (leading edge up), leaving the remainder of the aileron undamaged.
"(2) The wing tips flew up and aft, clearing the tail by a considerable margin.
"(3) The left wing tip failed approximately .02 sec. before the right wing.
"(4) The elapsed time between the first visible wing tip deflection and complete separation was approximately .05 sec.
"(5) The wing tips failed first on the lower surface, the failure traveling progressively around the leading edge back to the rear beam.
"(6) Final separation of the lower surface occurred at 5.8g and 250 miles per hour approximately 1 second after the start of the pull-out. At this time the acceleration was increasing at the rate of 1g per tenth of a second. Assuming the failure started at 4 1/2g (based on static test results), this would indicate an elapsed time of .15 sec. between start of failure and separation of the lower surface.
"(7) The acceleration leveled off rapidly after the wing tips failed. The pilot noticed

only a slight jar when the wing tips failed. The entire pull-out took place in 3 seconds.
"(8) A landing without wing tips was satisfactorily accomplished, the wind velocity during the landing varying from 10 to 25 miles per hour (gusty) approximately 60 degrees across the runway.
"(9) Although the design of the F4F airplane was sufficiently successful to warrant incorporation of the breakable wing tips in the XF8F-1 airplane, the design of the XF8F-1 tips will present additional complications. Mach number effects at higher XF8F speeds will complicate the problem of calculating the design load distribution. Because of the high rate of roll, the possibility of a single type of failure occurring in unsymmetrical flight will have to be considered and the controllability of the airplane with one tip off established. (In this connection, the rate of roll for the F4F was not high enough to be critical for the wing tip joint.)

"However, it is confidently expected that the XF8F breakable wing tips will prove to be satisfactory, and that the significant weight savings to be achieved thereby will amply justify the effort expended in developing a successful design."

Commander RAMSEY: "Than you very much, Commander Conlon. I hope you could see those slides. Due to the fact that compressibility was mentioned in his talk, I think it would be a good time to take up that subject now. What are the various companies doing about it?"

Mr. PARKER: "I think we have licked our compressibility troubles to the extent of being able to recover from compressibility dives. However, we are still working on the control of the airplane in compressibility. On later designs, such as Woolam's company is contemplating. I think it is very important to have smooth flowing lines of the plane for immediate high speeds in level flight. Possibly that will run into some new information on compressibility. I still consider compressibility as a phenomenon. Too many people do not know too much about it, and I'm one of those who don't know."

Commander RAMSEY: "I am in the same category. Just the word itself is hard enough for me to spell. Has Vought got a representative who can tell what they are doing?"

Mr. W.C. SCHOOLFIELD (Chance-Vought Engineer): "We at Vought are not doing anything that is very remarkable about it except our damnedest to find out more about it and understand it better. I might mention a point in connection with the curves that Commander Conlon showed in his slides on the buffet boundary for the F4U1. We have been able to extend that boundary at high lift coefficients. We are going into it gradually and trying to find out more about it at the same time. We are preparing an airplane with dive recovery, flaps similar to those that have been operated successfully on the P-47 and also on the P-38. We haven't anything very remarkable to contribute at this time."

Commander RAMSEY: "Has Grumman done anything or does it contemplate anything in the near future on compressibility study?"

Mr. HALL: "We also have been going into this business very slowly and we don't know very much about it either. We are doing some instrumented flight research with a completely instrumented airplane in order to find out what does happen to our airplanes. When we find out what happens, we will endeavor to fix it. We are also preparing an airplane with dive recovery flaps but to my knowledge we haven't gotten into a situation where we needed to recover from high-speed dives. Our trouble has been with a buffet boundary which perhaps isn't compressibility at all, although it may be linked up with it. That in our case has been a structures problem as well as an aerodynamic problem. As Joe says, I don't know much about it either."

Mr. STEPPE: "Along with the rest of them, we have done but very little on pure compressibility investigation. We have done a little investigating on fabric-covered control surfaces and high-speed dives. We have taken movies and the results are rather shocking. We got pictures that show terrific bulges during straight dives and also during pull-ups and we have been playing with metal-covered surfaces and experimenting with changed angles of incidence on the fixed surfaces. We don't have much of a picture put together on that now, except that we had quite enough to dictate that the 51 coming out shortly will have all metal surfaces. Wright Field has recently run high speed dives on a P-51. I don't have all the details. Perhaps Colonel Terhune can tell us about that."

Lieutenant Colonel TERHUNE: "I can't add much to the picture. What we are trying to do is establish a base line on the whole picture and NACA, the Army and Navy are all working on it. One of our first projects, which was mentioned before, was the dive test on the P-80 in which we are trying to instrument the ship completely and get some sensible results. I am sorry I don't know anything about the results of the P-51 dive tests."

Commander RAMSEY: "I am sure that the Bell Company has had some trouble with compressibility. Jack Williams, will you say something relative to that?"

Mr. WILLIAMS: "We don't know very much about it either. We have conducted dive tests on the P-63 and got an approximate Mach number of 0.8 at which the plane started to porpoise slightly and if you go faster it porpoises more and more. At that point we decided we had gone far enough. We evidently get an alternate change in flow over the wings, changes due to compressibility. So far as the P-59 is concerned, on the dive test on the compressibility investigation, we would take it up to about 37 or 39,000 feet and point it straight down with the Mach number instrument in there and it finally got so we were coming down at about 10,000 feet almost vertically and hit a constant Mach number. Evidently we are running into compressibility of some sort because we didn't accelerate any further and we were getting down near to terminal velocity. We assume that compressibility occurred at various parts of the airplane and was one of the major factors in holding the airplane to a constant Mach number. We did not have any control difficulties whatsoever. I have pulled 6 G's in the 59 at a Mach number of 0.8 with no indication of any abnormal control force. It is a possibility that compressibility may be very much affected by slip stream. We will know more about it when we get more jet airplanes flying. We will be absolutely amazed at the lack of effect on the controls of high Mach numbers, as on the P-59. I think that Colonel Terhune may be able to shed some light in view of the P-80 dive tests. When we get more data on jet airplanes, we will find out more than we know now about compressibility."

Commander RAMSEY: "Due to the fact that we have had trouble with the Curtiss Helldiver, is Lloyd Child familiar with that phase of Curtiss development?"

Mr. CHILD (Curtiss-Wright pilot): "We have been working on it for the last year and have instrumented the SB2C very thoroughly. It was a more thorough instrumentation than anything we had before. Strain gauges all over it to measure pressure all over it and cameras all over it too. We are able to get up to speed and demonstrate the plane. We wouldn't have been able to do it without the use of these instruments. We learned from them what we could do and we're going ahead with that. We are also building a new wind tunnel which will be used to investigate it. I would like to ask Mr. Von Valkenberg, of our aerodynamics department, if he would amplify my statements."

Mr. Von VALKENBERG: "I am afraid I don't know too much about the SB2C, Lloyd, but I will say from the standpoint of aerodynamic design of new airplanes that we are trying our damnedest to permit the highest possible dive speeds by having the least possible compressibility effects from the standpoint of the data we know. Now, how that is going to work out, we are going to find out as soon as possible, I think by means of instrumented flight tests in conjunction with wind tunnel tests."

Mr. CHILD: "I would like to add that the moving pictures which we took of the controls of the vertical surface of the tail are very alarming to the pilots. It really amazed us how you couldn't feel oscillations and variations. The fabric was bulging terribly and we wouldn't have known it without the moving pictures. But it was the strain gage recordings that were constructive in solving the problem. The movies alarmed the pilots and it is better that they didn't see them."

Commander RAMSEY: "I agree with Lloyd on that scaring of pilots. I was out there when they showed those films at Columbus and Mr. Harrison of Curtiss swore he would never get into another Helldiver. Since then I have never gone over 250 myself, although I do know in the fleet they are going much higher than 250 to get those Japs."

Mr. GUYTON: "I might say that we took pictures of the same thing and got photographs of the fabric bulging. The pilot was not as scared as the engineers when we told them about it."

Mr. R.W. FOOTE (Eastern Aircraft pilot): "Our new ship will be our first experience with compressibility. The FM-2 just about ran into it in our terminal tests. It just about hit the border, so we haven't had to worry about it. On our new ship the main thing we are doing is planning as complete an instrumentation as we can and we're beefing up the tail considerably. It has been designed to operate in compressibility ranges and we are following along the conventional lines. We are going to install dive flaps."

Mr. B.O. MALMBERG (Eastern Aircraft engineer): "On that new airplane, we are providing strength in the tail for pull outs. On that same ship we are putting dive flaps but the intention is not to use them unless it is absolutely necessary. We are going to instrument the airplane's tail with strain gauges and recording equipment, and approach compressibility ranges very slowly."

Mr. MARTIN: "Except for the dive recovery flap on the P-38, I think the only thing we have developed along that line is a handsome respect for compressibility. I think the cleanliness of the P-80 will keep it out of trouble for quite a while. As one of the gentlemen pointed out here, the absence of a propeller may be the thing that will save us on that airplane."

Mr. HALL: "I would like to ask if anyone knows the results of the dive test on the P-51 without propeller along the lines of Jack Woolam's comments."

Mr. STEPPE: "I don't know anything about the details of the dives. I know that NACA at Ames was making them. After about five dives they had an accident but they were towing a 51 up without a prop and diving it. But they didn't get it worked up to high enough Mach numbers to get anything definite before the thing happened."

Mr. BURROUGHS: "It seems to me that there is too much or at least enough about what lift coefficients and Mach movements you can go to in each airplane that has been investigated. But there is not quite enough accent on the altitude at which you get these Mach numbers and lift coefficients. I don't think it is appreciated by service pilots in particular that the so-called compressibility effects and sudden increases in stability you get, the nosing down tendencies, are a function of not only Mach numbers, but also of lift coefficient. You find that if you dive very steeply you get into a vicious circle because if you are on a steep angle and attempt to pull out, the harder you pull, the higher lift-coefficient you get, the worse are your pitching moments. On the other hand, if you don't pull hard on a steep altitude dive you may not come out in time. Speaking from a piloting technical standpoint for investigating Mach numbers and G, we have found that it is most important to get the Mach number you will shoot for at the minimum angle at which you can get it. With power and a lot of altitude loss don't get to any angle steeper than you need to get to the Mach number. That means that at the time you get the Mach number you are at terminal velocity for the angle at which you are diving, and when you pull the least bit of G over that which you had in the dive, you are going to lose speed. If you go to a steeper angle than that required for the Mach number you will shoot for when you start to pull, you are not able to change your flight path and you go to a higher Mach number and become worse off than when you started to pull. I have never seen written up for fleet consumption or Army consumption any such ideas as these. But I think the reason you can get away with investigations that would sometimes prove very dangerous if done in the fleet is that most test pilots use that technique in their dive. There is one other comment which I think is important with regard to pull at high G's at the lower Mach numbers which Commander Conlon was referring to. For that condition, the tail loads are most critical although the Mach number is low, much lower than it is on the outer edge of the boundary at which you are going to get the highest balancing tail loads. On one occasion when I was pulling a high-balancing tail load – a high G at relatively low Mach number, a so-called accelerated stall point – I found that in trying to get the G, if you do not pull it rather suddenly you do not get it because the airplane will develop a high speed stall and all you get is a rough ride, but not enough acceleration to get the contract specifications. So I settled upon a procedure of snapping the stick back even more suddenly than I had done before, and in the process I pulled too hard and realized it the moment I pulled. So I immediately kicked the stick forward again to prevent over-shooting the G. That might have been a good idea hadn't we developed the records afterward and found that the tail load obtained in that dive was much worse than tail loads we had gotten at the same Mach numbers and lift coefficients. So when we tried to figure out how it could have happened, we found that by kicking the stick forward and attempting not to overshoot the G, the tail load caused by the pitch and acceleration of the airplane or at least the load required to put the nose back down again, augmented the balancing load and could have caused a failure. That is another remark I would like to pass on regarding the technicalities of investigating high dive speeds. If you ever think you are overshooting when you are pulling a high lift coefficient at a low Mach number, it is better to overshoot a little than it is to kick the stick forward suddenly. It is always important to ease the stick forward whether you overshoot or not. You prevent then the high loads on the tail to get the pitching acceleration."

Commander RAMSEY: "Does anybody have anything to supplement Dick Burroughs' remarks? The next question is, do we want Mach number indicators incorporated on the instrument panel of fighters? If so, when – during the test stage, during use in combat or what? I'll call on the British delegation."

Lieutenant Commander CALLINGHAM: "It has been some considerable time since I was in contact with the question of these Mach meters. About a year ago the tendency toward the idea of installing them in high speed fighters and even twin-engine high altitude interceptor fighters was definitely for. You have to make a quick mental calculation to find if you will get into trouble or not. At that time RAE had produced quite a useful Mach meter incorporating the altimeter and air-speed indicator and I know at that time very definitely that we were in favor of a Mach meter as opposed to an air-speed indicator for airplanes which are liable to run into compressibility trouble."

Col. Lee B. COATS (AAF PGC, Eglin): "I don't think that we can add anything to what Joe Parker has already said. So far as our testing at the proving ground is concerned, we haven't been using any Mach meters and I don't know that we shall go to them. However, it has been our experience that we have been able to wring these airplanes out and with the proper indoctrination and techniques of wringing it out, we have saved a great many pilots. In talking to the people from the English theater who have been through compressibility. I learned that they didn't know what it was until they got back to the states, and they had ridden it right out when they got on Jerry's tail and had come out all right when they got down to lower altitudes and could pull out."

Commander BOOTH: "I might estimate that the air-speed system on which the Mach meter itself depends is as susceptible to compressibility as any other part of the airplane, and I am wondering what the accuracy of the indicator is at the speeds you have to worry about."

Mr. VIRGIN: "I happened to have the opportunity of talking with Major Barsody of Wright Field following the dives he made in a P-51 in which he used a Mach meter and on the series of dives he claimed by the use of this instrument he had perfect control in hitting the Mach number he was shooting for. It aided him immensely in conducting the series of dives. One statement he made was that at the very high Mach numbers the instrument lagged at 0.03 just a small amount, but he claimed it was a material advantage in making those tests."

Lieutenant Colonel TERHUNE: "The thing I want to say is more in the form of a question. It is possible to put another hand on the air-speed indicator which could be set for each individual airplane for the limiting Mach number of that plane at sea level and all the way up to altitude if you calibrate it right and I wonder if it would be a useful gadget to have in the cockpit of a fighter airplane. This might be a small hand which, say at sea level, would point to 550 in some particular airplane and at 30,000 feet would be adjusted according to the flight tests that were made by the contractor. That would give the pilot an indication not only of the Mach number he was flying, but of the limiting Mach number at all altitudes. He could see it in just one simple glance at the instrument."

Commander RAMSEY: "I don't get just exactly what you mean."

Lieutenant Colonel TERHUNE: "You could add some kind of mechanism to the air-speed indicator which would be controlled by the density of the air and which would give you a reasonable indication at all altitudes of the limiting speeds of that airplane."

Colonel COATS: "I would just like to say that such an instrument will be tested very shortly by the Army Air Force. For each airplane tested it will be stopped at the maximum indicated air speed. At sea level then if the top air speed that the manufacturer puts on the airplane is 500, it will stop right on 500 and decrease as you gain altitude."

Commander RAMSEY: "Do any contractors wish to comment?"

Mr. WILLIAMS: "I was going to suggest the same thing and to clarify it further this Mach air speed meter is just another air-speed meter with a red hand on it, and as your Mach number changes with altitude so does the position of this red hand. All you have to worry about is that you don't change the air speed meter past the setting at your critical Mach number. I don't think it is necessary with airplanes which have gone through dive tests and show that there are no bad compressibility problems. In other airplanes which do have a critical Mach number I think it would be a big help. The Mach number instrument is very good for test purposes and it would be vastly superior to the regular Mach meter because it would combine everything into one instrument. That is about as far as it should go."

Mr. BAKER: "In the absence of a developed Mach meter number instrument, we at the Propeller Division in conducting propeller efficiency tests with an F2A2 airplane approached the Mach number problem by designing three rings around the air speed indicator which would give the Mach number for the speeds desired at 25,000, 20,000, and 18,000 feet in the speeds we wish to investigate, we found that we would reach a Mach number of 0.072. To evaluate this speed on camera film with which this plane was equipped we went into the dive series at 25 miles per hour increments and used this home-made method of staying within the critical Mach numbers of the particular airplane."

Lieutenant Commander CONLON: "I have wondered just a little about the use of this air speed indicator with the second hand, in that you will recall the load factors and Mach number of any airplane is a curve, you can get the hand at one point, but if you set it on the limiting condition of the airplane and you hit that point, the only way to slow down is to pull out at a higher load factor which means that you are going outside of the buffet boundary."

Mr. WILLIAMS: "You can make allowances for that by setting your meter at a lower indication than the critical indication and allow yourself a little bit of leeway for that."

Mr. Von VALKENBERG: "I would like to point out that I think it is possible to devise an instrument that will take care of that difficulty by combining the accelerometer and some sort of density, or it would be a pressure altitude capsule really and then you get a hand which would follow a calibration that would follow the curve of cl versus Mach number."

Mr. PARKER: "My experience with compressibility is very limited. I have made about 100 dives with compressibility but it seems to me that a combat pilot, when he gets into compressibility, he's in. He's not on the boundary line like we are discussing here. This instrument and instrumentation is very fine provided you are doing tests and shooting for certain Mach numbers like the Vought pilot has just remarked here. But there wasn't a damned thing I could do about it when I got up to indicated speeds. I just brought the damned thing down and I don't think an instrument would have done any good at that particular time. You must note now that going into compressibility from the vertical position is quite a bit different than from an angular position. You have also got to take into account that a terminal velocity dive is 112 degrees and not 90 degrees from where the pilot is sitting. We have made many dives at about 45 degrees and we were able to stay right on the edge of compressibility and have full recovery. But steeper dives than that seem to penetrate the wall or so-called wall to the point that elevator compression just throws you into compressibility. At 35,000 feet with power off, with the airplane trimmed straight and level, I could make vertical dives, simulated attack or necessary attack with the vertical dive condition, but I had to leave the power off so I wouldn't gain that speed. That may be of interest to North American in diving without the propeller. I think compressibility is the coefficient of lift and speed and also propeller effects."

Commander RAMSEY: "Thank you, Joe. I agree with all your remarks except the first one. I think you are modest in saying you have not had much experience."

Mr. WILLIAMS: "I believe that wing loading has quite a bit to do with getting into compressibility troubles. Ships with lower wing loadings do not have to go to such a high lift coefficient in order to get the required G's and therefore don't get into trouble so easily. That is one variable on this P-59 situation and we don't know whether the reason we didn't have trouble was because we have a wing loading of only 26 or whether it is because of a lack of propellers. I believe there is a definite correlation there."

Commander RAMSEY: "Any other remarks on the Mach number indicator?

(Speaker not recorded.): "I would like to comment on the question that Commander Booth raised on the use and the accuracy of the pitot static tube and also the Mach meter. The Mach meter will work or will operate directly off a pilot static tube. It has been pretty well demonstrated that a pitot static tube is quite accurate up to a Mach number of 0.9 or 0.95 which is about as high as we would want to take an airplane at the present time. In this connection, I think that means that if the pitot static tube is well forward of the wing the results will be fairly accurate. The use of flush type static orifices are going to throw some doubt on the reading of any Mach number indicators and also the reading of any air-speed indicator at very high Mach numbers."

Commander RAMSEY: "While I was in England, I saw many interesting features undergoing tests and I would like to ask if our manufacturers are doing anything along these lines. One was an adjustable pitch wing angle of attack with a horizontal stabilizer. The horizontal stabilizer being adjustable to postpone compressibility effects."

Mr. Von VALKENBERG: "We designed one for an airplane of ours and we decided we could not use it because it was a Navy airplane and it needed an extremely high down load on the tail to land it. The stabilizer would stall on the undersurface on approach and that feature could not be used. We could get the tail down with the stabilizer anyway. From the standpoint of dive recovery, since we don't need it for landing, we are not going to use it anyway."

Mr. BURKE: "In line with this use of the adjustable stabilizer for dive recovery, some of the work that I did earlier in the year leads me to believe as a personal conclusion that it is a very fruitful line of endeavor. We had a large airplane with a very low Mach number. The compressibility effect we got was a tail heavy wing change and the airplane was uncontrollable at about a 0.71 Mach number, but at the same time it trimmed at about 3 1/2 g. We had enough testing to prove it was a complete compressibility problem because in making the final pull out at between 5 and 3,000 feet we got the required speed and acceleration where we could not get them pulling out at 10 or 15 or 20,000 feet. I believe some good work on the use of adjustable stabilizers might be done to control to somewhat higher Mach numbers than we have had with airplanes with fixed stabilizers."

Mr. WILLIAMS: "We have an adjustable stabilizer on a new fighter which is quite a large airplane and is intended to be a very high-speed airplane. The reason we have this stabilizer is to be sure that we have adequate control until we learn more about the airplane. It entails a considerable amount of extra weight which does not do the CG any good. We are going to try it if we can and then get rid of it. We will put a fixed stabilizer on and have a trim tab on the stabilizer."

Mr. Von VALKENBERG: "I doubt the effectiveness of a horizontal stabilizer change for dive recovery. I think it would be as dangerous as using your trim tabs and I think it would be quite possible that having gotten into an attitude where the tail might be ineffective due to weight and then having the compressibility cease, the pilot would be unable to overcome the down load that the adjustable stabilizer would have gotten."

Lieutenant Commander CALLINGHAM: "I saw a P-40 at Langley Field recently with a moveable stabilizer and no elevators at all. I would like to ask if anybody knows anything about that along this line."

Mr. BURROUGHS: "We have a similar project at Vought wherein we use instead of an adjustable stabilizer and an elevator a single all-moving surface with link leading tabs to get the full control stability. We have had pretty good success with the control arrangement, but we haven't had any high Mach numbers with it. The P-42 at NACA is equipped with a horizontal all-moving tail. They have had very good success with it recently at fairly high Mach numbers. The thought at NACA is that while you can build dive recovery flaps that will successfully recover an airplane from a compressibility dive with something like the P-47's flaps, it is nothing but a temporary measure because soon planes will be flying at level flight at compressibility speeds. Naturally you are not going to put a dive flap down because if you do you won't have the speed. Whether we like the idea of adjustable stabilizers or all

P-38L

P-47M

P-51D

P-61

269421

269421

YP-59A

FM-2

F6F-5

FG-1

XF2G-1

FG-1A

XF8F-1

16

MOSQUITO

SEAFIRE

ZEKE

moving tail surfaces, we might as well accustom ourselves to the idea because we are going to have them before long. We're going some day soon to go to some such arrangement because it boils down to the fact that you need a certain amount of pitching moment and you want to get the pitching moment without a lot of drag."

Commander RAMSEY: "Gentlemen, we will continue for about 1 more hour and then go to the oyster roast. To continue on the compressibility phase of our talks, I would like to go into the question, having seen the slide on the breakable wing tips, of pilots' and engineers' likes and dislikes relative to that feature to be incorporated on present and future designs."

Lieutenant Commander CALLINGHAM: "I would like to know a little bit more about the features of this wing tip. One in particular: if you happen to be approaching the critical "g" factor, to have these wings come up and one wing strike a bump, for instance, which temporarily leaves the loading on that wing, what happens then?"

Lieutenant Commander CONLON: "I think really that should be answered by Grumman, but the answer is you can pull one wing tip off, especially in present-day fighters where the roll produces a large bending moment in the asymmetrical flight condition. With one tip off you still have part aileron on that side and full aileron on the other side. Calculations show and the wind tunnel tests show it to be perfectly satisfactory and rather akin to the DC-2 1/2 which flew out of China with a DC-2 wing on one side and a DC-3 wing on the other."

Commander RAMSEY: "I'd like to get comments from engineers and pilots relative to the incorporation of the breakable wing tip feature in either present day fighters or future designs of fighters."

Mr. Don W. FINLAY (Boeing engineer): "Quite honestly, the feature is so novel to us that I would like to get a little more experience or knowledge of the company's practical experience of these tips before even attempting to answer that question."

Commander RAMSEY: "Colonel Coats, I'd like to have the Army's ideas on this subject."

Colonel Lee B. COATS (AAFPGC, Eglin): "I feel much the same as Commander Callingham. I would like to take a look at one of these things twice before committing myself. However, it does seem like a rather novel idea for reducing weight by reducing load factors to such an extent that you can break the tips off and give the effect of a 12g pull-out without having the 12g's there. As the commander said I would like to have another look at one of these before saying more."

Commander RAMSEY: "I would like to have the flight test officer of the Navy give us his ideas on the relative merits of such a design to present day Navy fighters."

Commander BOOTH: "I flew the airplane for the first time a few days ago. We are interested in seeing the contractor demonstrate the plane and I believe he has to demonstrate that part of the structural strength of the airplane. Isn't that right? I think it is an excellent way of getting the wing bending moments down so you can get the weight out of the airplane and make a maneuverable plane, yet keep it within reasonable limits if you want to and exercise it to its fullest if you have to."

Commander CALLINGHAM: "I would like to ask the Grumman people just exactly what percentage of weight reduction they have achieved by this method."

Mr. HALL: "The primary object in this novel design was not to save weight particularly, although obviously it does save weight. I believe the amount of weight we saved on this particular plane was on the order of 125-150 pounds on the wing structure. But we believe in the idea of a safety valve on the place where the plane would break and, if it broke, the effects of which would be known in advance. We all know that on any plane there are times when emergencies arise where the plane can and has been broken. But nobody knows where it is going to break. Usually a break occurs where the wing bending moments are highest, in other words right next to the body, and certainly such breakages are disastrous. This same principle can be applied without thought of weight-saving. If the number we have picked (8.6g) for that wing tip to break off is not considered quite enough, it can be adjusted upward so that the ultimate strength can be put in the rest of the wing to allow for upping the 8.6 to some other desired value but still having a spot that will be known to break first and thereby eliminating losses in combat or training from overstressing the plane either purposely or carelessly or inadvertently. That was our idea and it came to us because we had so many failures in the tails of the 6F where people lost half of the stabilizer and elevator on either one or both and were able to come back and tell about it. We didn't think the tail was the best place for it to break. We thought about this breakable joint in the wing. As regards demonstration, we are required to show that the airplane will fly with left tip off, right tip off, or both tips off. I don't think we are required actually to break them in the air. So we will have to take the plane off with the left tip off. If we can get it off the ground, we can certainly get it back again.

"I think that will answer the questions about one tip coming off and one staying on which is certainly possible in rolling pull-outs, but we believe that the plane will be satisfactorily controllable under those conditions."

Commander RAMSEY: "There seems to be very few who know a whole lot about those breakable wing tips except Grumman people, so we will go on to the next question. I would like to revert back to the bubble canopy and the wind screen discussion, which we had this morning relative to defogging and degreasing. Of course, this doesn't apply very much to twin engine aircraft, but it does to single engine aircraft for carrier work at night. I would like to call on a pilot first, one who has had quite a lot of experience in the Pacific. We'll see what his experiences and ideas are, his ideas on the need for degreasers and what in his experience has been done. Lieutenant Runyan."

Lieutenant RUNYAN: "Well, we have had quite a bit of experience with windshield fogging in the Pacific. I think it is very discouraging to quite a few of the pilots. The humidity of the air out there is quite high and fogging of the windshields very prominent. The defogging devices in the planes which we had most of the time were insufficient and I think that we found out after we got the planes with no windshield outside the bulletproof glass that this became more prominent. Out there we just had to put up with anything. The greasing of windshields and oil on the windshields were a definite factor not only with carrier landings but morning strafing and air combat work in the early hours of the morning. It is the thing that most people felt was causing us not to carry out our job as we should have carried it out and I think the contractors should look into it pretty thoroughly."

Commander RAMSEY: "Have you any remarks relative to the defogging device used in the Army Air Force and on the subject of degreasing, Colonel Terhune?"

Lieutenant Colonel TERHUNE: "I personally know very little about the subject. The only thing I do know is that we are running into fogging trouble in our high altitude flying on every portion of the canopy ahead of the armor plate, with the exception of the bulletproof glass, and I think Jack Woolams will back me up on that. That also happened on the P-80 and they are busy trying to fix up a system of hot air which will defog that part of the windshield other than the thick glass; they are trying to fix up a system that will defog the part of thick glass. I haven't heard the results."

Mr. WOOLAMS: "We have had fogging trouble on the P-59 in particular coming down from very high altitudes. The entire cabin has fogged up many times with the exception of the front glass, and every once in a while under very severe conditions on a very humid day we have had the front windshield fog up also and had to open a hatch in order to land the plane. We have had quite a bit of trouble on the P-63 with the front windshield fogging up but not so much with the rest of the canopy fogging. To alleviate that on one of our airplanes we put a fish-tailed arrangement on the windshield to try to circulate the air sufficiently to keep it from fogging up. However, this has led to the point that when we put enough air on the inside of the windshield it blows up into the pilot's face, which is very objectionable, causing the eyes to water. We haven't had enough installations to decide whether or not we have the right angle on this. We may have to change the velocity of the air or the direction in which it flows. I believe that it is the way to fix it, but it requires quite a bit of fooling around to get the proper combination so that you can get rid of the fog."

Commander RAMSEY: "How are you going to do that in pressurized cabins, Jack?"

Mr. WOOLAMS: "You've sort of got me there. I don't know. We can divert some of the pressurizing air against the windshield. We could take a take-off from our pressurizing duet and might be able to do it that way. We are going to wait until our next ship comes out before we worry about it."

Lieutenant HALABY: "I would like to ask a couple of questions of Woolams of Bell and I think they will be of interest because I have already heard of one proposed cabin pressure ship being developed. In a sense they cover almost all the topics we have mentioned this morning. One of them is how do you feel about the prospects of developing a cabin that will pressurize and hold pressure at some medium level like 20,000 feet? How do you propose to heat and cool that cabin? Now, for example, on the P-59 you not only have very bad fogging but it's impossible to wind that cabin open at any speed above about 120 miles per hour indicated, and it is also quite likely if you could wind it open and jettison it, because of its extreme size and weight you would carry away the vertical fin or stabilizer and I am wondering if we are not going to run into problems like that on all airplanes that require a heavy cabin system. I don't mean to put it on Bell in particular, because they have a very good cabin, but I think everyone is going to run into troubles. Perhaps part of the troubles could be prevented if Woolams would give us some ideas on it right now."

Mr. WOOLAMS: "I will answer the questions one at a time. In regard to holding pressure, I believe the main reason in the first place that we have been losing pressure has been the seal tube that seals the cabin hatch. We had better start at the beginning and experiment with how we get pressure. We have a rubber rim running all around the bottom of the hatch and at the juncture of that hatch with the main windshield frame, and when we pressurize the cabin we take air from the compressor into this rubber-sealed tube. This air is taken from the engine compressors and as the plane goes up in altitude, the amount of compressed air available decreases until eventually the point is reached where there is insufficient air to keep the sealed tube blown up enough to hold the pressure in the cabin. Once we get the air in there and get the proper pressure we will just have a valve which the pilot will have to turn to lock the pressure in. It's a rather simple matter and we have worked it rather successfully on the P-59 for special test. We intend to use the same type of sealed tube arrangements for the bubble canopy that we have used for this other canopy. The canopy on the P-59 was our first attempt, and one of the first in the country, in getting a successful pressure cabin and as such it is not by any means perfect. It is heavier than necessary and there are several things about it which are undesirable and which we know about and believe we can fix when we put the bubble canopy on. The bubble canopy by its very free-blown nature has quite a bit of strength in itself. We are not going to pressurize our next plane as much as we have pressurized the P-59. We don't think it is necessary to pressurize it as highly, so we will save weight and save structural difficulties, and we don't think we'll have much difficulty because of the data that we already have from the P-59. As regards jettisoning, the canopy pressurization is ideal from that viewpoint because you have high pressure inside the cabin tending to blow the canopy, as well as your low pressure on top of the cabin, so we feel that the thing will fly off much more readily than it will in a nonpressurized cabin. As regards opening the canopy in flight, when you turn the release handle, it automatically trips a lever which lets the air out of the sealed tube so that the canopy will have no binding tendency, and then it will open just like any other canopy, so we don't think that is any problem at all. The reason the P-59 canopy does not open is because the radio doors have to be collapsed in order to slide the canopy back; this is purely a mechanical difficulty as a very high bursting will collapse them. If we did not build the canopy in the first place so we could collapse the doors, we would not have had the problem of opening it so we don't think there will be any problem in that respect. The problem that we look for is the matter of providing adequate circulation of air in the cockpit and adequate control of temperature. We can get plenty of hot air from the compressors, and in fact, the air that comes in to pressurize the cabin is hot – you have to put it in an intercooler if you want to cool it for warm weather conditions. In addition to that, as has already been indicated, the bubble canopy itself makes for a very hot oven and we are looking forward to having some trouble circulating the air and providing enough cool air. We think that we have the answer, which is to circulate with these fans that we put in the P-59. It is working pretty well and if we put a larger fan in for cooling, inasmuch as this fan will have to be in there anyway in order to circulate the air, we will probably install a little canister of dry ice, put it in front of the fan and have ourselves a refrigeration system. So I don't think the bubble canopy will give us any trouble. It is a matter of providing circulation and proper temperature in the cockpit."

Commander RAMSEY: "I would like to hear from contractors on problems of degreasing, especially the single engine manufacturers. Is Republic doing anything along the degreasing line? Or do you have any trouble from that engine?"

Mr. PARKER: "I think Pratt and Whitney has the trouble with degreasing. Their rocker box covers are not properly designed and they habitually leak and we have tried to put the maximum number of gaskets on them. But I can say that some of the activities that I have been with have stolen the rocker box covers from the 1830 engine, which are of heavier material, and in using one gasket of the sealed type has found it very satisfactory. I believe Pratt and Whitney is going into heavier and more durable rocker box caps that do not warp and will seal off that engine oil. Maybe Mr. Connally will make some comments on that."

Commander RAMSEY: "He says negative. Lloyd (Child), do you know what Curtiss is doing? Or would you like to say what they are doing on degreasing?"

Mr. CHILD: "No; we have no solution. We would like to learn how to do it. We did have the problem that came up this morning about jettisoning the cabin. One of our pilots was killed and another one was hurt when the cabin came off and carried away half the rudder. After that the Army got us to make a jettison rig so that the cabin would come off and yet be caught in the net outside. We let it go at different air speeds but they didn't ask us to jettison it completely. I agree with Mr. Virgil that there are too many altitudes at which it could be let go, and therefore we should not have to demonstrate that the cabin comes off."

Commander RAMSEY: "I disagree with you Lloyd; I believe it should be demonstrated once to see that the jettisoning works. I still think that the manufacturer should have to make sure that his hood does jettison, that the pins pull and it will come off. Outside of that it will prove very little. Do you agree?"

Mr. CHILD: "Yes sir; I agree. The thing was that it was always in a yaw when it came off and that just swept it across the cabin. If it had been straight flight, I am pretty sure it would have come off."

Lieutenant Commander MILNER: "I want to say that I agree that it should be demonstrated because in the case of the SB2C it wraps around the pilot's ears before it goes overboard in normal flight."

Mr. BURKE: "From my own personal viewpoint in flying, if I want to get out in a hurry and pull the jettison and the canopy comes off and knocks off the vertical tail, that is just as well too because it's one less thing I can hit when I go back there. So I think that trying to build a jettison canopy that will not damage the rest of the plane when you are only going to throw away the whole plane is a more or less immaterial consideration."

Commander RAMSEY: "I expected remarks like that and more."

Lieutenant Commander OWEN: "There might be a case when you have to jettison the canopy and land aboard a ship. In case you have to land aboard and your canopy is damaged or struck by gunfire so that you can't open it, it's rather uncomfortable to make an approach with the hood closed."

Lieutenant RUNYAN: "One thing that I believe some people are not considering is, as Mr. Owen said, when you return to the carrier and want to land in the water, you want to get rid of the hood because you don't want to sink with the plane and you still want enough rudder to land in the water."

Commander RAMSEY: "That sounds like a vicious circle to me. Has anyone a solution to that?"

Mr. WOOLAMS: "One thing you can do to check whether the canopy will jettison. We did differ on a P-63 that ran a bunch of spin tests to see if the doors would come off while we were spinning. We just fastened the doors to the fuselage with cables that were about 1 1/2" long. These cables kept the doors from doing any damage. They just flopped out and hung there and you could do the same thing with a canopy to see whether or not it would come off."

Commander RAMSEY: "I'd hate to do that – at 300, Jack.

"I would like to ask Major Lamphier, before he gets out of our hands again, to say a few words about this P-38 which flew minus part of its wing."

Major LAMPHIER: "That was not I flying that P-38. It was a cohort of mine. We were strafing a destroyer and went in a little too closely. It tore the star off the wing back into the aileron and it bothered him so little he was about to make another pass before we quit. When he came to land several people raced ahead to tell them to prepare for an emergency landing, but he just came in and landed it and so I guess they are stable enough with half the wing on."

Commander RAMSEY: "Bud, you can go back and tell your factory about that. Has someone else a comment to make? Major Lamphier."

Major LAMPHIER: "I might add that the Marines took umbrage on that, and 10 days later one of them strafing Munda got part of his wing shot off. He came in to land and he wrapped his up – a Vought Corsair. He didn't get more of his wing shot off either."

Commander RAMSEY: "It's too bad there aren't more Marine pilots here to stand up for that."

Lieutenant Colonel CANOVAN: "I would like to go back to windshield defogging. As you probably know, in the Pacific area the dive bomber people have had a lot of trouble with their windshields fogging while coming down from high altitudes with a change of temperature. And the Marines conducted separate investigations. They tried several defogging combinations but didn't arrive at any final conclusion. Finally, a Marine F4U squadron on Midway lent their services to the project and they came to the conclusion that all you had to do was to get sufficient air circulation around the cockpit particularly around the windshield area – warm air. Just how they accomplished that, I don't know, but I do know that they dove from around 35,000 feet down to 5,600 and never had a case of windshield fogging after they got that circulation in the cockpit."

Commander RAMSEY: "I believe some degreasing work has gone on down there."

Commander PALMER: "We had a project assigned here attempting to overcome grease on the windshields of night fighters. We assigned six materials for test. Of the six materials tested paint thinner turned out to be the best and is applied to the windshield of the F6F3, by an inverted T-shaped tube with a hand-operated plunger. As I recall, the quantity of fluid required was about 4 pints. After some redesigning of the tube holes, it was found to be effective up to 200 knots, that is, clearing the entire windshield top to bottom. It cleared very effectively at any speeds lower than that and it cleared three fourths of the windshield at any speeds above 200 knots. That design was turned over to Grumman I believe to be incorporated into his windshield for night fighter aircraft."

Mr. GERDAN: "I would like to make a suggestion here and that is that the aircraft manufacturers contact some of the larger automobile companies who have played with air conditioned cars. Fogging was a terrific problem in those cars and was finally whipped with air circulation. I believe both Packard and General Motors have worked out the problem."

Commander RAMSEY: "Unless there are further comments, I think we have covered this subject pretty thoroughly. Anybody else like to offer comments? Mr. Virgin?"

Mr. VIRGIN: "One more thought on degreasing. That trouble is not foreign to a P-51 either. We had a considerable amount of difficulty with it due to several causes, the highest of which was the engine breather which was fixed by us. Right at the front of the engine cowl there is a gap of about 1" between the spinner and the nose of the airplane. The governor is located in that particular spot. We had a little trouble at one time with governor pads being of not sufficient strength and they would warp and any leak at that spot would carry out to the rear end of the spinner, be thrown back and eventually wind up on the windshield. That condition was corrected. We had trouble also, since our engine cowl parts right in front of the windshield on top side, of any leakage around the engine which normally occurs trickling down this parting line along the skin and eventually winding up on the windshield. We cured that by putting a rubber strip the entire length of the cowl. The last problem which we do not have as much trouble with as we used to, although we do have occasional reports from the service activities is that the propellers leak oil. The seals sometimes leak and there again the oil comes out and is thrown back on the windshield. There has been quite an improvement in those seals, but I think there is room for more development."

Lieutenant Commander CALLINGHAM: "On the question of defogging or deicing a canopy, I would like to just mention the Spitfire high altitude airplane with the single thickness plexiglass canopy playing hot air from nozzles on the interior of the canopy which was covered with ice. It was very effective. It just left three small holes where the hot air was actually playing directly on the canopy through which the pilot could see. That problem was overcome; by having a double thickness sandwich with dried and heated air circulating inside which proved very effective indeed. The air was passed from the cabin pressurization system and it was at a fairly high temperature having gone through the blower system and through a dehydrating agent such as silica-jel and it definitely kept the cold from the outside from coming in contact with the warm air and probably damp air inside the cockpit and it definitely worked. Some of you may have seen the latest version of the Spitfire 7; I believe there is one at Langley Field. I don't know whether it has the single or double thickness, but having the double arrangement was very successful indeed."

Mr. GUYTON: "We are rather happy to hear what we just heard about the hot air not being blown directly on the surface, but circulated, and doing a lot of good. On the new Corsair, the F4U4, we take heated air from the right hand oil cooler and pass it up to the windshield through a filter and it circulated through the entire windshield system and the two side panels. We think this is the answer and we have checked it so far in dives: but as anyone knows, the final test is out on the proving ground in combat. What we did find was that coming down on the flight the cabin was cracked slightly open and we couldn't close it up and lock it. So I happened to notice this little oil cooler cooker and I just turned it on, and it kept me warm during the flight down here. I was very surprised that the draft came completely around the bulletproofing and around the windshield and cabin and I think we might have the answer in this system. It is very simple. It does without any of the former Stewart Warner job that we used to have so much fun with and we now take a simple device to get heated air through the cabin."

Commander RAMSEY: "The next subject is safety equipment in the cockpit. The first is crash harness. As you all know the Army and Navy have got together pretty well on the type of safety equipment for crashes, and that is a regular safety belt with the two shoulder straps. However, the British have a system which has its merits also and that is a four-strap system which is secured by one pin in front of the pilot's chest. The several Army pilots have commented so far in the meeting that they liked the British system very much as it gives freedom of movement and still the safety feature involved for crashes. As that is open to discussion, I would like to have someone comment on it. Lieutenant Eluhow."

Lieutenant ELUHOW: "I was talking to Mr. Parker about the American type of shoulder straps. They are not so comfortable and not so good as far as flying with them on because although the bubble canopy insures you great visibility to the rear, you cannot comfortably look to the rear, with the shoulder harness on, because you can't twist your body sideways it doesn't give you that freedom of movement to look sideways to see what is behind you. Usually most pilots slip their harness off after taking off and it is quite a bother to slip it off especially while in close formation. There should be some other method where it gives you more movement and also gives you safety because many times you forget to slip back into it."

Lieutenant Colonel TERHUNE: "The first time I ever flew a Navy airplane was yesterday but I was very impressed by the feature you have in your shoulder harness which allows you to draw it up very quickly and I believe that maybe if that were properly used after take-off it would not bother the pilot too much and it could always be in a position where a quick tug would give the proper degree of tightness. I think it is a very good arrangement and we ought to do something about that."

Mr. GERDAN: "We have incorporated that also in the F4U-4, inasmuch as the reel behind the seat can be let out and it can be locked at any distance up to, I believe, 18 inches. You can lock the shoulder harness, either full forward at 18 inches, or you can lock it halfway. You can lock it full aft, and anytime you go back you simply go back easily and come up to the 18-inch point. We like it because you can lean forward to what you think the give will allow you without hitting the panel and you can lock it there. You are comfortable throughout the flight because you can reach anything in the cockpit."

Commander RAMSEY: "I think there might be some merits above those which we already know about the English system, but as far as I'm concerned the Navy system is good for carrier operation. We must take our parachutes off before landing on board ship for safety reasons and the present harness allows for ease in releasing the tightness and taking your parachute harness off and really pulling down the straps again. The British system however, I believe, does not lend itself to that as quickly as the American system."

Mr. BURKE: "I should like to say another word in defense of the Army-Navy system of a tight seat belt and shoulder straps as against the four straps of the Sutton harness and that is that in the event of a violent land crash such as a barrier crash, or running off the end of a runway and into a ditch with the Sutton harness the lower part of the body is not restrained. The upper part of the body is restrained but the head is not and as a result with a sudden deceleration the pilot is bent into a V-shape and there have been one or two fatal crashes resulting from broken necks and where the plane was not seriously damaged and the pilot was wearing the Sutton harness. In the Army-Navy type the body is fairly uniformly restrained and will probably stand a much greater crash with safety to the pilot."

Lieutenant Commander CALLINGHAM: "Answering Mr. Burke's point. We have recently modified the Sutton harness and brought the lower straps down much closer to the waistline of the pilot. We found in one or two cases that had occurred where planes stood on their nose and pilot had slipped through since there was no restraint on the lower part of the body and that has now been taken care of. Answering Commander Ramsey's point, with the British type parachute we have a quick release box which does not necessitate slacking the harness at all. It merely consists of a small box and by giving it a smart tap, your parachute harness is completely free and you have only to slip your shoulders out of a couple suspenders."

Lieutenant Colonel TOUBMAN: "One modification I would like to see in the shoulder harness which I fully approve is a few stiffeners which would permit the harness to stand by itself so you wouldn't have to throw your shoulder out of joint hunting for a shoulder harness that has slipped behind a seat or one which requires the presence of a crew chief. Maybe spring steel stiffeners that would not jam into your shoulders could be wound into the harness itself. I wonder what could be done on that."

Commander RAMSEY: "Has any contractor present given any thought to that? In any plane out here I know I have had and have watched other people have, a terrific scramble without the aid of a crew chief. We should avoid that scramble to get into those things for quick manning of planes and taking off. Any contractors wish to comment on that?"

Lieutenant Commander MILNER: "It might be interesting to know that the Navy Department is now conducting a statistical study of just about how many G's the seat installation plus the safety strap can stand for safety purposes. They have access to the records in operational training units and they have some remarkable crashes, sometimes where there is nothing left of the plane and the pilot walks away. There is some argument in heavier aircraft as to how strong fittings should be that secure the safety straps and how many G's they should withstand."

Commander RAMSEY: "If there are no further remarks along this line, I will go on to padded sights."

Mr. WOOLAMS: "I want to agree with Colonel Terhune that the Navy shoulder harness, I believe, is vastly superior to the Army harness in regard to the adjustment. In regard to the British harness, I personally do not like that harness at all when it comes to aerobatics. In order to prevent yourself from dropping on your head when you get upside down, it is necessary to tighten the harness up which throws you against the back of the seat so you can't reach forward to shove the stick forward properly on inverted maneuvers, slow rolls and things like that and I always have a feeling of claustrophobia, as if I'm tied up and can't move. I have heard a lot of the other pilots make the same remark."

Commander RAMSEY: "Gentlemen, again I reiterate that I hope you will all feel free to attend the oyster roast this evening and extend a cordial invitation to you to eat as many oysters as you can. The bus will leave here to go to BOQ for you to freshen up a bit or change into old clothes or vice versa and will leave there at 1745 to take you to the barn.

"Tomorrow we will have an 0830 meeting, concerning the topics we have just had under discussion. At 1030 we will have a discussion on armor and armament by Commander Monroe from the Armament Section of BuAer and in the afternoon, if the aerologists are as correct as we hope, we will continue flying, with further discussions on Sunday and Monday. Are there any points anyone wishes to bring up?"

NINTH MEETING
21 October 1944.
0930-1230

Commander RAMSEY: "This morning I would like to start out on pilot comfort again. We have finished that item, but there are a couple of phases which we have not touched upon – one is the electrically heated suits. There are several different types in the Navy. Unfortunately I have not used either the British or the Army type and probably some of the contractors also have a different type. I would like to call on Lieutenant Commander Owen for any remarks he has relative to electrically heated suits."

Lieutenant Commander OWEN: "The question of electrically heated suits I do not believe is quite as important considering Pacific operations as it might be in the European theater or the North Atlantic. The various electrically heated suits that we have seen around here in the past year or so have been very well received and very comfortable and I don't think there is any question but that an electrically heated suit would be highly desirable considering the high-altitude work that long range fighters are going to be called upon to do. I really do not feel I am very well qualified because I do not know too much about those suits. I haven't seen the very latest and have actually used only two of them. One suit that we do have here now, which is a sort of nylon fur-lined affair with electric leads is very comfortable. I think I can ask Mr. Halaby and Mr. Brown to speak a little more on that because they have used this particular suit. The main trouble in the past, I believe, with electrically heated suits has been that they are so bulky and stiff that they are uncomfortable. This new one I referred to was very comfortable, and it is much more pliable than the previous ones we have seen."

Lieutenant HALABY: "We have two suits here – one which is a standard Navy issue, which is made of leather covering similar to our flight jackets which you can see in the room. It weighs about 12 to 14 pounds and is fairly stiff. Because of its stiffness it is difficult to move around in the cockpit with it on. The latest experimental suit that we have been playing with is one made of a material similar to nylon, weighs about 10 pounds, is much more pliable as Mr. Owens has said, and feels more comfortable because you can move around in the cockpit. It apparently has sufficient wearability and durability to be satisfactory. There are a couple of considerations however; one is when you take off at 100 degrees from a deck or from an airfield in an electrically heated suit, you are pretty well perspiring by the time you get to an altitude where you will need it, and the possibility of pneumonia is as great going up as it is coming down. Another item is what to do with an electrically heated suit in case of bailing out over water and whether or not the manufacturers of such a suit could incorporate features that would make it buoyant enough to use as a means of sustentation. We feel that there is a lot of work to be done on them yet, on those latter two items in particular."

Commander RAMSEY: "Captain Kessler, have you anything relative to the electrically heated suits used in the Army or the African theater?"

Captain KESSLER: "Commander, all last winter we did most of our work about 31,000 and we found that they were superfluous in a single seat fighter. I really don't think you need them, they are just extra. Of course the P-38 boys that had their engines at the side didn't get that heat off the exhaust manifold and they got pretty cold, but we really didn't need them."

Lieutenant C.C. ANDREWS (NAS Patuxent): "I would like to add just a little to what Captain Kessler said. I believe they won't be necessary in single-engine planes. We made two cruises into the Arctic where it was just about as cold as you can get in low altitude flying. They furnished us with electric boots, electric gloves, and electric suits. I believe it is the same suit that Halaby spoke of, and so far I have never seen anyone plug one in. We liked them because they looked nice and were windproof."

Lieutenant ELUHOW: "I would like to add to what Captain Kessler and Lieutenant Andrews said, that we are not concerned about the heating proposition, but the cooling proposition is up in the air."

Commander RAMSEY: "Jack Woolams here this morning? Buffalo gets pretty cold. How about Lloyd Child? Lloyd? Have you got anything to say about test flying in these electrically heated suits?"

Mr. CHILD: "We have never used them and we have never found the cockpits too cold. It is only in getting in and out, as was said before, that we get cold."

Commander RAMSEY: "Mr. Virgin, have you anything to add to that?"

Mr. VIRGIN: "In California we use only the summer flying suits and we go as high as 12,000 feet and minus 65 degrees, and we don't have any trouble at all. The majority of the test flights are only for 1 hour to 1 1/2 hours' duration. For longer periods, of course, it would be necessary at least to wear a heavy flying suit, a winter flying suit, but we only use a summer jacket."

Commander RAMSEY: "Joe, are you also a member of the Chamber of Commerce of California? Or have you anything to add to it?"

Mr. PARKER: "I'm a member of the Chamber of Commerce of Long Island, I guess. We have never had any trouble with pilot comfort as far as heat. We have a 2,000 horsepower heat engine out front there that really keeps us warm. However, we do use a medium winter jacket to go to and from the airplane on the ground, and we are a little bit concerned in case we would have to jump out at 35,000 feet."

Commander RAMSEY: "Mr. Burke, we are talking about electrically heated suits, the need for them, etc. Have you anything to add to that question?"

Mr. BURKE: "Personally, I would like very much to have a good electrically heated suit. I haven't had the opportunity to try any of the latest service suits, but I think that one of the principal problems that is facing us in pilot comfort is the amount of drapery that the pilot is required to hang on himself before he takes off, and if the electrically heated suit can be made light enough and close fitting enough to reduce that effect of a Christmas tree, I think we should very definitely go in for it. I would like to suggest that in addition to having a suit that we try to combine as much as possible of the pilot's flight gear and equipment in one item. In other words, I would like to see work done along the lines of when the pilot gets into one piece of equipment having on his parachute, his flying suit and his life jacket, and anything else we can think of that he has to take with him."

Commander RAMSEY: "I think that is an excellent suggestion, Burke. I know that on a carrier, by the time you sit around in the ready room for some period of time, you are pretty damn well worn out before you get into the plane because of the gear that you have on over you. If we can incorporate that in one piece of apparel, I think it would be very well worth while. Mr. Foote, have you anything to add so far as Linden, N.J. is concerned? Mr. Foote."

Mr. FOOTE: I think the subject has been covered very well. In our particular case where you have these jet exhausts and are more or less enclosed in a can of sterno I have found even past 30,000 feet that an ordinary jumper over just ordinary underwear will suffice. There is one thing I would like to add. For going to and from the planes in addition to this problem of bailing, it would be highly advisable if something could be developed in the line of a battery, or two small batteries to fit in the knee compartment of your flying suit so that the whole suit would be electrically hooked up and would not need any outside power. If this could be done I think we could dispense with these connections in the cockpit. We wouldn't have to worry about when we bail out and it might solve the problem of pilot comfort outside the plane."

Commander RAMSEY: "Lieutenant Commander Callingham, will you speak on the British use of electrically heated suits? You have probably had operations in cold weather."

Lieutenant Commander CALLINGHAM: "I personally have had no experience with electrically heated clothing myself and it was not in general operation at the beginning of the war, but I believe that our aeromedical authorities are rather more keenly interested in heating the extremities of the body – electrically heated gloves and boots – rather than concentrating too much on a complete suit. We have those suits, of course, but as I said before, they are keen on just the hands and feet. I think we could do away with a lot of the bulk if we could pay attention to having just those two separate items."

Lieutenant Colonel RENNER: "Well, I wanted to bring up a point. Although it sounds great to take a flying suit and put your life jacket, your zoot suit, your bladders, to restrict the blood flow toward your pedal extremities and so on, we attempted to do that and the first thing we ran into were objections from the pilots because they did not want to wear their life jackets when they were in the ready room – they wanted to be able to take them off. They didn't want to wear a pistol with 15 or 20 rounds of ammunition hanging around their necks all the time they were sitting in the ready room. So there are two sides to the problem and it is very simple to build a life jacket if you are going to have it on you all the time. But if you want a flying suit that you can sit around in you are going to have to take it just as thin and comfortable as you can get for the tropics, and hang the rest of this material on you some place.

"We have developed, although it hasn't been put out yet, a type of vest that fits over your head with crossed straps and has about, I think, 14 pockets to carry miscellaneous gear. You can carry various pistol cartridges, and the compass, a first-aid kit, can of water, and this and that and the other thing in this vest. But other people have objected that they don't want all that stuff hanging on the front of them. So the idea of building a hunter's coat that you can stuff 20 birds in and carry your ammunition is fine, but we have got to get together and make up our minds that we are going to wear them inside and outside and in the ready room or else it is going to have to be in pieces that you can take off. Our new life jacket is now packed with the one-man raft. We have pockets below the knee, on the shin and behind the calf of the leg so it will balance. Put the heavy stuff in the back and the light stuff in the front. We also have a thigh pocket. And we have a cigarette pocket on the chest and two pencil pockets on the left shoulder. We have cut off the tabs so it doesn't flutter when you make a carrier approach with the hood open."

Mr. BURKE: "In answer to Colonel Renner, one of the main things I wanted to get at in suggesting the universal piece of equipment was the possibility of not wearing it in the ready room but of having all of this equipment in one item, made in such a way as to get into it quickly enough so you wouldn't have to sit around with all this stuff on. If it could be made so that you could jump into it as quickly as you can into the three-strap parachute harness, and I believe that if sufficient thought were put on it, it could be possible, then you would get away from the problem of having to wear all this stuff while on the ground and could just jump into it before you jump into the airplane of after you jump into the airplane."

Lieutenant DAVENPORT: "As far as having everything in one suit, if you had it all built intricately into the one suit, when you went into the water you wouldn't be able to detach it with the present type of life jacket that we have. If you get too much weight on you the life jacket isn't worth a damn until you get into your raft. With the present paraphernalia that we now have if you are too heavy you can get rid of some of it but if you have it all built in to one suit you won't be able to."

Commander RAMSEY: "Any further comment along that line?"

Lieutenant ELUHOW: "We had a regular flying suit that we always wore around the ready room – it didn't have anything in it, and some of the pilots at the last minute, put their miscellaneous articles in their lower pockets, but most of us had leather flying jackets and a few extra pockets added on the inside of it, plus the two pockets on the outside, and we usually kept all of our miscellaneous stuff in there. That was hanging up in our locker, and when we were ready to take off, we just slipped into that, put our Mae West on top and off we went. I thought that was pretty simple. I mean compared to some of these suggestions, and no one had to wear them, it wasn't hanging too many things on, and it did seem like a pretty good idea."

Commander RAMSEY: "Joe, do you want to refute anything that Lieutenant Eluhow said about your suggested suit?"

Lieutenant Colonel RENNER: "To get back to the argument about putting everything in one suit, we had the zoot suit problem come up just recently and the anti-G suit, and we had two types advanced. One was a pair of chaps that had merely a belt and a couple of skeleton legs dangling down. That was supposed to be worn inside your regular flying suit and it appeared to be a pretty good solution. But everybody that tried the suit – we had one built into a flying suit and one that was just the bladder separated – wanted to wear the suit that had the bladders built into it. However, we did rig up an arrangement whereby the legs could be ripped loose and snapped up under your armpits to give you added buoyancy, because we realize the biggest objection to the zoot suit is that people are afraid they will drown if they have to get out in the water in the thing, and we have tested it in the water and found out it adds buoyancy rather than detracts.

"I would like to get back to Mr. Burke on this parachute business. I will say this, we are certainly open to suggestions. We have tried the British quick-release harness, we have tried the British obsolete harness that they have thrown out, we have had correspondence with the naval attaché' at London because we didn't use the right harness the first time, we have tried every Army harness, we have tried all the Navy harnesses, we have had some of Grumman's engineers design a new snap that zips loose if you push a button, we have had Pioneer design 2 new quick-release gadgets to make it loose and come off with grabbing 1 pin and pulling it. At Lakehurst last month we tried 57 different harnesses. We have had 2 parachute men from BuAer that do nothing but try to design new harnesses and they are both qualified parachute men. They take them up and jump them and try to get out of them, and by God, we haven't got the answer."

Commander RAMSEY: "Looks like that is still an open question gentlemen. Any other remarks before we go on to the next item?"

Lieutenant HALABY: "I would like to go back to the heated suit just 1 second, and emphasize that the Navy wasn't contemplating the use of that suit at medium altitudes, but at very high altitude in excess of 35 to 40,000 feet. I don't think we can just stop with the jumper or skivvy shirt in this situation. I think most of you are familiar with what happened to Colonel Lovelace, when he jumped out of the Boeing ship at 42,000 feet. He lost 1 of the 2 pairs of gloves he had on and the hand almost had to be amputated because of frostbite, and I think if anyone goes above 40,000 feet, with the prospect of fire in the airplane always present, without being completely clothed even to the clothing over his eyes, he is just taking an unnecessary risk. The bail out from altitude in excess of 40,000 feet, without all the covering you can possibly get on, has been demonstrated to be very dangerous."

Lieutenant Colonel MENG: "I just wondered if Colonel Renner could tell us what this vest they are going to put on would weight when we had everything everybody wanted to put in it and what they finally pared it down to, if they did."

Lieutenant Colonel RENNER: "This vest is a direct development of the Central Pacific. Commander E.R. Sanders flew a PV there and he had seen some Marine squadrons and Army squadrons using the vest. The vest we have developed is built of nylon, and when empty weighs less than 4 ounces. It has all these pockets and is not intended that any equipment be assigned to this thing at all. It is given to you as a piece of equipment. You can use it or not as you see fit. If you don't feel there is enough water in your jungle kit you can carry additional water in this vest. If you don't think there are enough Very pistol cartridges or that there is something in the jungle kit that is useless you can throw the jungle kit away and put the necessary parts in the vest so that you can do without the jungle kit. It fits over your shoulders somewhat like a housekeeper's apron for clothespins or a carpenter's apron for nails, only it just hangs down slightly below the waist and has crossed straps that come across and tack on to it under your arms. It has a place in it for a pistol and an extra clip of cartridges and the usual things that you might want to carry."

Lieutenant Colonel MENG: "I asked that because the Army at the present time has one. I believe they call it the C-1 jungle vest. It is going to be a jungle vest. Someone here might know more about it than I do. The first one I believe weighed, when the man had everything in it, 20 pounds with equipment. The one they are putting out now is very large. I wondered if anyone knows what's in it and how much it weighs."

Lieutenant Colonel TOUBMAN: "We tested a vest similar to what Colonel Renner was describing, about 6 months ago, and we had about a dozen such articles. The general consensus of pilots flying it was that though there were a few minor changes to be made, it was quite the answer to what everyone wants. If I remember offhand now, I believe it was in the neighborhood of 24 pounds. If you are tramping through the jungle, you want to stick all these necessary articles in your pockets, in your hands and every place else. There is really an amazing amount of equipment you can carry and still get away with it. And I think to the boys down in the jungle every one of these articles was really a necessity rather than a convenience. The last answer I had was that the Army threw it out and I would like to know if they are going to build it or not."

Commander RAMSEY: "Can anybody answer Colonel Toubman's question? If not, I guess we will have to let that stand on the record as a question. If there are no more comments on the electrically heated suits, I would like to go on to the subject of oxygen masks."

Mr. E.J. SPECHT (General Electric): "We have had a suit, probably most of you have seen it, designed so that over your electrically heated suit you place another suit. If you are going to jump you are pretty well insulated, and you will stay insulated. I couldn't make any time estimate, but if you are heated when you jump you are going to stay pretty warm all the way down."

Mr. WOOLAMS: "I don't know whether it is exactly appropriate to bring this up at this time, but I thought I might pass on some interesting information regarding jumping at high altitudes. I recently talked to Colonel Lovelace concerning his jump. He jumped from an altitude in excess of 40,000 feet, using a static cord to pull his rip cord immediately after clearing the airplane. His airplane was slowed down to 110 miles per hour indicated and the chute opened when he was about 100 feet from the plane so he was traveling at a very low rate of speed when the chute opened. However, when it opened, it opened so suddenly that it knocked him out. The motto of this little story is that at high altitudes, although your indicated speed can be very slow, your chute evidently opens much more suddenly that it does at low altitudes. It opens with an explosive violence and he recommends that if you jump out do not pull the rip cord, don't even consider pulling the rip cord, until you are down to at least 18,000 feet, or you are liable to wrench your back, or break your back, or knock yourself unconscious causing grogginess which will result in severe injury. I just bring this up for the information of anybody that may have to jump out up there."

Lieutenant Colonel RENNER: "Along the same line of what Jack Woolams has said, the Russians have done considerable experimenting and they find that the G involved on the jumper above 30,000 feet for some unknown reason increases with the square of the altitude so that you will really get a 10 or 15 G pull and rip all the panels out of your chute. It is an extremely good idea to wait awhile after you jump before you pull the rip cord."

Lieutenant ELUHOW: "If I'm correct, fighter planes don't have these oxygen bottles when you bail out of a ship, so you couldn't very well pull a rip cord at that altitude because you would be a dead man by the time you got out. So besides the G it is very unwise to pull the rip cord immediately after bailing out at high altitude because you would not be able to survive the jump due to lack of oxygen."

Mr. WOOLAMS: "I would say that if fighter pilots are not equipped with bail-out bottles at high altitudes, it is high time that they were. It is extremely dangerous to make even a delayed jump from any high altitude without one. In our test work, we would not think of going over 30,000 feet at Bell without bailing out bottles."

Commander RAMSEY: "We have something here in bail-out bottles now passing under test."

Lieutenant Commander CALLINGHAM: "I was just going to mention, on that question of bail-out bottles, that we have a small bottle attached to the life raft on our single-seat dinghy that has a capacity of about 12 minutes duration at the height of 40,000 feet. It is hooked into a regular oxygen system and with a little T.P. all you must do is bail out and crack the valve and you have a full oxygen supply for 12 minutes. It is attached to the side of the dinghy."

Lieutenant HALABY: "I think the Navy has a pretty good rig for high altitude bail-out and I would like to discuss it. It is still in the experimental stage, but in brief outline, we combined a pressure breathing mask, similar to the latest type of pressure breathing mask that the Army developed at Wright Field, with a bail-out bottle which is attached to the electrically heated flying suit or any flying suit that you are wearing. There is a T fitting, one line coming up from the bail-out bottle, the other from the ship's pressure oxygen system, and this T fitting has two distinct plugs: one plug can disconnect the bail-out bottle or connect it and the other plug can disconnect the ship's system. That movement is rapid and requires only about a second, so that the pilot who is going about 35,000 feet has pressure breathing oxygen all the way up. If he wishes to bail out, he disconnects the ship's system and is then automatically on his bail-out bottle and from there on, he has positive pressure breathing from the bail-out bottle, which will last, depending on the altitude at which he jumps, for 5 minutes or more. I think that gives his as much as he could possibly need. It seems to me if we go above 40,000 feet, these things are really important and we cannot go blithely flying in skivvy shirts any more."

Commander RAMSEY: "I think that is a very important item. I have a convert for throat mikes, which ties in here. Are there any comments on throat mikes versus hand mikes or oxygen mikes?"

Lieutenant Colonel RENNER: "You people are probably wondering why I have so much to say about all these subjects, but I happen to be a Marine and the only Marine in the Military Requirements Division of the Bureau of Aeronautics. I catch all the odd jobs: oxygen, parachutes, life jackets, life rafts, and all that stuff is poured on to the Marine in the end. Along the line of the throat mike versus the hand mike versus the mustache mike and of course the rest of them, our radio and electric people in the Bureau of Aeronautics spend about 10 billion dollars a year trying to figure up new gadgets and we are continually bombarded with a new type of mike that is absolutely the answer. We have at present on order 250,000 noise-canceling microphones that you wear under your nose like a mustache and, needless to say, when that order arrived and we sent them out for test, they found out that unless you wear a mustache, it irritated your upper lip. So, the mustache mike has a double meaning: It not only looks like one, but you need a mustache to wear it. This proved to be so uncomfortable and there was some kind of chemical in the mike that when you perspired a little, which was normal with wearing this mustache mike, it wasn't long before you had a tremendous itching and you tried to push the mike away with your tongue, your teeth, etc., and strange noises were uttered during all speeches. Don't let that worry you, because the same brainstorms that figured this one up have a better one. They have a microphone now that they call the boom mike. The boom microphone is fastened to the helmet on a wire that comes out in the shape of a loop and it is pivoted on to the solid car pad (the doughnut pad). On the outside you have this pivot and two little wires run down and, believe it or not, they were able to use the same microphone they hung under the nose as a mustache on this installation. It's just a different method of suspension. This boom microphone is curved so that the mike fits in front of your mouth when pulled down into position. If you are fighting at low altitudes such as the Navy has been doing (10,000 feet and below), you can use this boom microphone swung into place. We have a mike in the A-14 mask, as you know; if you have to use the mask, you can move the mike and swing it up on top of your head, or you can swing it down so it is poking you in the chest – either way you prefer. But then, you can put on your oxygen mask and use the mike in the mask. So, we really have the answer now."

Commander RAMSEY: "It is regrettable that Richardson is not here to defend himself. Are there any other comments?"

Lieutenant ELUHOW: "I would like to add to what Colonel Renner said – that he sort of found the answer. We only had a microphone inside the mask; that oxygen mask we're using, that A-14 I believe, is sort of hot and you begin to perspire, especially at low altitudes. We had to have access to a mike at all times. We perspired pretty freely and had a pool of water below our chin in the mask. Some fellow solved that problem by having the parachute man put in a chamois on the inside; it kept the rubber from sticking to our face. After the mask got hot, the rubber stuck to our face and hurt us each time we moved it. The chamois prevented that and absorbed the perspiration."

Lieutenant Colonel MENG: "There is one thing that is always brought up at every Fighter Conference. It is not very important, but it's this: Having your oxygen tube and your earphone lines and your microphone lines all together. We really like this idea and we're going to do something about it. Every fighter airplane you get in – Army or Navy – you grab a long electric line and pull and pull and you say "That is the mike"; then you grab another line and you pull and you pull and say, "That's the ear phones." I wonder if the Navy has worked on that. We got into it, but there was too much objection to what we were doing. I am wondering if the Navy has done anything to combine your oxygen tube, ear phones, and your mike into one line or a quick connect and disconnect."

Lieutenant Colonel RENNER: "We have been working along that line. The main problem, as you probably know, is to get a quick disconnect – the oxygen disconnect has been solved. I won't say completely satisfactorily, but we do have a male and female plug now that you can break and disconnect your oxygen, but the main problem of getting the head phones and the microphone on to the same set of lines which was the thing we started out to do has made it almost impossible to find a type of disconnect that will fit the electrical connection and still break with the same movement as the oxygen line breaks. So, as yet, we haven't found a type of contact or disconnect of the electrical contact that will break with the same disconnect that the oxygen line takes to break it. We have tried all three lines taped together. If you're willing to break each line individually, you can tape them all together and have them running up to your face. However, if you take the three lines and tie them all together, you must arrange your cockpit so that the oxygen and electric lines come from the same side. I must confess that not much has been done mainly because we haven't found out how to disconnect the electrical lines with the same movement that the oxygen disconnects."

Lieutenant Colonel TOUBMAN: "I have had a pet theory for a long time and have not been very successful in persuading the Army. My idea was to bring the oxygen tube, the radio disconnect and the throat mike disconnect right between your legs where it will have only 2 feet of travel instead of the present 6 feet. Use the same connectors, use the same force to disconnect them and I don't think there will be any difficulty as far as bailing out. Then also you must wear the three connectors or you don't have the bulky oxygen tube dangling across your legs and half way on the front of the cockpit. I tried that on a P-47 and it worked pretty successfully. I noticed on the F8 and the new design of your stick does not make it quite practical, but on several other types of aircraft, it would make a clean installation. I would like to know if any others have tried it or if they have anything against that type of connection."

Mr. BURKE: "I have tried that on a test cockpit, bringing all the lines up and clipping them to the front of the seat between my legs and that it seems to me was the least likely way to get tangled up. I think it is quite possible that if that line were followed we would get the equivalent of the single attachment without the mechanical difficulties which were blocking it. As you move around a cockpit, if you watch carefully, you will find that the distance from that point to your head remains more nearly constant than any other point in the cockpit."

Commander RAMSEY: "It sounds like an excellent suggestion. Any other comments?"

Mr. HELLER: "Going back to Colonel Lovelace's dive where he left the Boeing at 42,000 feet, he reported that it was the G's caused him to black out at that altitude. The pictures of that jump show the severe oscillations of the chute that caused most of the difficulty, except his hands freezing. On oxygen systems in our present equipment, I believe pressure breathing should be incorporated and the suggestion of the bail-out bottle so you would not have to remove the mask, would be excellent. If you take the demand type mask off and go to a small mouth piece and use the emergency bottle with about 12 minutes of oxygen in the period of time you would need to get out of the altitude you would find frostbite on your face. Incorporating pressure breathing is a must in our program. We are at the present time conducting endurance tests of propeller equipment at 30,000 feet. We are attempting to get 200 hours at that altitude and we have 36 hours. I have never been able to keep any of the men in the equipment after 3 hours. We have been able to hold that 3 hours – 3 hours, 18 minutes, was the longest at that altitude. We have used food to help us. We have had orange juice and milk aboard, which is now discontinued because it causes gas pains, and we are going to try plain hot soups. However, plain chocolate bars have helped a great deal. We turn the emergency supply on and move the mask aside quickly and take several small pieces of the candy and then put the mask back on, leaving it on emergency for an additional minute. I offer that as a suggestion for high altitude work. Dr. Boothby of Mayo Clinic maintains that the period of consciousness at 34,000 feet without oxygen is 7 seconds, so it's practically a must to keep the mask on. Dr. BuBullion, who makes the BLB mask and who always has kept forward on this 14 mask of the demand series, is at the present time trying to incorporate a full feature having the mask microphone and head sets and the mask incorporated into one unit with one connection. That is still in the experimental stage."

Mr. WOOLAMS: "I might refresh the memory of some people, or perhaps maybe a lot of these people don't know what the Army's newest oxygen mask system is. It is no longer necessary to disconnect your mask in order to connect your bail-out bottle. We have a fitting that goes on either the A-13 (pressure demand mask) or the A-14 (demand mask) and this fitting takes one line from the main oxygen supply and another line from the bail-out bottle. In order to bail-out all you do is pull a rip cord on your bail-out bottle, disconnect your oxygen line and go over the side. It is as simple as it could possibly be and that is standard equipment now, although it possibly hasn't been circulated throughout the Air Corps yet."

Mr. LINDBERGH: "I was just going to bring up that at altitudes above 40,000 feet there is not time to change from the ship's oxygen system to the ordinary bail-out bottle mouth piece. That is inadequate support. If you have to go through any exertion to get out of the cockpit, that is the reason for changing to the Army's system Mr. Woolams speaks of. So it is not a question of desire at these high altitudes – it is essential to use a system that would not require the removal of the regular oxygen mask. However, as far as eating bits of food is concerned, there is no problem in that, because you can remove your mask for a few seconds at a time even at 40,000 feet. I think for about 10 to 15 seconds. After 15 seconds, there is a very noticeable effect of the lack of oxygen which sets in. That depends, to a large extent, on the condition of the pilot at the time he removed the mask. If he is well oxygenated and in good condition he will last considerably longer at 40,000 feet without a mask than if he had been up for an hour or more."

Lieutenant Colonel TERHUNE: "I would like to add one word on his new gadget that Jack Woolams is talking about. I may not be up on the latest improvement, but we had a paratrooper officer down at Wilmington, which is a glider base about 30 miles south of Dayton, who made a delayed jump from about 40,000 feet and used the same ship which Colonel Lovelace used, the only difference being that he never did open his parachute. The point there is that he used this equipment Jack was talking about, but it seems that when you jump out of the plane and disconnect the bottom of the oxygen tube, that bottom is still open and attachment for this bottom fits into the tube some place below the oxygen mask. In falling down that free end flaps in the breeze and pulls the oxygen out of the mask and into the tube. That is a conclusion we have come to, as I remember it, because they took a B-17 up shortly after this accident happened and had a man standing back by the rear gun window who stuck the end of the oxygen tube out there and he couldn't breathe against the suction that resulted from that. They are pretty much disturbed over it. But there is something wrong and I think they are going to have to close up the end of the tube, and get some automatic closing device for the end of the tube so that that won't happen again."

Commander RAMSEY: "Lieutenant Halaby has been playing around with that sort of thing."

Lieutenant HALABY: "We have that very device on our T fitting which I described earlier.

The quick disconnect which you pull out to disconnect the ship's system, also closes a relief valve which blocks off the open end. It is a very simple installation. Another thing I would like to bring up, but hesitate to bring up in detail because I am not sufficiently familiar with it, is being issued by the Bureau of Aeronautics, I understand, as a technical directive describing a method of voluntary pressure breathing. Now it is not universally accepted as a good idea, but I believe in one situation it is something that all test pilots and high altitude combat pilots should be familiar with. Vaguely and with the caution that you should consult your own aero-medical experts on the thing, it is this: You take a very deep breath and compress your chest and abdominal muscles (and you should be instructed or indoctrinated in how to use it before even attempting it at altitudes), but when you do compress the partial pressure of oxygen available in your lungs, by osmosis it can get through these little tissues and carry away the carbon dioxide. It enables you to stand 35,000 feet without oxygen for periods up to about 10 minutes. Now that is the rare case, but in the average case, I believe they have been able to stand 35,000 feet in the chamber for at least three to five times the period at which you could ordinarily stand it without oxygen. That could be a very useful bit of information, if all these other devices failed and you were without any oxygen. I think it would be interesting if we could get some information on the subject and use it for the emergency bail-out situation."

Lieutenant Colonel RENNER: "This isn't directly in line with what Lieutenant Halaby is talking about, but just recently Dr. Gell, who is our medico in Military Requirements, has discussed the high-altitude problem with me and we have decided to build 100 sets of the Navy pressure breathing gear with the T fitting and the bail-out bottle, and letters should have been received prior to the time by all the Navy contractors for fighter aircraft offering them any number of sets up to 5 of his high altitude pressure breathing gear combined with the bail-out bottle, explaining that it is necessary for the test pilots that are going to use it to receive a course of instruction in the type of breathing required. It is an unusual thing – it is something in line with what Lieutenant Halaby is saying – you have to regulate and use certain breathing techniques even though the oxygen is coming. And under pressure, you just can't breath it like you normally would in an A-14 mask. So if any of the contractors are interested, they might obtain sets from the Navy and I wish to caution them again that if they intend to use this particular type of pressure breathing that it is safe for the normal pilot up to 45,000 feet and in cases of people who are especially adapted to high altitude, they can go up to 48,000. Be sure that your pilots go to Philadelphia and get trained in the use of it, otherwise they are likely to make their last flight when they first try it."

Lt. W.C. HOLMES: "In line with this voluntary pressure breathing Lieutenant Halaby mentioned, we have just had report to Tactical Test the doctor who developed this thing. He had a great deal of trouble convincing anybody that it worked. He took an Army pilot up to 40,000 feet and removed his oxygen – of course after training him in the use of the system. The pilot flew for 15 minutes at 40,000 and then flew down to 15,000 feet and suffered no bad effects. As Mr. Halaby mentioned, the system takes some training in its use, but you just use the voluntary pressure you can work up by your diaphragm muscles and it amounts to just an emergency pressure system which uses the pressure in your lungs to such an extent that it becomes phenomenal. In fact, this doctor has done it so much that he can sit in this room without smoking any cigars and he can blow smoke rings by building up so much pressure. We have him here on the station. When he lets this air out, I might mention, it sounds like a .45 going off. He is in Tactical Test and if the conference thinks it is worthwhile, I am sure he could be persuaded to come up here to give us a demonstration on this emergency system."

Commander RAMSEY: "Extend an invitation to him for about 1230 before we go to lunch. Let's go on to the next item under pilot comfort. The Navy feels the mock-up of the F7F cockpit down on the hangar deck is for an ideal VF. That may not follow the Army's viewpoint and some contractors may have some other ideas on it. It is important that you put down your ideas on the questionnaire.

"Arrangements of instruments and controls is the next item for discussion under pilot comfort and we will go into that now."

Lieutenant Colonel RENNER: "I don't know how much introductory speaking has been done on this mock-up cockpit that is down on the hangar deck, but I want to cover some of the things that we have attempted to do and tell you why. This is the outgrowth of a Joint Army-Navy-NDRC combination where they all got together and wanted to adopt a standard cockpit for single-engine fighters for the Army, Navy, British, and everybody. Anybody who sat in on this standardization committee meeting knows that the road is long and rough and when it finally finishes up, everybody hoes home and says, "Damn it, we'll do it to suit ourselves anyway." So this is not an attempt to force a standard cockpit on anybody by the Navy.

"We feel that there are certain things to be gained by always having throttle, propeller, supercharger, and the mixture controls in the same arrangement and having the stick in the same place in the cockpit. That is about the only thing remaining constant over the years. We do not intend to send out the drawings and dimensions and photographs of this cockpit and say it must be just like this. But we intend to say that this is the general arrangement of the Navy fighter cockpit. If you have a single- engine plane, you just leave out the duplicate controls that are required for a twin-engine plane. We started out on the F6F cockpit. We tried to combine safety also, in burying the knobs and switches that usually poke holes in aviators' heads when they crack up. The doctors are pretty worried about this. The aviators themselves are not bothered by it, because they never intend to crack up. We have tried to bury the switch handles and remove the obstructions that are liable to poke holes in you when you crack up and when your shoulder straps break. The safety idea, the standardization idea and in a long-range airplane the comfort idea for long hops have all been attempted in this one cockpit. I want to call attention to the barber-chair foot pedals we have in there. A lot of people have been in the cockpit and didn't even know about them. I must confess that the idea as demonstrated here is not as good as we had hoped in that the foot pedals are merely wooden mock-up pedals and we could not hinge them as low as we wanted to. The brake pedals fold back toward the pilot so that he can stick his feet through the stirrup formed by the rudder pedal and he can stick his foot straight out and rest on the back of the leg some place around the ankle or farther up; it is my personal experience that you can fly a plane in loose formation on the way out merely adjusting the rudder with the tab or with a slight deflection of the pedal by pulling on it with your feet as they stick through the stirrups. Another thing tried here is to get all the switches for night lighting on a separate panel that can be put out of the way for daytime use.

"Those are just some of the things on which we would like to hear comments.

"The gunnery panel is out of sight and out of the way so that you won't leave your brains scattered on it in case your shoulder straps break.

"The mock-up for the zoot suit is in there. The disconnect, or plug-in, is at such an angle that it is not necessary to disconnect your zoot suit if you have to bail out. The mere act of climbing out of the cockpit will break the disconnect to your zoot suit so that is one less thing to worry about. We may be able to do something along that line on the oxygen mask to effect a sidewise pull that will break the disconnect so that you won't have to stop to think about that.

"Those are some of the things we have tried to accomplish in this cockpit and we would appreciate any remarks on it."

Mr. STEPPE: "The main comment I want to make about that cockpit is that the grouping of the flight instruments in it rather puzzled me and it is not the way I would personally prefer, but I would like to know what the reasons are for putting them the way they are. There is undoubtedly some very definite reason for doing it, but I will be darned if I can figure it out."

Lieutenant Colonel RENNER: "The fight instrument set-up has long been a bone of contention. The Army, as I pointed out, has had one system whereby they want the directional gyro on the same level and to the side of the gyro horizon. The British have wanted their directional gyro under the gyro horizon. The Navy has always wanted the air-speed indicator to be the upper left hand instrument so that they could glance from the signal officer and down to the air-speed indicator and back with the least movement of the eyes. Climb indicators, according to most pilots, can be thrown out of the cockpit. The turn and bank indicator the Army wants on the center line of the airplane because of the parallax and they want to use it as a stop. I guess they are using the attitude system. They feel it should be on the center line.

"The British have had a standardized cockpit with the air-speed indicator to the upper left; the altimeter lower left; the gyro horizon upper center; directional gyro lower center; rate of climb upper right; and turn and bank indicator lower right; I understand, for the past 15 years. We knew that wasn't as good as ours, so we set out to establish a standard Navy panel inasmuch as we disagreed so violently with the Army that we couldn't get together with them and we arrived at the same conclusion that we wanted the air-speed indicator on the upper left for our carrier approach; the altimeter lower left, because it is the second best spot to look into when you are looking out at the signal officer. We wanted the gyro horizon using the stick and we wanted the directional gyro lower center because we used our feet on the rudder so that you operate the upper instrument with the stick and the lower instrument with the feet and we disagreed on the turn and bank and the rate of climb. However, we decided that the best thing to do was give in to the British system and of the two instruments we didn't think the rate of climb was very important, and we didn't care where we put it. We thought we could probably budge a point on that inasmuch as it was an instrument we didn't care about so we adopted the British system.

"In the cockpit that you find down there, you'll find that the standard instrument layout of the six primary flight instruments follows that general configuration except that the air-speed indicator was displaced off to the side a little and also the rate of climb has been moved out to the side a little, but the six primary flight instruments are in that arrangement."

Commander RAMSEY: "We will call a recess for 10 minutes while we set up the movie and the slide machine."

(10 minute recess)

Commander RAMSEY: "Gentlemen, this morning we have the pleasure of having Commander J.P. Monroe, head of the Armament Branch of the Bureau of Aeronautics, who will lead discussion on armament and armor."

Commander MONROE: "I have some films here of the Navy 12-inch rocket which I thought you might be interested in seeing before I get started on my talk. These films were taken out at Naval Ordnance Test Station and show the very early attempts to launch this rocket. I thought you would be interested in it because it's a very promising weapon. I think it can be launched from any type of aircraft, from the fighter right on up.

"The first film shows the SB2C dive bomber trying two methods of launching. We at first thought we would shoot it off like a regular 5-inch rocket – that is, right off the zero length rails. But we found that the pilot had so much powder smell and so much turbulence that he was allergic to that method of launching. So we went over to what we call the lanyard method. We drop the missile free and it drops about 4 feet, then circuits set off by the lanyard just complete an electrical circuit. The third method we have doped out for launching this rocket, which weighs about 1,300 pounds, is to displace it on a device similar to the regular dive bomber displacement gear. It is primarily in use in the bomb bay to get it out of the way of the propeller.

(Film showing with comments follow.)

"This is what we call a zero length launcher. Those slots that you have there are the whole guided travel of the missile just the same as the way we launch our 5-inch projectiles. This is the system that does not work. It puts too much stress on the airplane surfaces, and we have already killed one pilot doing this by the plane's being thrown into a violent dive and before he could get out, he crashed.

"This is the projectile with a dummy head on it. It weighs 1,300 pounds, has a velocity of 800 feet per second plus the speed of the airplane. That normally brings it up to about 1,200 feet per second when it hits the target. It is far superior to a gun in that respect, in that the bullet starts out with its maximum velocity at the muzzle.

"These are some platform shots for primary testing before we went into the air. You will notice they have two of them on the wings there.

"This is an SB2C, Curtiss Dive Bomber.

"Most of that disturbance is dust. This is out in the desert and it is extremely dusty.

"These are embedded particles of putty. We had quite a bit of trouble with CalTech, which designs these rockets. Apparently for some time they had a lot of junk down in their basement which they had no use for, and they were sticking it in the rockets. We had a large variety of stuff coming out of the tail end.

"We have to abandon this method due to the damage to the airplane. It also extensively damaged the pilots' nerves.

"This rocket is 11.75 inches in diameter. It uses a regular 500 pound bomb as the head, or of course it can have any special head you want to design for it.

"I was out at CalTech the other day and they said they had a 22 inch one they are going to try out. The behavior of the plane is almost normal with only one rocket on one wing.

"This is zero length launching. You can see pretty much why the pilot doesn't like it. He flies through quite a bit of smoke, and he complains of the smell of powder and the airplane jumping around.

"That was especially trying on the pilots' nerves. When he came down after that one, we gave up the idea of launching them any more like that.

"This was the same type of airplane that later crashed. We had a fixed tab; you can see it on the corner of this picture, on the side of the elevator, and the flash from the rocket knocked the tab up about 20 degrees and flew the plane into a violent dive and there is a piece of metal in there, so that he couldn't get it straightened out; he went right on in.

"That was due principally to the igniter blast which sets off the charge. It was found to be too strong. We reduced the charge from 1,200 to 235 grams, and we're getting away from all this damage now.

"That is some of the stuff out of CalTech's basement that makes this. There are some unidentified objects here. This lanyard drop appears to be the dish for getting rid of this projectile.

"Use a regular Mark 51 bomb rack with a spring loaded reel to hold the piano wire lanyard. When the reel gets down to about 4 feet, it closes a micro switch and the weapon fires.

"This is not it. This is one of our first micro switches.

"This pulls the lanyard back up and gets rid of it.

"This weapon can go on any 1,000 pound bomb station, except for flight restrictions, because it actually weighs about 1,300 pounds now.

"Notice how much easier it is on the airplane.

"I have just had the question asked: Was the accuracy affected by the free drop? As a matter of fact, the first three shots they fired, on different flights, of course, fell in the same hole.

"The accuracy there was miraculous and probably quite accidental, but the nose of the projectile falls down about 3 degrees during its free fall and the accuracy is excellent because the action of the projectile is always the same and you can allow 15 to 20 mils in up correction in your sight and you'll be right on.

"I have another short film showing this thing going off a fighter that I think will be rather interesting to you.

"That is Assistant Secretary for Air Gates.

"There was some little doubt whether this concrete was the right type or not. A lot of people claimed it was adobe.

"This is the result of firing directed shots at it on the ground – that damage you saw there.

"That's the big boy there, the others were 5 inch.

"It burns I think for around seven-tenths of a second.

"This TBF fires it with a displacing gear.

"That head on there was a 500 pound semiarmor-piercing bomb.

"That's Admiral Horne and Admiral Schuirmann. Admiral Schuirmann sent us a dispatch saying, "This is the most revolutionary weapon I have ever witnessed."

(End of film)

"That gave you a little idea of what may be coming in the future. Now we can get down to some of the things we have at present. I feel nervous in front of all these experts, but I trust it will pass away.

"Here is a slight discussion of the armament that exists in our aircraft, from an armament point of view, and please remember that all my remarks are from an armament point of view, and if you're thinking about something else like oxygen, electrically-heated suits, you may disagree with it. In general, our planes fall into two classes as regards armament – that is, the four-gun fighter and the six-gun fighter. Primarily, of course, we are interested in the six-gun fighter. Among the former types you have the FM-2, the little CVE escort carrier fighter. The F8F is coming out as a four-gun fighter. We hope to get that up to six guns. We have already got Grumman working on that. (I am an exponent of heavy fire-power naturally, and also from a quite logical point of view, I think.) Our experimental types are also of the same two general classes; that is, the four-gun for CVE's, of which the Ryan fighter is a typical example, and the F8F is another one, and the six-gun type, the F6J, the F2G, F15C. We have the F7F, which has eight 50's, or four 50's and four 20's. Except that we haven't got the bugs worked out. I think rather highly of that armament. It seems to be a pretty good dish for the night fighters. Of course our fighters also carry a large assortment of bombs, ranging from 100-pounders to 2,000-pounders. We have one fighter – the F6F – which can carry a torpedo. We acquired sets for torpedo carrying and sent them out to the Fleet, but from investigations about 6 months later, we found that the Fleet had no interest in them at all. They had lost all the gear; in fact, ComAirPac had never heard of the F6F carrying a torpedo. So apparently they don't want to do that, but it can be done. Our fighters are authorized to dive up to 85 degrees. Of course they have no displacing gear. Careful investigation down here shows absolutely no danger of the bomb hitting the propeller. At least the airplane and the bomb keep their relative pressures fore and aft, and the bomb drops away from the airplane, which was a great relief to everybody.

"I am, of course, very interested in increasing the fire-power of naval aircraft, and I would like to give my ideas of what is coming.

"As it is now, we have the 50-cal. gun which has reached its peak. The only improvements will be minor. The only good increase is to increase the number of guns. So it seems to be just about the right time to look for a better weapon. There are two possibilities here – the one we have and the one we might get shortly. The one we have is a 20-mm. gun. I think very highly of it. It is a fact, it is one we have here, and it is one in hand. It won't do what the 60 will do, but we haven't got the 60, and we won't have it for a year. So we are gradually working into all our aircraft the 20-mm. gun. To give you some idea of the 50 versus the 20 and dispel a lot of ideas that have bothered us, I would like to give you a comparison. When somebody goes from four 50's to two 20's, to the layman that means a decrease in fire power. Actually, quite the reverse is true. In the horsepower of the gun, one 20 is equal to three .50-calibers. In the actual rate of fire delivered at the target, one 20 equals three 50's; in kinetic energy at 500 yards, one 20 equals two and one-half 50's.

"That adds up to four 20's equaling twelve 50 calibers, judged by those standards. Of course you have other advantages of the 20. You have the much greater penetration of armor. The 20 will go through 3/4 inch of armor at 500 yards, while the .50 cal. will go through only .43. In addition to that you have one more great advantage – that is, you can have longer and more frequent bursts without damage to the gun with the 20 than you can have from the .50 cal. That is important for the strafing airplane, because they are burning up their barrels and ruining their guns on one flight. Sometimes it is long before that one flight is over. They will come down with screaming barrels and get trigger-happy, and then all the barrels are gone in one flight. It should not happen in a 20-mm. Of course, you have disadvantages. You have a heavier installation, one-half as much ammunition for the same weight. Our standard ammunition in the Navy is 400 round in one gun. The Fleet has set up 30 seconds of fire as a minimum requirement for the .50 cal. gun. We can't do that with the 20, so we give them 200 rounds. The 20 is lethal enough to get far more results out of that 200 rounds than the .50 ever will out of the 400 rounds.

"With the 20 you are putting out a new weapon. Fortunately, we are over the headaches to a great extent in the Navy. The SB2C has led the way with the 20-mm. We had an awful lot of headaches getting the new ammunition, the new lengths, etc., and getting the ordinance men to learn how to use the gun and get around the temperamental characteristics. We are over that now, and the majority opinion is very enthusiastic about the 20-mm. gun. We have at the present time 200 Corsairs going out, at the rate of 50 a month, which will go into action as soon as we can get them aboard the carriers to get an evaluation of the 20-mm. gun in the Navy fighter. I am personally very anxious for the first report on that, and I think the first time they open up on a Jap fighter, it's going to fly into a million pieces. We have a great cry for the 20-mm. gun.

"Another disadvantage of the 20 is the time of flight. Out to 500 yards you've got three-quarters of a second as against a .62 second for the 50. These airplanes go 450 to 500 feet per second, and in one-tenth of a second 35 to 40 feet.

"It also hurts when you try to mix the batteries. I am personally very much against mixed batteries, with the guns at the present ranges. If the Mark 23 sight does what we want it to do, it is going to push the hitting range out so far that the 50 cal. will get to the target a helluva' lot ahead of the 20. If our fire control that is coming is as good as we think it is, it is very unwise to mix the batteries. For present battle ranges it is perfectly all right. Of course the Navy is fighting in the Pacific. We have a rather inferior bunch of enemy airplanes. The Japs have played ball with us very nicely all during the war so far and they have refused to improve their armament or their planes as fast as we could expect them to. I have a WAVE in my office who keeps a chart on all Japanese aircraft trying to figure out what the trend is and it is very definitely towards larger armament, heavier guns, more guns and leak proofing and much better performance airplanes. In view of that, it behooves us to get in ahead of them. I feel it is much better to put the 20 mm. in now, although the 50's are doing a perfectly acceptable job and we could probably finish the war with them. It doesn't make the pilots feel very good to be shooting 50's when the enemy is shooting 20 mms., however. Their 20 mm. is a somewhat inferior weapon, of course. We've got hold now of an experimental Jap fighter. It has four 20's and two 30-caliber guns. The guns in that airplane are of German design copied by the Japs and they are pretty good guns. They're worth anybody's respect and it is not going to be any fun to come up against an airplane like that with inferior armament.

"We're finding all the time 20-mm. guns, 37 mm. and 50 calibers in increasing numbers, in Jap aircraft. I don't know why the Japs stick to the 30 caliber. It is a completely ineffective weapon particularly the way they use it. It might be all right in the fighter if you had 16 of them and got right on somebody else's tail. You might saw him apart. The day of the 30 has long since passed. The Japs stick to it and it's fine from our point of view. In the Fleet we have the SB2C which originally came out with a single .50 turret. due to the marginal performance, they took the turret out and put twin 30 cal. guns in its place. That was a very fine solution as far as the Fleet was concerned. They were delighted with it and have continued to be so. The same comparison holds true for the 30's and 50's as has held for the 50's and the 20's. Two 30's are approximately 1/3 to 1/2 as effective as one 50. The two hand-held 30's can't compare to the effectiveness of the power turret, as far as getting on the target and holding your aim and following the enemy around. The Japs are very easily frightened off by those two streams of 30 caliber tracers going out there. The boys are able to scare most Jap pilots by hosing a stream of bullets out there in their general direction. However, it is my firm belief that if the Jap pilot would get over his awe of those tracers and just plow right in, he would have no trouble at all taking the SB2C any time he wanted to with his 20-mm. guns.

"In armor our goal is protection against the Jap 50. Our problem is this: We consider the Jap 50 to be a 2,500 foot/second gun. We figure the Jap will overtake us at 100 miles per hour. He will shoot at us at 200 yards at 20,000 feet, so we figured out armor that would take that. It is 1/2 inch, if you had no other protection in the airplane, and you do have a lot of other protection, you have the skin and the radio and all kinds of members in the airplane which give protection. We took a lot of fuselages down to Dahlgren and fired them and discovered a lot of interesting facts – that you could reduce your armor and thickness, the total weight of the armor carried, by simply taking the radio and putting it in the solid wall behind the pilot, or that you could substitute a sheet of two of steel or dural possibly and decrease your weight tremendously and I wish the engineers would start thinking of that in their new designs. Just make a wall back there and put your radio right behind the pilot as close to him as you can. You can take a helluva lot of weight out of armor.

"Another study is the trend in the way our ships are getting shot down in the Pacific. The trend is now that for every one we're losing by enemy aircraft, we're losing five by enemy antiaircraft, by shore or ship-based guns. So our problem is taking another turn in armor. We get to forward protection again. You have nice forward protection in the engine. The oil cooler needs it. The pilot has to have something to sit on. He has some very valuable and vulnerable material there that he does not like to get hurt. So we're going to have to think very seriously of forward fire power particularly now that we are getting into large Jap land masses. We presume that the Japanese Fleet is doing the same things we just covered. We have developed some wonderful strafing techniques. They are sinking destroyers and they have stopped cruisers dead in the water with 50-cal. gun fire. It's a tremendous weapon. When we get these 20 mms. out there, it is going to be even tougher. They can put a pretty good-sized ship out of commission with them. So, for that you need some kind of forward protection for the pilot.

"A new item that has come into the armament picture particularly for fighters, and it applies to all types, is the rocket in the Navy. It is called a 5-inch HVAR. You will see a lot more of them in the future. HVAR means High Velocity Aircraft Rocket. It is a 5-inch head with a 5-inch motor driving it. It gets a velocity of 1,200 feet per second plus the airplane's velocity, which brings it up in the neighborhood of 1,700 foot/seconds when it hits. It is actually just about a 5-inch shell. Of course, we have the older two-type rockets – the 3 1/2 inch rocket with the 3 1/2 inch head and the 3 1/2 inch rocket with the 5-inch head. They have lower hitting power. I suppose you are all very familiar with the launching picture. I will give it to you chiefly. We started out with 7-foot rails. They were much too cumbersome and handicapped the planes too much so we were very fortunate in having some people on the West Coast develop the zero length launching, which is really only two studs in the wing with two lug bands on the rockets. They have about 1 inch of guided travel until they are free in the air. This installation on the F6F costs 3 miles per hour at V Max. That installation will launch all the rockets except this big 12-inch job. We put that on in general as an after thought. It is primarily designed for aircraft already in the fire in that it is a retroactive installation and it is just screwed on the wing of the airplane at a strong point. It makes a bump on the plane and for that reason we are trying to improve it. We have down here at Patuxent a single pylon launcher which is nothing more or less than a flat plane with its end pointed forward which goes in the airplane wing and has two little zero rails about 1 inch long and 12 inches apart. That will make the contractor's job a lot easier because instead of having two posts for the launcher, he can just put this launcher in the wing or any strong point. We feel that the accuracy will be just as good, the drag will be lower – it's a production-line installation. We will have that ready in a few months, I hope. Our firing of the rockets is done by the bomb button on the pilot's control stick. After trying several other places in the cockpit, we decided on that, which gives the pilot a selector switch so that he can decide whether he wants to shoot rockets or drop bombs. He is probably used to firing with the right hand on the stick and this is going to work out much better. Our present launcher has a shear wire which holds the rocket in. When the rocket is fired, it has to cut that shear wire to get free. It has been a problem to get a shear wire which did not tear the launcher up when it tried to get away, but would be strong enough to stand carrier landings. The contractors will be delighted to know here that we have done away with the shear wire and have a spring loaded latch which needs less such paraphernalia. It is positive and will allow the rocket to be fired in the air with no danger. You can also take carrier landings without the rockets falling all over the decks. We have had a lot of trouble getting a sight for the rockets. Our solutions are the tilting head sight or the regular Mary VIII sight. That had a lot of vibration and the pilots felt that they did not want to bother setting up this sight head at different angles because they were too busy. So, instead of that, we changed the needle in the sight and gave them a ladder reticle. This reticle is a series of short horizontal lines in the vertical plane every 10 mils right up the center of the sight. The pilot finds out soon how many mils to allow for the drop of the rocket. Of course the trajectory of the rocket is not nearly as great in flight as the machine gun bullet. He finds that point on his ladder reticle. There is no adjustment. He just takes the 30 mils lead and uses it. It is not a perfect sight. He's still got to change his range and dive angle. We have found that those are the two controlling things in firing rockets. If you could get something to solve the ranging and the dive angles, you could fire rockets very, very accurately. We are working to a great big radial sight or a big 50-pound object that will do that because it is too much and doesn't have any relation to the gun firing problem or the bomb dropping problem. At the moment, we have settled down on the ladder reticle on the sighting problems.

"We are trying to connect up the rocket launchers by the nose. The reason for that is the rockets are tearing up control surfaces. The blast from the ignitors that sets off the blast – the initial blast, is causing it. It tears holes in the control surfaces. The slip stream does the rest and tears them to ribbons. The assorted trash that comes out of the end of the rocket is very hard on control surfaces. It tears holes in them and particularly the plastic plug which seals up the end of the rocket. We are trying to get that changed to a more frangible type. If we could get rid of the shock of the ignitor firing which breaks the ribs in our control surfaces, we would probably have the answer to an acceptable rocket installation. You contractors' representatives are going to have trouble with that – the ribs in your flaps and the ribs in your elevators. The after ends of them are just going to give way. Finally, you will just have your after edge which will be flapping in the wind. That is particularly a shock wave phenomenon and the answer may be metal-covered surfaces or doubling the number of ribs or putting gussets in there. It carries the stress up some place where it will do less harm and it is a subject that you all ought to give a lot of thought to. Our structures gang has not got the answer. They are setting up a project at NAMC, with dummy rockets flying at standard control surfaces to try to solve this problem. We are getting a large supply of 5-inch rockets in the fleet now and I'm very much afraid that we are going to be deluged by reports from the fleet of damage to the surfaces for which we have no ready answer except to change the damn things when they give way. Another advantage of this up angle to the rocket is that it will fall more nearly in your machine gun pattern at a given range. So far we have found that a good up angle for the rocket is 3 degrees. On the F6F set at 1,000 yards, that puts the rocket right in with the machine gun, and makes it a nice aiming problem. Of course, this 11.75 rocket is really tough on control surfaces; it just tears it all to pieces. We have had to attack this problem in a different way. We have had to instrument the airplane with pressure gages and fire it on the ground; find out what the pressure is so we can find out how to use it with the lanyard drop. Then we go up in the air with the same instrumented airplane and drop it in the air and see if we are right or not. At the moment the only outfit that is liable to get in combat with this thing in the near future is using the displacement type launchers, which apparently hold the rounds far enough away to insure that there will be no damage. This displacement type launcher by the way does not hold the round down when you make your firing run. When you press your launcher down somewhere during its travel, when the rocket gets to a suitable point it fires and leaves. That makes quite an aiming problem too. During one of our lanyard drops was when we lost the only plane in the process of testing this weapon. That was quite a shock to us, not only from losing the airplane and the pilot, but from the fact that it looks from those pictures that the airplane is close and above the blast and should have no effect on it. But we found by later experiments that the ignition blast was so terrific, something like 1,200 grams of powder going off, that it just raked the controls until they gave way; particularly the elevator – the rudder shielded it. What we had to do about that is beef up the surfaces and drop the weapon farther away from the airplane and make the lanyard longer. If we had to go to a lanyard drop for the 5-inch we would be on the spot. The answer to the 5-inch is to tilt it up about 3 degrees and try to direct the blast away from the control and to get this ignitor charge cut down. BuOrd is on expedited program to cut down the ignitor charge and cut down this blast and to make the sealing caps out of a soft material which won't damage the airplane.

"Our current ideas on charging guns in our fighters is to charge them mechanically with hydraulic chargers. We have two reasons for that. One is that all the pilots insist that we give them charging. We question them very closely when they come back from combat areas and they state that they have to have charging and they started up their guns many, many times with charging. I think Joe Foss said if he didn't have charging he would be dead; that he charged his gun, got back in the battle and saved his skin. There is one more reason peculiar to the Navy in charge guns. That is on carrier operations. When an airplane lands on a carrier the guns are pointed at just the right angle normally to rake the bridge. The admiral would of course be allergic to bullets flying around his ears so he insists that something be done about it, and we have to have a hold-back feature in the charging. You have to have a hold-back so that nothing will happen. For these two reasons, recharging your gun in the air and for carrier operations, we are going wholeheartedly for the charging. Now there was a period when the Army started to abandon charging in the air and we started to follow suit. All it has gotten us is a hell of a lot of trouble for the contractors to get charging back in the airplanes that were never designed for it. A typical example of that is in the F8F. It was laid down with no charging put in to save weight, save complexity requirements, so we are asking Grumman to put it in and of course it does not do it any good. You can't find any place for control valves in the cockpit and it's a narrow squeeze to even get the chargers in at all. The Navy has always gone for hydraulic charging. I know the Army shifted over to pneumatic charging and it is damned good. But the logistics problem of changing over to the pneumatic charging has meant that we would have to put charging equipment in bottles and compressors on all our ships all over the world and every time we come in from a hop, unless we used the Cornelins compressor or some such arrangement as that, we would have to look at the bottles and see if they could stand another hop. For that reason and because we have the hydraulic systems available there, we are sticking to hydraulic charging and I trust that we will continue to do so. We would like very much to get all electrical charging if anybody can dope out what is reliable and will go in wing guns. We have a pretty good charger now, that is suitable for turrets. It mounts on top of the guns and out for wing installations. Made by Eureka, it was an automatic device; we have changed it now to a manual device where you have to push a button and it charges every time you push the button.

"The subject of blast tubes is getting increasingly important here. They have the F7F down here with 7-foot blast tubes and the bullets occasionally behave very strangely in these tubes. They all of a sudden will go off at some wild angle, one or two bullets will. We don't know what it is unless it is a pressure phenomenon that builds up in those narrow blast tubes and that long length of fire. But we don't like the long blast tubes; unfortunately we are stuck with them in the F7F if you want those nose guns. Another thing that has bothered us is that our blast tubes are not tied down tight enough. Securing mains for blast tubes could bear a lot of investigation. They've got to be tied down so that they cannot move any in flight. We lost an F7F down here and there is a strong suspicion that the reason for losing the F7F was that a blast tube became loose in the nose and deflected the bullets into some hydraulic lines and some gasoline which set it on fire. The cause of the crash was fire but probably the underlying cause was that the blast tube got loose. Another thing that is coming up is the cooling of guns in the wings. That is a very serious problem with strafers. The boys quite often burn out every gun they have got in one dive and then come back and raise hell. While it is initially hard to tell a pilot that you are absolutely limited to the number of rounds you can get out and still save your rifling, you still have your barrel in a fit condition to fly. When he thinks he wants to shoot 200 rounds, he wants to shoot 200 rounds. The fact that it is an impossibility – the gun won't do it – sometimes doesn't seem to make any difference to him. You are held down to 75 rounds for your initial burst and 25 rounds thereafter in your 50 caliber but in the 20mm. it doesn't hold true. We have lately run across a pretty slick idea for cooling which apparently might have merit. It consists of running the blast tube clear back of the gun completely enclosing the barrel, taking an air lead into the outside air right by the muzzle and an air exit back. It does not require much fighter speed and doing that blows a continuous stream over the gun barrel. Apparently it doesn't make any difference how hot it gets, but your cooling does affect the recuperating powers of the barrel in between bursts. That is where your cooling is important and where enclosing the barrel with that stream of air might be the solution. We have one example we are trying out on the F7F. It may be the answer. We are just about to put out via BuOrd permission for the fleet to take the oil out of their buffer plates and thereby increase the rate of fire of their guns, and the result of taking the oil out of the buffer plates increases your rate of fire around 150 rounds per minute. The investigation down at Dahlgren indicates that that will double the wear on the gun. If they want more rate of fire than that, bad enough to take additional breakage, maybe they will decide to go to it. We at least would like to get a large service evaluation of this method. Everybody seems to have a different experience with that but I believe we are going to put it out to the fleet, and tell them to do it if they accept that breakage. Dahlgren assures us that you will get an absolutely reliable gun for 5,000 rounds, and if you take care of a gun you might get 20,000 rounds out of it. Of course, 20,000 is quite adequate. We have one more way to increase fire which can be used immediately as against more theoretical ways that are not so close. It is putting these boosters on there. We simply screw them on the gun barrel and restrict the passage of gases and they give you about the same increase in fire as taking the oil out of the buffer. As a matter of fact when I said you can get only twice as much wear out of the buffer, you get 3 times as much wear on taking the oil out of the buffer where you use the adapter. The Army has the adapter which is just about to go in production and the Navy also developed one. Captain Chin, down here at armament test has developed one. They both do about the same kind of a job. They are both called good. I think the adapter is about 2 ounces lighter. It appears right off that the adapter would be the easiest way of increasing your fire. We are going into production on adapters and dishing them out to the fleet.

"That is about all I have to say on that rate of fire business. It is a controversial subject and I wonder if we will ever settle it until we give it to the fleet and they make up their minds very shortly whether they want it or not.

"Now we come to some new developments here. One is a Mark 23 gun sight called by the Army K15. It is a lead computing gyro gun sight which everybody has gone to except the Navy. The British are apparently using it every chance they get. The Army is for it and the Navy is kind of diddling around on one foot or another and can't make up their minds. Actually, they want to put it in but they hate to go ahead and tell the contractor to get rid of that nice old Mark 8 which is doing a perfectly swell job in the Pacific. It has done an absolutely satisfactory job and there are a lot of people that argue with us about what is the use of getting rid of the Mark 8 sight, that it is doing such a good job, and put a bigger sight, which requires greater maintenance and is harder to see around. Of course, the answer is that the Mark 23 is probably the greatest single improvement in fire control that we have had recently. It gives the average pilot terrific increase in hits. I think that has been proved all around the world and the reports on it from every theater of enemy activity except in some test activities, are highly enthusiastic. It is one of the few things I have seen about which all the reports are enthusiastic for the sight. Due to the fact that it is almost impossible to evaluate this sight on a tow sleeve, it has fallen on its nose in some of our tests, particularly down in operational training, where they took a bunch of people that were used to the Mark 8 sight and went out on gunnery runs and have been shooting on it a long time and knew exactly where to put the pip on the Mark 8 sight to get hits. They got plenty of hits with the Mark 8 even under those conditions where the target was not big enough for the Mark 23 to range on. The Mark 23 sometimes did better that the Mark 8 but normally the Mark 8 has a tremendous advantage there.

"Does everybody know what the Mark 23 sight is? Briefly, it is a lead computing gyro stabilized sight which has a fixed reticle and a moving reticle. You turn the thing on and you range by means of rotating your throttle. You have about a 50 degree throw in your throttle. You have a circle of lights in which you are supposed to outline the wings of the enemy aircraft. You first set your enemy's wing span on a counter and then if you will keep your circle of lights on the wing span you will have a correct range. It goes into the sight and the sight computes its own lead. All the pilot has to do is put the pipper on the point he wants to hit. Well, that is very attractive and if he sets up the problem right and the sight is accurate, that is where the bullets will go. He has no more mental lead to put in and can make a full deflection shot theoretically at 800 yards. The sight is supposed to be good for 800 yards. Actually the airplane is so small out there you can't range accurately, but you can hit far in excess of the range that you used to.

"Paul Ramsey said you have one set up back there which should be very interesting to look at. We are sending an advanced set of kids to try it out in combat in the Pacific. Everybody else in the world tried it out, the Army and the British were delighted with it, but the Navy is not as yet convinced that it is good. Our training people think very highly of their training method and claim they have the best deflection shots in the world coming out of their schools, so possibly, they argue, the lead computing sight won't help the Navy gunner as much. I think that by the first of the year we will have the sight going in the first airplane.

"Another development that is getting a little bit old is taking the reflector off the Mark 8 sight and shining it on the bullet deflector glass. That takes some garbage off the sight, allows the sight to be moved considerably away from the pilot and is a solution as to the normal reflector plate on the Mark 8 sight. It is a strange thing that practically all of our combat airplanes were laid down with a telescope sight. Then the Mark 8 was finally put in. It is in combat now but it was a hell of a job to get it in. It was a mental hazard and an actual hazard to pilots on crashes, on water landings or any kind of rough landings particularly before we put the shoulder straps on.

"We probably have as lousy sight installations in naval aircraft as you will find anywhere in the world simply for the reason that the space on that instrument board was hard to rearrange and to get rid of the old telescope and get the Mark 8 in as an afterthought. It is much harder to get the 23 in because it is a lot bigger and I think it's going to scare the fleet to see the Mark 23 come out at first and it's got to be good to get by.

"I have already discussed the ladder reticle versus the tilting head and the fleet opinion was that the tilting head was not as desirable as the ladder reticle. Douglas is coming out with the power type method of dropping bombs. As you know, we have been plagued all our lives in the dive bombers by having to displace the bomb past the propeller; at 90 degree dives with a flapped airplane the bomb has a very bad habit of going right square into the prop unless you forcibly keep it from doing so.

"Douglas has doped out what we call a power-displacing gear, which consists of a bomb rack which he has invented himself, and a starter cartridge which kicks the bomb out far enough to make sure that it doesn't go into the displacing gear. When you drop your bomb, you fire the cartridge and it will kick a 2,000 pound bomb approximately 4 feet, and get it clear of the prop. It looks promising, and we are trying it out. He's also cooked up another way to get rid of the bomb without displacing gear. That is an automatic pull-out device. This device, when you normally press the tit to drop the bomb, and in order to drop the bomb, you energize this pull-out device. A fraction of a second after you start your pull-out, the bomb is dropped off the conventional bomb rack but the airplane has started to pull-out. The thing is figured out very finely and it is only supposed to displace the bomb about 8 or 10 feet off its normal point of impact. The result is you need no displacing gear – the plane actually pulls itself away from the bomb. We are putting that in a plane and trying it out.

"We are now putting lugs on the torpedo so you can get away from the old suspension bands. The old method of suspending torpedoes to aircraft was by means of a couple of wires. The lugs consist simply of a big channel on top of the torpedo, which has points for hooking on to the 14-inch rack and the 30-inch rack. You can choose either one of them. At the present time it has a little draw-back of quite large deflection of the channel on top of the torpedo. It is only secured at either end so, under G's, the thing deflects. I am afraid it is going to make sway bracing a rather difficult problem.

"The idea of putting a band around the torpedo and stopping the deflection came to mind, and they decided to remove the channel and just use a band around the torpedo, which, according to water tunnel tests at the model basin, hardly increases the drag of the torpedo at all. You can move the bands either to 14 or 30 inches. All our late airplanes are mocked up for the lug suspension and getting away from the sling suspension.

"We're coming out with a line of gun packages now. I mean by that, a package of guns attached to the bomb rack as against the regular bathtub that you permanently screw on the side of an airplane underneath the wing. We have in production now the twin .50-caliber package with something over 300 rounds per gun, weight 346 pounds, and it will go on the Mark 51 bomb rack. It will be jettisonable in the air if you want to get rid of it. We are going into production on the single 50 and single 20. Both of the latter two have passed their tests and are ready for production. The single 20 is a helluva nice package – runs well and will increase your fire power. The next two steps are the twin 20, which is being worked out and should be ready shortly, and we're only waiting for the .60-caliber to go into the package on it. We have some predesign studies made on it. You cannot synchronize these packages. If you get in a tight spot and want to get rid of them, you can simply release the rack. I am about talked out here. That is about all I had to say.

"Are there any questions or any discussion? I apparently put everybody to sleep."

Lieutenant Colonel TOUBMAN: "You spoke of the .60-caliber gun. Can you give us a little more dope on that gun?"

Commander MONROE: "Yes; I can give you a brief outline of it. It was a development that was started in May 1941 by the Army, which, of course, develops all guns. After 3 1/2 years we have about a dozen firing models. The gun has not gone too far along. From my experience with guns, the short time I have been in the Bureau, it is at least 1 year away. Even if we had a gun now that was satisfactory that would give us 1,000 rounds with no breakages, which is a sort of thumb rule used, I believe it would be practically a year before we could get them in production. The .60-caliber gun is so attractive a weapon, and so lethal, that we are trying to change our contracts to put a hole in the plane that a .60-caliber will go in, and to make provision for it. Of course, we cannot guarantee what the final shape of the gun is going to be. I cannot guarantee anything except that we can show now what a wooden model looks like. I believe new production models are coming out at the moment. But it is too far away.

"I don't know whether the Army has any objections to my discussing the gun or not. If not, I can give you figures on what the gun will do. We're dishing the guns out to contractors. I suppose everybody here has been cleared. Well, it is a 3,600-foot-second gun, weighs 140 pounds, and has a rather large bullet, that is, the powder chamber is large and a conical shape, which means it is going to be hard to stack them. That has been reduced recently, so it's not quite as conical as it was. Rate of fire is variously reported, anywhere from 500 to 700, depending on who is using it. This chart has 725 revolutions per minute. I think that is high. It is probably in the neighborhood of 600. The armor penetration is very spectacular. It will go through 1 1/4 inches of armor at 600 yards at a 20 degree obliquity, which means that you can't arm against it, you can't carry that much armor. Its time of flying to 500 yards is .045 seconds, against 0.62 for the conventional .50-caliber gun. It has a very satisfactory effect on leak-proof tanks. When the leak-proof tank is full the hydraulic shock wave set up is so severe, the exit side of the tank is ruptured with every shot. You can guarantee that if you hit a full or semifull leak-proof tank, you will have a fire in every case. The German leak-proof tank with this projectile did not have a fire. In addition to not being able to armor your airplane, you cannot use the conventional leak-proof tanks. One hit, one fire. It is a very handy weapon to spring on your enemy at any time. I believe if you can get that gun in the air you could knock out any air force in the world immediately. There is absolutely nothing that can take it and these are not fanciful theories. They can be proved on the ground and have been proved on the ground. So when the gun gets here we will really have something. Of course, by that time, the other fellow may have something too. The weight of the projectile is about the same as a 20-mm., that is double .50-caliber, or 60 pounds per 100 rounds. That is a pretty good picture of it. I would like the contractors to keep in mind until it does come out, put in a hole for it and it will lead the field. I want to impress that on all the Navy contractors, and due to the fact that it will almost fit into the 20-mm. hole, you should not have too much trouble; it will be a good thing if the fighter contractor could bear that in mind. Reserve a space for it."

Lieutenant Colonel RENNER: "Jack, I had some questions here. The first one is in moving the sight forward and reflecting against the bullet-proof glass. We have found that the pilot eye level, if sitting in a normal position, is such that he can see the sight, but if he leans forward to get within the 6 inches that's necessary to use the sight, the distance from the lens of the sight to the reflecting glass and the pilot's eyes in the Mark 8, I think, is 18 or 16 inches, and in the 23 sight, as I recall, you have to fix your eyes 6 inches from the sight. Well, if you are sitting in a normal position and you want to get forward where you can use all your deflection, you move your seat, because the movement of the body forward shortens the relative height of the pilot from his duff to his eye and he is too low.

"So it is going to be and unless we get a hydraulic jack to run the sight up and down every time we want to use it, we are going to have trouble. I wonder if anything has been done to increase the distance that the pilot can keep his eye from it over the present Mark 8 or 23 so we may be able to get back 16 to 24 inches from the sight which will allow the pilot a slight forward movement and still use the sight. Another question: Has anything been done on putting the sway bracing on the bomb instead of on the bomb rack so that we can drop the garbage when we drop the bomb and will anything happen to the trajectory of the bomb at the short distance that the fighter is dropping the bomb? I know that internal racks have been put into the F7F and I wonder if they are satisfactory – if we could go to all internal racks to get away from some more garbage. Has anything been done to build us a high-velocity .50-caliber gun, something that will give us 3,500 to 3,600 feet per second? The .60-caliber gun sounds wonderful, but if we load our wonderful planes down with so much armament that the enemy will be sitting on our tails, we will have to have it so that it will shoot 360 degrees or we are not going to get any shots at him."

Commander MONROE: "That sight location is a problem that has continued ever since we have had illuminated sights. Unfortunately, it is controlled by the size of the sight. In other words, the distance from the reticle to your eye in the Mark 8 sight is somewhere around 16 inches, and there is nothing you can do about that. The same thing is true of the Mark 23, where you have a much smaller outlet for the light, and you have to be much closer to it. In order to get farther away from that sight, you've got to make a bigger optical system and a bigger sight. That is our present experience. In order to get away from it in the Mark 23, you would have to redesign it and make it larger.

"At the present time, the reticle in the Mark 23 is a 35-mil reticle while that in the Mark 8 is a 50-mil reticle – the smallest of the reticles. There is nothing that can be done about it. The sight just has to be put back closer to the pilot in the case of the 23 and pilots of varying stature will have to adjust their seats to see the sight the best way they can at the present moment. There may be some better answers to that, but I don't know what they are.

"As to the sway bracing on the bomb, there are a number of ways to get around that. One of them is used on the Grumman. The sway brace falls away with the bomb and you neither have it on the bomb nor on the airplane after you come out of the dive. That is a very attractive method if it doesn't hit your tail surfaces and do more harm than good. In general, I don't believe, in fact I know, we have no plans for putting sway braces on bombs as far as the Navy is concerned because of the deleterious effect on the aerodynamical qualities. I don't know what the result would be, but they would certainly deflect the bomb in a way you did not want it to be deflected. It appears if you want to take the logistic problem and the expense of furnishing a sway brace for every bomb the Grumman method is satisfactory. It is simply a fork that fits into a hole in the airplane's wing and it is held in there by compression and when the bomb leaves the sway brace leaves with it. As far as this internal rack business goes, we have a beautiful internal rack arrangement in the F7F wing. You could walk by that plane and look it over and never know that there was a bomb rack on it. All you do is open up a door and slap the Mark II hoist in there and run the bomb up. It has this disadvantage, though, they couldn't get the rack down to the lower edge of the skin, so they had to put a lug in there to reach from the lugs on the bomb to the hooks on the rack. That means to use that internal stowage of the rack, you've got to have two links with every bomb that is dropped. And again, you've another piece of ordnance material. The plane can't carry a bomb if you don't happen to have the links. You have to have a barrel of links on every station and every carrier. You would probably have to deliver links with the airplane. It is somewhat doubtful whether that's wise or not. Personally, it's such a nice installation that I am all for it. And if we can hold down similar installations so that everybody would have to use the same link it would be no problem at all. But if we have a different link for every plane, it would be quite a headache and be absolutely unacceptable. A lot of the planes of course could bring that bomb rack down further so the lugs protrude slightly and the hooks can get at them without a link. I foresee a great future for buried bomb racks and getting away from the pylons. It is cheaper in every way – the drag is completely removed and there is very little built-in weight and I think Grumman has demonstrated that you can hoist a bomb up there in no time at all.

"The high velocity 50-caliber you know is limited by the outline of your bullet. However, there is such a thing as a double base powder which will give you a helluva lot more kick than a single base powder, but it has the disadvantage of being very hard to control. It is not stable enough at the moment to be used, I believe. The last time I heard from it, it wasn't. The outline of the 50-caliber bullet is fixed, so you've got to find a much higher brand of propellant to kick the bullet ahead faster. With the redesign of the gun and a bigger bullet, it is doubtful whether it would be worthwhile to try to get muzzle velocity in that way. But that higher muzzle velocity is a very desirable feature."

Colonel COATS: "Very little has been said about the 37-mm. high -velocity M-9 cannon. We had the XA-41 up for a brief evaluation which Eglin Field had for a considerable length of time. I have seen this weapon fire and it is really terrific. The contractor's pilot has assured me that this particular installation is very good. In fact, he said that he had only three stoppages in several thousand rounds of firing. He said he thought it was more dependable than a 50-caliber. It seems to me this 37 mm. is a better gun by the same proportion as the 20-mm. is over the 50-caliber. Does anyone have any information on this weapon?"

Commander MONROE: "From Armament's point of view in the Navy, you have a lot of draw backs on the 37 mm. Of course, its rate of fire is only at 140 revolutions per minute. While it has a high velocity it is questionable whether that rate of fire is desirable for air-to-air combat. You also, of course, have a large amount of built-in weight. You have a heavy gun installation. You have a very small number of rounds you can carry in the plane. We have had very little experience with the 37 mm. in the Navy except for a couple of installations for submarine warfare where we want to get a lot of penetrating power. Possibly Colonel Meng can give us some further dope on this."

Lieutenant Colonel MENG: "They did fire that gun at Eglin. Some people liked it. One thing against it was slow rate of fire and the small amount of ammunition you could carry. It was tested down there and they had only one or two stoppages. We have an old ordnance colonel down there who became quite excited when I quoted a lot of figures. He said, "Yes, son; that is right, but I can prove the same things with a 16-inch gun." I believe the Army specifications last year were for 20-mm. alternate mounts and I think they have stopped that and are now going into 60's."

Colonel COATS: "We have a requirement for a minimum of six 50's on all our fighter airplanes and we try to get 300 rounds per gun. We do not have any requirement for 37's at the moment. Those we do have are supposed to be interchangeable with 20 mm. and 60-caliber if possible."

Commander MONROE: "I wonder if somebody in the Army could explain why the Army is not interested in the 20-mm. gun. They developed it but apparently have no requirements for it while the Navy feels quite differently about the gun. We are going to it in a large way, I trust, in that we are putting it in the Fleet to let them try it out. I personally have a tremendous amount of confidence in the gun and believe the requirements will be very great. Anybody in the Army who can speak on that?"

Colonel COATS: "I'll try to answer that in this way. I believe the feeling in the Army generally is that we would like to have a lethal density pattern. The most bullets going across one place at a given instance. We would like to have the smallest caliber gun that can do the job. If it takes a 22-mm. to tear a Messerschmitt or a Mitsubishi apart, we want 20's, but as long as a 50 will do the job we feel that if we can carry a greater number of guns and a greater amount of ammunition with the same weight, with an equal or greater fire power, that is the gun we want. If you are strafing an airdrome you can put out more bullets. A Jap doesn't care whether he gets killed by 20 mm's or a 50 caliber. We can put out more bullets and we have more weight covering the same area. Another thing that comes into this matter of sighting is the training of the personnel. I believe that with more guns, you can put out bigger density pattern for the training of your personnel. When we get sights to the point where we can pull the trigger just once and hit a fellow, then we can go to the bigger calibers. It is a matter of training of pilots. The Mark 14, the gyro sight, we found didn't increase our accuracy for our control gunner to any great extent. However, it did bring the people in the middle and lower brackets up as much as 5 or 6 times better than they had shot before. I think we in the aircraft game should be worrying about the people in the middle third or the bottom half, that we have to make better sights, better cockpit arrangements, easier planes to fly for those people. We don't need to worry about our top shot or our best pilot. He can get along in any kind of a rig. That is the reason – we feel we can get a bigger density pattern.

"I would also like to point out, I won't go into an argument with 20's versus 50's, but I think a lot of it has to do with the arrangement in the plane. For instance, in a P-47 or F4U, you have the guns in the wings. Of necessity you must cross the fire pattern at some fixed distance from the plane. With all your guns over one fixed point at a given number of yards, you have a great X forming out there. At 600 you are wasting a great amount of your bullets. If you close up on a fellow at 200 yards, you are also wasting bullets. In the F7F or the P-38 you can put all your guns in the nose; firing parallel streams of lead, your bullets all going out forming a lethal density pattern as far as the bullets go. In an installation like that you could possibly be better off firing four 20's than you would be firing six 50's. In the P-47 with four guns in each wing, we recommend that they cross the first two guns at 250 yards, the next at 350, at 450 and 550. That gives you a density pattern in depth as well as width for about 200 yards, which in turn gives the mediocre pilot a better opportunity to hit an airplane in flight."

Commander MONROE: "We have put out a lot of bore sight patterns. Right at the moment, I might say something about our policy on that. We got all hepped up about our pattern bore sight and went into it in a big way for all our fighters. We put it out to the fleet and they grabbed it and went to work with it and immediately practically discarded it. They claimed that they could not get the hits with the pattern bore sight that they could with the point bore sight. We made some extensive tests and found that in some of our planes the guns jumped around so much the point bore sight actually gave you a pattern bore sight at the hidden ranges and when you got a pattern bore sight you were really scattering lead all over the sky. We went back to work on a point bore sight and advised them to converge them at 300 yards. They disregarded all our instructions on that and I don't think you will find any two squadrons that have the same bore sight. One of the sovereign principles of the squadron commander is that he figures out his own bore sight and uses it. That is a point on which it does little good for the Bureau to put out anything except information.

"I think that was a very good presentation of the arguments against the 20 mm., and it shows the Army's point of view, which I have been wondering about for a long time. We are going to let our case for the 20 rest entirely on the way the Fleet wants it. If the 20's are wanted, they will get them; if not, we will stick with the 50's. I think the 20's are coming in here very shortly."

Colonel COATS: "Commander, I wonder if the Navy is doing anything about automatic ranging for this lead computing sight."

Commander MONROE: "We have a project set up at M.I.T. to hook in the lead computing sight with the radar so you will have an automatic input on the sight which will take away the most objectionable and most sensitive feature of the present Mark 23 sight, which is range. If you don't range accurately, you lose a large percentage of the advantages of the sight. If you get an automatic radar tie-in, you've really got something. There are several discouraging limitations about this radar tie-in, in that in order to have an automatic radar tie-in, you've got to have a locking device on the radar in order to get the radar to lock on the right target, particularly if somebody else flies across your pattern. The radar might decide he is in the baby and go on a wild goose chase. So, until we lick a few of those problems, while the radar automatic ranging looks very attractive, it's not quite here yet."

Colonel COATS: "Commander, I would like to ask if in the interim before you get the radar ranging you have given any thought to electrical ranging – that is getting away from the cable drives to electrical motors or some device to get your range and get away from the cables. I have found the cable control rather objectionable in smooth ranging."

Commander MONROE: "The English have gone into the electric control and we are following their lead and investigating it. It is a sort of step-by-step affair. You are either all on or all off. You've got the maximum and the minimum rate and apparently that stepping method works very well. We haven't any examples of it yet for test. The English are fond of it. We have, however, on the FM-2 a hydraulic control for it whereby when you move the throttle it changes the sight hydraulically. That is very attractive and it all probably boils down to mean that the hand ranging is a limited interim method and it will probably disappear in favor of these more automatic methods."

Lieutenant Commander MILNER: "In connection with the radar tie-in and lead computing sights, when both fighters and bombers use jet propulsion, manual ranging with the lead computing sight is going to be extremely difficult. With both planes making high speeds, the rate of closing will be great. Also flying the so-called pursuit curve may be impossible. I believe at Englin a P-51 and a P-47 using lead computing sights made runs on a 29 flying at high speed at 30,000 feet. Both planes stalled out before completing the pursuit curve. Fast and accurate ranging will soon be a requirement for lead computing sights and manual ranging will not give it. Maybe Colonel Toubman or Captain Turner will say a few words on the test made at Eglin."

Lieutenant Colonel TOUBMAN: "I am sorry. I wasn't in on those runs there, but the pilots, I understand, did have considerable difficulty making a pursuit curve at altitudes and it will be a very serious problem with fast aircraft at very high altitude. On these lead computing sights, they haven't decided what they are going to do with bullet drop – that is bullet drop at 30,000 feet and bullet drop at 5 to 10,000 feet. That is also a problem they are facing. I would like to know the Navy attitude regarding all-electrical bomb shackles – are you going to give them mechanical and electrical, or just a good electrical one, or what is the attitude at present?"

Commander MONROE: "That is a highly controversial subject inside the Navy itself. Armament's feeling in the matter is that the all-electrical is of course the answer. However, we have been unable to convince the Fleet of that and as we are governed entirely by the Fleet, we have to give them some mechanical means of both arming and releasing the bombs. This really puts a severe strain on the bomb rack, and the contractors to get it in the planes. The reason is that the Fleet was severely disappointed with the early electrical arming and electrical releasing. They had a lot of trouble with it and they haven't forgotten it. In getting this improved arming and release method out electrically we are gradually dissipating that but there are a lot of die-hards out there who haven't forgotten it. In addition to that, it has been customary in the Navy to duplicate everything. It is done aboard ship and you have two methods of doing everything so if one is shot away, you still have the other one, and there are a lot of pilots who still like that. The Bureau is endeavoring to try to get all-electrical systems sold and approved by the Fleet. As yet, we haven't arrived at it. Our Mark 8 shackle is just about as advanced a shackle as we have at the moment. As a matter of fact, it consists simply of a lever that can be hit either electrically by a release box or hit mechanically by any kind of arm that pushes it. So you've got either one there."

Group Captain DEAN: "I would like to raise a few points which you might like to hear of our reactions to Commander Monroe's talk. In the first place, our policy on caliber is pretty well in line with the Navy. We have made a lot of study on the ideal size of gun and theoretically it comes out to be about .76. That is based on statistics for weight of gun, weight of ammunition, velocity, armor penetration, amount of H. E. you can carry and a host of factors like that. The 20 mm. is about a .8 so we're fairly happy on theoretical grounds. The 60 caliber, of course, is a little different in its conception. It is twice as heavy, for instance, as the 20 and there is going to be difficulty about packing the rounds, etc. We are extremely interested, but right now, and since the beginning of the war, we have had the 20 and we like it.

"Regarding some of the detailed points that came up. Gun charging – we have abandoned gun charging in our 20-mm. installations because it is a gun you cannot charge in the air with any hope of clearing a stoppage. The Navy point there is interesting in regard to safeguarding the captain on the bridge. I don't know what our Navy's policy is on that. They feel the same about their captains. So if we did put in 50 calibers, I think we would agree that it is a good thing to have a charger and certainly in turrets, which is not the concern of this Conference; we have a requirement for gun charging.

"Blast tubes – we have tried to standardize our telescopic form of tube that sits in behind the leading edge of the wing and sits on a little ring around the rear of the casing so that the whole thing is forced apart by a spring and if you care to examine the Mosquito down below, you will see that type of blast tube. We have never had any trouble with those tubes coming adrift but when I was doing tests sometime ago on American planes, I found much the same things you found; these tubes would come adrift and deflect the bullets to most awkward places. That blast tube is pretty well standardized now.

"The point about the cooling of the gun – I think the reason we have not run into much trouble with the 20 mm. is because of the principle of the gun. It has a driving band on the bullet and it hasn't got quite the same heat in the powder and there is much more metal there to absorb the heat generated.

"There has been a lot said about this Mark 23 sight. We have felt that sighting policy must keep in line with gun policy and so far the sights have lagged behind tremendously. We have been an awfully long time bringing that sight out. This sight really begins to pay off at long ranges out to about 800 yards and is really best at small angles of deflection; that is, not a full beam shot or quarter attack, something about 10 degrees in to direct astern. It's under those conditions I think that the pilot does not recognize that there should be any lead given. It is extremely difficult when you are looking at an aircraft from astern to see in exactly what direction he is flying and to give the correct lead, not only to give the right amount but also to give it in the right direction. It is quite a useful exercise to try to do that yourself and see how wrong you can be. The lead computing sight takes care of that, of course.

"Regarding the developments of that sight reflecting the reticle off the windscreen, we have tried that and had a certain amount of success at home, but the main limitation is in the aperture of the lenses and that in the sight is quite big enough already to almost prohibit an increase in the size of the lense system. The only way you can possibly increase the eye distance, which is the same as increasing the eye freedom, is put in a bigger lense; so we feel that that sight may be limited to its present shape until we can get some other scheme. We have also tried feeding in radar range and that's going ahead now. I haven't any details on how it is working out, but it's the obvious answer.

"It might be appropriate now to bring up two small points on the installation in the F6F. The ranging there on the throttle, I have found that to be a little awkward in the sense in which you have to use it. Admittedly, it is the same way we do it on most of ours. It might be worth thinking about – to change the system so that instead of rotating the throttle about a vertical axis, to make the throttle up in the form of a T and by holding the top of the T and rocking it is very much less awkward. It is just a point for consideration. The other point was, I found the reticle to be very uneven in its lighting. The center pip was very much too bright when the ranging pips were just visible at close range. Under those conditions of lighting the pipper actually hit the target at long range. You had to keep looking around it to make sure it is still there or to hold it on accurately. I think those are most of our remarks on our reactions to the Commander's talk. There was one point about the displacement gear for bombs – this cartridge fire device, Commander Walker wants to know what happens if in releasing the bomb the cartridge doesn't fire? Is there any danger of the bomb hanging up and wandering around inside the bomb bay?"

Commander MONROE: "I have been asked to terminate my remarks and give the doctor a chance. I will answer that last question; No, there is no danger at all. The bomb doesn't get away until the cartridge is fired. The bomb is not dropped off the rack in the conventional manner and just with a kick in the tail from the cartridge. The cartridge starts the whole business.

"I appreciate Group Captain Dean's remark and getting the British point of view on these matters. I enjoyed talking to you and regret we haven't more time to discuss these things."

Commander RAMSEY: "Gentlemen, I am sure we all appreciate Commander Monroe's coming down and giving the Navy's ideas and the bringing out of the Army and the British points of view on armament and armor.

"Today an F8F will simulate a typical carrier take-off with the jet assist. We will show the manufacturers what we want in take-off from a carrier deck. We will have a signal officer. In addition, we will have an exhibition P-59 flight by Lieutenant Halaby.

"I will turn the microphone over to Lieutenant Halaby who will introduce the next speaker."

Lieutenant HALABY: "I feel that we owe Lieutenant Barry Commoner a little introduction in that some might regard him as someone who is trying to put the cigarette manufacturers out of business. He is a Harvard PhD and is attached to Physiological Test, which is a unit of Tactical Test. I had the opportunity to try his method of voluntary pressure breathing at Corpus Christi. It is very efficacious for emergency bail-out and emergency cockpit procedures. At 28,000 feet I was able to take off the mask and operate normally using this system for about 15 minutes. He will explain just how far his technique has been used and I don't know whether he will get the smoke rings or not, but he was warned that a lot is expected of him. Dr. Commoner."

Dr. COMMONER: "I am sorry that my turban is nonregulation, so I didn't bring it. There has been a great deal of information passed about on this business and all I'm going to do is very briefly give the history of what we now call the Emergency Breathing Procedure. It was developed about a year and a half ago at Corpus Christi. It was developed for one and only one very specific purpose. We were trying to find a way to teach a man how to do better than with normal breathing in situations where he was without oxygen at altitudes. The philosophy behind it, since Mr. Halaby mentioned Harvard I will have to mention philosophy, the philosophy behind it is simply this; we have been paying a lot of attention to the equipment we gave the pilot, but we have not been paying much attention to the pilot himself. We have tried to find some way of adapting a man to high altitude quickly in cases of emergency.

"The procedure is extremely simple. All we try to do is find a way in which we could increase the total pressure in the lungs somewhat. It is very simple to do. We all know that the reason you do not get sufficient oxygen at high altitudes is not because of a reduction in the amount of oxygen, because there is still 20 percent, but because the total pressure is too low.

"This will tell you how to increase the pressure in your lungs and the procedure that was developed consists of taking a deep breath and at the peak of respiration closing off your wind system by sealing your lips and then squeezing and trying to blow out and not actually blowing out. It is just as though you took a football and hugged it. It just increases the pressure. You keep on doing that. I will do it for you now. There is nothing up my sleeves. You may be able to see a white cloud of vapor as I exhale and I'll explain what that is. The reduction in pressure is sometimes so severe that you get cooling of the air as you exhale, and the water vapor in your breath condenses to form a cloud. I can't do it. This is terrible. Let me first show you the procedure and then get against the blackboard so you can see the vapor if it does come out. You take a deep breath and close off. Also, if you want to, you can talk * * * not very fast * * * but you can say * * * a few words at a time.

"That is the procedure. Did any of you see the vapor? If any of you want individual demonstrations, I can give them to you later. The vapor is just a trick. The whole point is this; in the low-pressure chamber at Corpus Christi we tried several hundred people and here is the information we got. At 25 to 26,000 feet pressure altitude we could train a man in a relatively short time on the ground and perhaps 30 minutes in the chamber to get along for 15 to 30 minutes without oxygen. The first minutes are the worst. If he can get the first 6 minutes, he can go on for 15 or 20 and we have had people go as long as 45 minutes. None out of the several hundred had to be revived. We weren't interested in any longevity trials, just a short-time emergency procedure. Breathing air you can survive if you know how to do it. But you must be trained.

"Now, let no one please feel that he knows how to do the emergency breathing procedure. You don't, and I warn you not to try it unless you have been trained by someone who knows how to do it. I don't want to have anybody's fate on my shoulders. But that is what we are able to do in the chamber. We made a trial to see how it would work out in the air. We had the man not just sit still there, but do his regular job. We used the copilot and also the radio man. We had a copilot at 25,000 feet breathing air in control of the airplane instruments – there were bad flying conditions and the windshield was frozen over. He was in control for 18 minutes breathing air. At the end of that time he brought the plane down. It took 17 minutes to drop to 15,000 feet. That makes a total of 35 minutes he used this system without oxygen and then he put on his mask. The fact that we got back proves he was able to control the plane. In the same flight we had a radio operator using a hand-held microphone communicating by voice with the ground and he received and sent messages without error for 26 minutes and at the end of that time he was doing better than he had been doing for the first 5 minutes.

"In a Liberator at Miami, there was a man who handled it correctly for 18 minutes at 35,000 feet without oxygen.

"These men were trained. They didn't hear about it. They were trained. We don't intend, as far as the Bureau and Medicine and Surgery is concerned, and they are very interested in this, to give this breathing procedure general dissemination; because I think you know better than I do that it would lead to all kinds of misuse. It is not intended to substitute for oxygen equipment. It is not intended to substitute for tobacco – it is intended for a situation where you are faced with the absolute alternative of passing out. This is not easy to do. You are better off if you have oxygen, and you are much better off in front of the fireplace. If you are going to die if you don't use it, then it is nice to know how to do it. That is the attitude we have toward this thing. No one has tried using it for bail-out procedure except that Wright Field may be working on it now. I don't know. From 35,000 feet and perhaps a little higher, if you use the technique of taking a deep breath of oxygen and then bailing out and using the emergency breathing procedure when you have to start to breathe, you would be in pretty good shape when you got down. That has to be worked out.

"The data we have are from low-pressure chamber experiments. They have been confirmed by physiological measurement.

"There is a great controversy on the theoretical reasons why it works. Some people say it is deep breathing. We think the pressure is important, too. We are interested only in one thing – to provide a man with some means of getting through a situation which might otherwise leave him a dead cookie.

"It is all in showing the man how to do it. The training takes about 20 minutes on the ground and about 15 minutes in the chamber.

"I want to warn you again. Gentlemen, you are not now equipped to use the emergency breathing procedure."

Commander RAMSEY: "Thank you, doctor; and thank you, Commander Monroe."

TENTH MEETING
21 October, 1944
1430-1730

Commander RAMSEY: "You have all come in from having seen the jet-assisted take-off. That particular aero-jet installation is used in conjunction with F6F's, TBF's, FM's, and other aircraft. It is made by the Aero-Jet Company of Los Angeles, and proved very successful; the smoke is the main disadvantage to it now. Smoke level is very high, does not cause discomfort to pilots except as to visibility of other operations on carrier deck. However, the advantages to it of getting more planes on board, of taking off on shorter space, outweigh the disadvantages. Aero-Jet and other companies working along that line are making an endeavor towards removing the smoke aspect. So we hope in the very near future to have a jet assist without the disadvantage of great clouds of smoke.

"The F8F demonstration which you all saw was a good carrier-based fighter take-off in short distance. The first wave-off that Lt. Comdr. Owen took was to the right, which simulates wave-off conditions when landing off the bow of a carrier equipped with both stern and landing gear. It is very important that you have sufficient rudder in order to make the carrier wave-off to the right. Unfortunately, many of the planes we have operating on carriers in the present day do not have sufficient right rudder in the slow-speed landing condition of carrier approach to take a wave-off to the right. The BTD was a sample of that and that was one of the reasons the contract was cut down. Other planes which we have had to use still have a deficiency in right rudder control for carrier approach landing condition. The P59, as you all know, is the Bell product built for the Army, is really the first successful jet propulsion aircraft in the United States. We hope that the No. 12 plane, which is down in the hanger now, will be ready to be put on the flight schedule tomorrow for selected personnel. We particularly want those contractors' representatives who are interested in jet propulsion, for either the Army or the Navy, to have an opportunity in this aircraft. This particular plane, in case you wonder why it hasn't been on the flight schedule so far, is the plane we have highly instrumented for normal everyday checking out. So we hope to have the other plane in commission tomorrow by about 1000. Those pilots who are representing companies interested in jet propulsion and have projects either contemplated or underway in that field are invited to make application to Lt. Comdr. Milner or Lieutenant Shehan to be put on the schedule. I wish those that are interested in flying it merely from an academic point of view or from curiosity would refrain from flying it during this session. If the plane does get on the fight schedule we hope that those interested from the latter point of view will remain over past Monday evening and we will try then to give them an opportunity to fly it – that's Army pilots, Navy pilots, or contractors. But during the meet, we would like to restrict it to contractors I mentioned before.

"Mr. Milner thinks he is going to be swamped with requests. If he does have more than enough, we will have to weed them out – take out those who should fly it and hold others over until after Monday evening. It is unfortunate that we have so many requests and unfortunate that we have, also, only one plane to fly.

"This afternoon I would like to continue the discussion about power plants and accessories. Before we go into it, Mr. Gerdan had to depart and left some remarks which I'll ask Lieutenant Wilds to read into the record. They may or may not provide further comment."

Lieutenant WILDS: "These are comments left here by Mr. D. Gerdan, chief installation engineer of Allison:

"1. With increased complexity of power plants (turbo, two-stage, externally cooled, etc.), the problem of accessory disposition becomes more and more difficult to solve. There is not only the problem of finding a location, but also that of inspection and maintenance. The Services' demand for a multitude of accessories leads to the belief that in the interests of improved power plants it will be necessary to resort to accessory drive gear boxes. These, in turn, require space which is always at a premium in aircraft, and particularly so in fighters.

"2. I believe there is a tendency to use the demand for quick change power plants as a dodge for avoiding the hard work required for first-class engine installations. There is no disagreement with the thought of requiring quick change power plants, but there may be a tendency to use slap-dash methods of design.

"There must be care to avoid design restriction when demanding quick change plants.

"3. Single or coordinated engine and propeller controls will not be wholly practical until the spread in propeller governor variation is narrowed to not less than ± 3 percent.

The engine control is generally a precision instrument, but its usefulness is greatly impaired by governor variations.

"With the great variety in types of missions now demanded of fighters, it is very difficult to draw up a universal control schedule. The demand is for first, economy, second, ability to accelerate rapidly if jumped.

"4. We require much better coordination between engine and aircraft manufacturers."

Commander RAMSEY: "I think that is a very good point – the one about more coordination between contractors and the engine manufacturers. I know definitely that we have had a lot of trouble on cooling problems which should not have been possible on planes delivered for trial if the coordination had been there. I don't know if the Army has that same trouble or not. Many problems – stacks and so forth – should be ironed out before the trial plane gets here, we think. We touched on liquid cooled engines and air-cooled engines, jet engines, and combinations of those the other day. Are there any further comments along that line?

"Superchargers versus two-speed single stage and two-stage blowers were touched on. Is there any further comment?

"Operating limits were touched on. We had some pilots' viewpoints read into the record on what they would like. Along that line, I would like to suggest a point that might have merit. Take a little Cessna for instance; not a fighter but it's very simple for a pilot to get into a Cessna and fly it, due to the fact that even though you haven't read the pilots' handbook (I don't advise taking off without reading it), you can get in the cockpit and after a perusal of 2 or 3 minutes, fly the thing because all needles when in the air and operating correctly read toward the green painted line. If it goes up to red, something is wrong. I think that should be done by all contractors before delivering planes. I don't know if you noticed it during the fighter conference or not but I insisted all planes have it in the fighter conference, at least Navy planes, due to the fact that inadvertently you might forget what the limits are for cylinder head temperature, oil pressure and oil temperature, etc. I think that is an excellent point. Any comments for or against that? Mr. Gosselin."

Mr. H. Gosselin (Pratt and Whitney): Insofar as painting the limits on the gauges is concerned, I think we agree with Commander Ramsey. We would like to see all planes – particularly military planes – with the limits put right on the gauges. Unfortunately there is some regulation against that, I believe, by the Bureau. Just what the reasons are I'm not acquainted with. The Army has been painting limits on gauges for sometime and we, in cooperation with the aircraft manufacturers, have been doing it insofar as experimental airplanes are concerned, even on Navy ships. But there is some regulation prohibiting that on Navy production airplanes, I believe."

Lieutenant Commander MILNER: "I would like to hear from Colonel Renner."

Lt.Colonel RENNER: "As a matter of fact, I think the reason that we don't have it painted on the dials is pure inertia. I know that a memorandum from Military Requirements has gone down to Instruments Branch of the Bureau of Aeronautics at least 3 months ago requiring that our dials be painted. We can't mark the front of the A/N instruments because the instruments apply to so many different aircraft that it limits change, but we have put out a request that the dial face, the glass, be painted with different colored paint, and I think that the Navy will finally get on the boat along with everybody else."

Commander GREEN: "I'm all in favor of painting markings on the instruments but under the various systems of instrument lighting you have at night you will not be able to distinguish some of the colors as alternative. I think a good measure would, where practicable, be to have normal readings of all engine instruments either vertical or horizontal. You could then in one glance tell if anything was abnormal."

Mr. PARKER: "I have just recently had the opportunity of flying a P-47 at night and I don't know what the paint is made of or the consistency but it shows up beautifully under this fluorescent lighting."

Commander GREEN: "If the Navy is going to indirect red, under red light you will not be able to see some of those colors."

Mr. BURKE: "I don't think it makes much difference what the colors are and whether you can tell color or not at night. If you follow this system, that the highest line on the instrument is war emergency, next highest is normal rating, the next highest is maximum cruising, whether you can see the color or not doesn't make any difference. You just pick first, second or third line according to what condition you're flying at."

Commander RAMSEY: "I think we're all in agreement that instruments be painted. Anybody who doesn't want his instruments painted at all? The next item on the list is critical altitude – what we wanted. I think the other day the general consensus was that it depended on the theater in which the airplane was going to be used. In the Pacific Theater I think we haven't yet any need for planes over 25,000 feet and most of them are around 10,000 or below. In the European Theater, however, planes operate more efficiently at high altitudes. In the Pacific Theater and Far East, they use B29's for higher critical altitude. Anyone wishing to add further remarks than we had the other day on that? Don't quote me as saying we won't need more, because in the future I think we will."

Lieutenant Colonel RENNER: "Paul, I want to get back just a second – I had something to say on the instrument business. There are a great many pilots in the service that feel rather strongly about arranging dials so that all instruments point at 12 o'clock when reading normal or a 3 o'clock or 9 o'clock. We don't care where they point, but have them all point the same so that all you have to do is take a flash at the instrument panel to know everything is okay, or if one needle is out of place you can immediately spot that instrument. Along that line I had a discussion with Commander Campbell, R. N., the other night and he said it seems utterly foolish the way we have our dials marked now. The thing that brought it up was that we were discussing running a F4U4 at 70-inch manifold pressure and he couldn't boost that without quite a bit of calculating, so he said: "Well, what's the difference whether we run the pounds boost or manifold pressure. Why don't we just number from 1 to 12, like the face of a clock." I can't see why 12 o'clock couldn't be okay for oil pressure or fuel pressure, whatever it happens to be. I don't care if I'm running at 17 pounds fuel pressure or 20 pounds fuel pressure – 90 pounds oil pressure or 70 pounds oil pressure – just so it's normal, that's all I want to know."

Commander RAMSEY: "Commander Green talked to me quite a lot previous to this conference, and now Colonel Renner has voiced the same opinion. Any opinions to the contrary?"

Mr. BURKE: "I think that a number of the contractors pilots who have had an opportunity to fly the same airplane produced for a number of countries and marked in a number of different languages will concur with that because usually the same instruments were used and the face markings were the only thing changed. You more or less forgot all about what the quantities measured were but looked at the instrument to see if the hand was at the proper position, and it made it a lot simpler to go from an airplane marked in Finnish to one marked in French to one marked in English or something like that, and I believe that the same thing would carry through on military airplanes."

Commander RAMSEY: "Any further comments on that?"

Lieutenant Colonel MENG: "I'd like to get back on that critical altitude of engines we were talking about at lunch. One of the boys who had just come back said we had P-47's with turbos on them fighting with the IXth and they said they were always down on the deck. I might add a gem of wisdom here by saying; You can always go down but you can't always go up. This is particularly true, I think, in that you will have to realize that some of our airplanes, no matter if they are doing all our fighting on the deck, still have to be able to go up. I've heard that question brought up quite a number of times: Well, we're not fighting upstairs, why build an engine for that? But we've still got to remember that the minute you get your airplane downstairs, we've got to go up. That's just pure basic reasoning, of course, somebody may get stuck with that engine and make their airplane suffer in speed."

Lieutenant Colonel RENNER: "I'm surprised at Colonel Meng making such a statement. I think we're all agreed that we have to design an airplane to do a certain job. We've heard the argument this morning on putting 20-mm. and 60-cal. and 37-mm. guns in. Well, I can make the statement that the more guns you put in the more sure you are of knocking him down.

"In order to get the best performance possible at the altitude at which you wish to fight you have to choose the type of supercharging that will give you the lightest possible installation. There's the old fight about weight in an airplane and you cannot build the F8F out here with a 2800 turbo supercharged engine for the same gross weight that we built this one. You have to make up your mind at what altitude you want to fight. If you're going to fight from sea level to 10,000 feet, you can probably get away with a single speed engine and save yourself several pounds. If you're going to go up to 15,000 or 20,000 feet, you can probably get away with a single-stage, two-speed engine. If you're going up to 25,000 feet, you may want a turbo supercharger or two-stage engine, and so on. But the idea that you put enough performance in every airplane so that it can fight at 35,000 feet means that all those airplanes that do the majority of their fighting below that are penalized by carrying the added weight around."

Lieutenant Colonel MENG: "I didn't mean that all the airplanes should be that way, but some have to be that way. No doubt about it. Take the VIII AAF – they go ahead and have to escort the B17's but then as that part of the war wears out, they go down low. You send a particular airplane to the theater that can fly at all altitudes and that's maybe the reason you picked it, because it can do both. I'm not saying you should specialize, but some people listening to the airplane companies hear complaints because their airplane is penalized. My explanation is that you have to have some like that but can't build them all like that."

Commander RAMSEY: "Is that argument settled? Any more remarks about critical altitude?"

Lieutenant Colonel RENNER: "I have a theory about the altitude required in fighter aircraft and I'll just throw it on the table to let as many people tear apart as wish. In the first place, I'm speaking of a fighter as a pure fighter aircraft, whose mission is to destroy other aircraft. My theory is simply that the altitude of your fighter – the critical altitude of your fighter – is determined by the altitude of bombers. If you're on the defensive, it is determined by the altitude of the enemy bombers that are coming in and pounding you. You should have approximately 5,000 feet better critical than the enemy bombers that are making their attacks at, because the escort fighter will be coming in with the bombers and probably stick up to 5,000 to 10,000 feet above them.

"As you shift over to the offensive your altitude is determined by the altitude of your friendly bombers. In the case of the Army, the friendly bombers may be going over at 25,000 or 30,000 feet; that means you must have a critical of 35,000 or between 30,000 and 35,000. In the case of the Navy with a dive bomber, it goes at 15,000 and hardly ever attacks above 20,000. You should have from 20,000 to 25,000 feet."

Lieutenant ELUHOW: "I think I have a little bit to say on that. The colonel said you should have a plane built for one certain thing – knocking enemy planes out of the air alone. Well, I disagree with the colonel on that. For instance, the P-47 we used on escort work and while escorting, if we saw ground targets, a few of us left and went down and strafed them and then went back up and joined the escort again. It worked out swell because on missions like that we didn't meet any enemy opposition, but we were prepared for them and were doing two jobs.

"Now around D-day and after D-day, we weren't doing much escort work; we were doing area cover work – at the same time destroying all communications. We were carrying out two jobs and looking for enemy aircraft and going down on deck. That required two altitudes which, we in one plane, covered successfully. You can't use just one plane to knock out enemy aircraft alone. It wouldn't be worth it, a plane that could do two jobs like that one is much better."

Lieutenant Colonel RENNER: "I think that a plane that can do two jobs is a good airplane, but I think that you'll find that the airplane that did both jobs was designed to do only one of them and did the other incidentally. We find in the Navy that the F4U and the F6F are damn good dive bombers, but the fact that they were built as fighters and fighters first is what made them important and that they could carry a bomb incidentally and do a dive bombing run is something that we are very happy for and which we use it for. We also use it for strafing but those things must be incidental in the design and not primary."

Commander RAMSEY: "Any other comments along that line? Boone Guyton."

Mr. GUYTON: "I would like to ask Colonel Renner what critical altitude he thinks a fighter in the Navy should have for the next year or two."

Colonel RENNER: "20,000 feet."

Commander RAMSEY: "That's a good round figure. Any comments on that? We might get on the record at this time, now that we have an official viewpoint of the Navy, the viewpoint for the Army. Colonel Terhune, would you be willing to stick your neck out on that?"

Lieutenant Colonel TERHUNE: "Well if the B-29 is any indication of what they hope to run at later, it is going to have to be about 40,000 feet, maybe; I'd hate to guess because our requirements are set up in Washington, and this is way above my head. I don't know what is going on in the back of their heads."

Lieutenant Commander SUTHERLAND: "I would like to say a word for the poor aircraft manufacturers being required to shift their altitudes around like so many cards. It may have been all very well when we were on the defensive and afraid the enemy would go high and always want the altitude advantage, to grab for altitude. However, when we are on the offensive, it should be up to us to set our altitudes to operate at 12 to 18 months from now and if we don't do that, it is a defection of planning. I would like to give the poor manufacturer a break for a change and I think if we tell him a year or two in advance what we want and don't change seven or eight times in the middle of that, we would get a lot better airplanes built in the end."

Commander RAMSEY: "I would like to hear at least one comment from some contractor. Would that satisfy you, Boone, to get a round figure of 20,000 feet?"

Mr. GUYTON: "Will we be satisfied with 20,000 feet as a round figure, or are we going to have two types of fighters? We're a little bit puzzled on this deal. We went up in critical altitude and now it looks like we're going down in critical altitude. What do we want from the deck up?"

Commander RAMSEY: "Colonel Renner, can you give us an official figure from Washington on that?"

Lieutenant Colonel RENNER: "It goes back to my theory, I think, that you pick your altitude with the altitude of the bombers. When we were down in Guadalcanal we had F4F4's. The Japs started across the field at 9,000 – there weren't any fighters present at all. They were having a field day. Our AA wasn't too sharp. The AA began to register and he moved up to 12,000; then we got the fighters in there and he moved up to 27,000. Then the old F4F4 got in there – she was a little short on performance at 27,000 feet. As a matter of fact, for a squadron to scramble and rendezvous at 29,000 where we like to make an overhead pass, we usually allowed 45 minutes from the time the leader took off until the last guy could close up and get into position. Consequently, we were screaming "Give us 35,000 feet altitude to get up on top of the show and start giving them hell." Fortunately for us that phase of the battle did not last too long, and as soon as we swung to the offensive our bombers began to determine the altitude at which we would fight.

"When I say 20,000 feet, I say it with confidence that we will maintain the offensive from here on in. If ever we have to switch to defensive again, our altitude is going to be determined by the altitude of the enemy bombers. If they are going to come over at 40,000, you are going to hear us screaming 45,000."

Commander RAMSEY: "Boone, does that satisfy you further? Bob Hall, does that satisfy Grumman? Burke, does that satisfy McDonnell?"

Mr. BURKE: "I won't say that it satisfies us because we'd like to know what we are going to do. The only thing I can say is we'll just have to rely on the rest of the country to keep on the offensive and not have to be forced into the defensive so we would have to design a whole bunch of new airplanes. However, I do think that as far as the fighters are concerned, that the advent of the jet-propelled airplanes is going to have quite a bearing on this because the performance of jet-propeller airplane has very little in the way of variation of altitude as compared with the reciprocating propeller drive and that may be something that will work itself out in that direction as we go ahead with the development of that type of aircraft."

Lieutenant HALABY: "I'd like to take exception to that last statement. The efficiency of the jet-propelled aircraft varies as much with altitude as the reciprocating engine and especially in fuel consumption. And it seems to me that the point is well taken that if the Japs come out with one, and we have every reason to believe that the Nazis, in their last gasps, will give them all their plans in engineering that is necessary, it seems to me that 20,000 is not going to be half enough and the whole tactical situation governs what we need.

"Along that line, I would like to ask Allison two questions which seem to me are appropriate: first he brought up the point the other day during Captain Spangler's discussion about using the jet, in a composite airplane, for some supercharging. For example, this morning Williams said it took some of the boost off the compressor to pressure the cabin. Why can't we take some boost out of the compressor to get about 5,000 feet extra manifold pressure in the main engine, the reciprocating engine? I think perhaps Allison's done something about that.

"The other question is how are they coming with the variable speed hydromatic supercharger that has a similar straight line critical to the turbo supercharger?"

Commander RAMSEY: "The Allison representative is not present. Colonel Terhune, can you give us anything on that?"

Lieutenant Colonel TERHUNE: "I don't know what the Allison representative said – I don't believe I was here at that discussion. I don't see any reason to penalize the jet engine which is really coming into its own up at altitude to bleed out part of the compressor air to feed it into the regular engine. I think that the altitude that the Navy are mainly concerned with will read particularly lower than the Army is concerned with and therefore you're going into this composite arrangement to give you range as well as high speed and altitude, and I see no reason for penalizing the jet engine for altitude at all."

Mr. SPECHT: "I'd like to add something to Mr. Burke's comments. First, the JP engine, as you probably all know, is a thermodynamic engine which actually has no critical altitude. I think that is what Mr. Burke was trying to get across. In regard to the effect of bleed in our compressor in the jet unit, we can go to a small percentage of bleed with practically no effect in our performance. However, if we do start to bleed large amounts, such as what I think Lieutenant Halaby was driving at, we are of course going to affect performance due to weight flow and we'd just change our whole thermo cycle. I don't know whether that's the question Mr. Halaby made or not – to bleed air from the jet compressor back to the front engine for boost. Is that right?"

Lieutenant HALABY: Right."

Mr. SPECHT: "The figure we can go to is about 1 percent for bleed which is way above any demand for a cabin so far."

Mr. GOSSELIN: "I don't know just what that means in actual air consumption – the figures which Mr. Specht mentioned – but I think it comes back to what Captain Spangler was bringing out the other day. It is a matter of where we can get the most efficiency that is in performance – do you get it by putting it into the propeller or into a thrust from a jet unit, and it is something I don't think you can generalize on. It would depend on the particular application and the airplane and propeller combination; also the jet unit."

Commander RAMSEY: "Mr. Specht was under the impression 1 percent was the maximum you could bleed off without ruining the efficiency of the jet engine, or is that the amount which has been done up to date that you know of that doesn't affect its efficiency?"

Mr. SPECHT: "That's right. Our experience to date at the factory shows about 1 percent without any measurable effect on thrust. We haven't gone into the matter to the extent where we say for a certain percentage of airflow drop we're going to lose so much thrust and we've no curves plotted up for consumption for airplane design engineers. Along that line, I'd like to agree with the gentleman who just brought up this point. Perhaps you can bleed from the compressor and get increased power from your forward power plant as in the case of the FR-1, but it's going to make an awfully complicated installation, I think."

Lieutenant Colonel RENNER: "I think Lieutenant Halaby is a little bit worried about my statement of 20,000 feet, and wondering what the Navy is doing about this jet business and if we can't get more efficiency out of the combination by bleeding from the jet back into the conventional engine. I think probably Lieutenant Halaby is aware that the critical altitude of the conventional 1820 varies, depending on whether you have jet on or off. It may be that you'd get more efficiency by bleeding from your engine supercharger back into your jet just the same as bleeding it back the other way. I have an idea that the thing probably will work better with each engine running by itself. However, to give you a general picture of the Navy's attitude on jets, let me first say that the 20,000 foot altitude was for conventional aircraft. We're thoroughly aware that if the Jap comes out with jet-propelled aircraft that we're going to have to do something and do something damned quick, and if we find that carrier-based jet-propelled aircraft have to be catapulted in order to give them launching speeds because the jet-propelled aircraft is inefficient at low speeds, we're going to have to catapult them and we're proceeding along that line, designing aircraft that have to be launched by catapult and reducing our carrier approach, or rather increasing our carrier approach speed, so that the airplane can come back aboard. We're making a lot of allowances for a jet airplane that we do not allow our present conventional aircraft to have and we're doing that because we feel that even though we have to restrict the combat range radius of it to 100 miles, if the enemy comes out with fighters that will knock all our fighters out of the air, we're going to have to take the carrier into 100 miles or into 50 miles or any distance which it takes to launch aircraft that can still maintain control of the air. The enemy's equipment and tactics are going to determine how fast we go to this sort of thing. We feel that in leaving the carrier out at 75 to 150 miles, as some of them are doing, that we're still pretty safe because we have rather complete control of the air. But we won't have that control if the enemy comes out with jet-propelled aircraft. Does that answer his question?"

Mr. SPECHT: "I've another question then but before I go into it, there is one point I would like to make clear. Up at Lynn the boys feel that the jet-propulsioned airplane is not going to be limited to a range of 100 to 200 miles, and I know that several designs are being made where we will be able to get greater range. Probably the Navy is already aware of that, but one of the factors that must be considered is that a large saving in power plant weight is made by the installation of a jet engine and that can be taken up in additional fuel. That's only one of the points that the JP engine would save in weight. We're not under any delusions that we might be able to get a carrier-launched plane, but as the colonel stated, we are definitely interested in what catapult arrangements could be made for a JP unit.

"In talking about combining a conventional and the JP unit, we, of course, are interested in a straight jet-propulsioned airplane. You're throwing away one of your greatest advantages of a jet-propulsioned airplane when you put an assist unit in front. You have a propeller there and on a large – a relatively large – area which you have to buck against the breeze. The P-59 is one of the jet-propulsioned ships and was a very interesting ship to learn on. However, its speed is not that of the combat jet-propulsioned airplane, as we shall soon see. The limiting factor, as far as I have been able to see during these discussions, is that of a wave-off condition, which the colonel brought up. I was wondering if the Navy is interested or will be interested in the future in increasing carrier speed or putting, shall we say, bigger carriers out so we can land on them. Just what conditions would the jet-propulsioned engine have to go to? I have discussed this matter a little bit with some of the men here and I have a figure of approach power that is usually used by a service pilot when coming in for a carrier landing. It's approximately 70 percent, I understand. He can practically get instantaneous power with a conventional power plant. I was wondering just how instantaneous this would be. I have an estimate of from 2 to 5 seconds which the jet-propulsioned engine may be able to meet. You are familiar with the fact that the gas turbine has quite a bit of inertia, and we have to go through a thermo cycle to get acceleration. What I would like to do is get some comment from pilots and airplane designers as to what the limiting figure would be on approach power and what the critical time lag would be, if we could decide what this approach power would be, what would the pilots' reaction be to an assist unit to the jet-propulsion engine, such as what we saw this afternoon in the Aero-jet-rocket engine. I haven't the least idea as to what controls are used in the Aero-jet or to what complicated maneuvers the pilot has to go to start his engine and let it run. Does he have to watch temperatures, or are we dealing in secret information, now?"

Commander RAMSEY: "The Navy is not only interested in the jet propulsion from several standpoints but we do have, as you probably all know, a contract with McDonnell to build a carrier-based fighter with pure jet propulsion – a McDonnell FD. Actually, on the P-59 (I have had a lot of time in that) the difference between summer and winter test work is appalling. One of the features of the P-59 which is surprising is that on a day like today, here at Patuxent, you could take a carrier wave-off. However, in the summertime you can't. The one thing about a carrier wave-off is that you're coming around down wind, about 1,000 yards from the ship, you start making a turn when you get opposite the signal officer which was represented by Lieutenant Runyon out there today. You start making a turn and just as you get the cut you will just have come out of your turn or just be ready to bring the wing up and straighten out for the cut and landing. In the carrier wave-off, at the point where you might get cut and prior to that time if the signal officer waves his flags in front of his face, the plane which you fly, whether it be jet propulsion or combination engine and jet or just plain engine, should be able to take power and not lose any altitude and not spin in and miss the carrier deck, and depending on whether you are landing over the bow or the stern, either making a right or left wave-off. You are coming in at about 5 or 10 knots above stall, about 70 percent of power on normal engine and sufficient power on jet to maintain that flight speed with flaps and wheels down. If when you get the signal, and you apply power, you can start increasing altitude, practically instantaneously or at least maintain the altitude you have when you get the signal, and that is what we want for carrier wave-off. At the same time you do not get that by flying straight ahead; you either do it by making a slight left turn or slight right hand turn, depending upon the side the island of the carrier is.

"The question raised by Mr. Specht on the take-off of the jet propulsion has worried us somewhat. I'm afraid that the P-59, the FD and all pure jet propulsioned planes are going to have to have some assist for take-off. However it's unfeasible to have those bottles hanging on you during the whole hop so you can use it for carrier wave-off condition. There are a lot of things wrong with having an Aero-jet. Incidentally, its very simple – all you have to do is push a button or turn on a switch or you just ignite it. That is all there is to it. There is nothing complicated about it. The supply problems, the logistic problems and the spacing requirements on a carrier would go against having enough bottles there to hang on all planes which might be jet propelled above those required for take-off under heavy load conditions. I think some of the carrier pilots can elaborate on that. Mr. Burke."

Mr. BURKE: "In connection with this carrier wave-off, I would like to point this out that on our present carrier operations the fighter is not the critical wave-off airplane. The fighter by its very nature of very low power loading can take the wave-off because it can accelerate very rapidly. On the other hand the torpedo plane at maximum landing load or dive bomber at maximum landing load coming into the groove and coming on board is the critical wave-off airplane and in comparing the wave-off of a jet propelled airplane it should not be compared with the conventional fighter, but rather with the minimum operating performance that is obtained from the torpedo or dive bomber type of airplane. It does not make much difference whether you have a good wave-off or a medium wave-off, as long as you can take a wave-off. That is the one thing. It's just the yes-or-no question – can you or can you not take a wave-off so that the ability to jump up in the air which we have and I imagine the F8F will show it out here, from wave-off is not a necessary part of being able to take a wave-off."

Lieutenant Commander SUTHERLAND: "In setting up our requirements for a pure jet propelled airplane when we're allowed to catapult them as an habitual means of take off, we realized that very thing and took the worst overloaded airplane we could get hold of – two of them in fact – the TBF and the SBD, both of which are barely able to stagger when their wheels and flaps are down and by some rough calculations concluded they probably had a maximum of 500 feet a minute climb, wheels and flaps down and set that up as a requirement for jet propelled fighter. We don't expect, as you say, to have the thing jump into the air but they do have to have some climb, wheels and flaps down, and drag increase due to high lift flaps and tremendous wheel fairing doors is liable to be a more critical item than inefficiency of the power plant at low speed. The Ryan, as some of you may have heard, crashed a while back – don't know what happened but have some reason to suspect that he had conventional power plant and in pumping the wheels and flaps down by hand got so slow and low that it was conceivable that the airplane couldn't fly any longer."

Lieutenant Commander OWEN: "In connection with the wave-off of the jet propulsion airplane, this occurs to me that it will probably be some time before we see actual operation of the jet propulsioned airplane from carriers. And it's almost been admitted that the launching of these airplanes will be assisted by catapult or some other means.

"In that connection, assuming a catapult launching, the planes may not be launched necessarily from the flight deck but from a lower deck either directly forward or other side. If that's the case, it might be that the carriers will be modified so that the entire flight deck is available for landing. That is planes will not be parked on the flight deck as much as they are now. In that case, more deck will be available and wave-off becomes much less critical. That's just a thought. I still think we're making too much of a problem of the wave-off possibilities of jet propulsioned airplanes."

Lieutenant Colonel RENNER: "Mr. Specht of General Electric asked us about increasing the speed and the size of the carrier. Well, increasing the speed of the carrier would probably be a lot simpler if General Electric will give us some turbines that will drive it through the water at a speed of 55 knots. That will solve a lot of our problems. The size of the carrier, except as Commander Owen has brought out that you will not have as many planes parked on the deck so you can probably use the whole deck for landing area and strike them below may help some but I'm afraid it's a rather slim hope in that in order to have enough parking space on a carrier to carry as many planes as the task force commander wants to carry you have to park some of them on the deck. And the fact that you have 22 wires instead of 9 is not going to solve your wave-off problem. The increased area of the deck is not so much the problem, in that we feel that we can put the plane down on the area now that is available, but at the higher landing speeds which will not be solved unless we get higher carrier speed the arresting gear does not have the ability to absorb the energy of the airplane coming aboard. So we have to get speed and weight of airplane down so that the arresting gear can handle it. Just building a bigger carrier with the same speed, I do not believe, will be the answer.

"Another question on the use of jet propelled fighter in the Navy. What good does it do us if we are on the offensive, which we are, and we have jet propelled fighters on the carrier when we have nothing to escort? We can race out to a target, shoot it up and race home, but we would have to take off an hour after the rest of the carrier air group gets out there, dash out to rendezvous and escort during strike, dash back and land – and land before the rest get back. In other words, will Mr. Specht build a fire under those people and give us gas turbines for our dive bombers and torpedo planes so that the whole group can go out at high speed?"

Commander RAMSEY: "Lieutenant Commander Callingham, we haven't heard from you yet."

Lieutenant Commander CALLINGHAM: "I'd just like to ask Mr. Specht if any work has been done on the use of ducted fans on turbines to increase take-off characteristics."

Mr. SPECHT: "I'm afraid I don't understand – you mean ducted fans to the compressor inlet?

"Oh no, we have not done anything on that."

Mr. WOOLAMS: "I think that a rather dark picture has been painted of the possibilities of landing jet propulsioned airplanes on carriers. Possibly so, on a basis of observation of the P-59 which is the only jet fighter we've had a chance to get a good look at so far. In the first place, it appears obvious that it will be necessary to catapult jet airplanes off or get them off with assist rockets or some means other than what the airplane normally carries, for two reasons; first of all, in order to build high-speed airplanes the wing loading necessarily has to be pretty high, and the other reason is that jet airplanes carry a much greater weight of fuel in proportion to total weight of the airplane than the conventional airplane does and at the present stage of the game this fuel weighs approximately 2 pounds per gallon more than gasoline. That means that the airplane is pretty heavy when it takes off and acceleration characteristics at low speeds are poor. However, it must be remembered that after this airplane has been in the air and completed its mission, used most of this fuel and the weight has been decreased terrifically, the landing speeds themselves are not going to be such a bad factor as is believed due to the fact that wing loading decreases appreciably during flight. It must also be remembered the P59 is relatively an underpowered jet airplane and future jet airplanes will have better take-off and wave-off characteristics.

"As regards wave-off, the P59 is extremely marginal but I don't believe that the later jet airplanes will be anywhere near as bad as the P-59 is.

"One other factor is that there isn't any change in rudder trim or gyroscopic effects which complicate the matter of suddenly pouring full power to the airplane which will make it easier in that respect. Due to the fact that the power which we have on the jet airplane, or jet engine, is a function of your static thrust times your velocity, the airplane accelerates at any increasingly rapid rate as speed is increased and it may be necessary to use a slightly different technique on wave-off. It may be found better, instead of maintaining altitude to turn the airplane sharply enough to miss a carrier and drop below the level of the carrier deck. I don't know much about Navy operations, myself, but it may be necessary to do so. You get a heck of a lot more acceleration once you do pick up a little bit of extra speed on that type of power plant installation."

Commander RAMSEY: "That's good, but personally, I don't want to decrease altitude from the position where I'm going to get a wave off. It's pretty bad to make a turn or do anything because when 55 feet of bulk is staring you in the face dead ahead for the first appreciable distance, say 50 to 75 feet, you have to go straight ahead and must at least maintain altitude if not increase it slightly. That's a wave-off.

"Incidentally, speaking of jet propulsion, this afternoon I believe we will have the Ryan down here. Unfortunately it suffered a 9 1/2G landing at Philadelphia so it will make the hop down here but will not be available for flight. Tomorrow morning there will be here also the F2G – either this afternoon or tomorrow morning. The P59 we hope will be in commission at 0815 tomorrow morning and the following pilots are picked to fly it. Those who feel they have been slighted, don't take offense but these particular pilots are representing companies who are interested in jet propulsion either on paper or building and those whom we have picked to fly. As I say, those pilots whom I did miss are interested I know, but they will have to wait until later on in the week after the fighter conference. I don't know in what order Lieutenant Shehan has put them but they will be posted on the bulletin board outside right after this meeting and you can see them then. Please try to live up to the schedule and not stay out longer, though the fuel supply will limit you somewhat. Make your hop from wheels to wheels – wheels off the ground to wheels on the ground – 30 minutes. That gives plenty of safety factor and also gives every one a chance to get a hop in tomorrow.

"Anyone wish to talk – Mr. Gosselin."

Mr. GOSSELIN: "I have just one question – I don't know whether it's in order to ask it or not – and that is, is the available space for fuel supply on a carrier usually the limiting thing insofar as operation or is it some other factor?"

Mr. WOOLAMS: "With regards to the fuel consumption for low altitude operation it is considerably higher with the jet type unit than with the conventional engine, and I was wondering if the available space for fuel in a carrier was an important factor or whether it was something else such as ammunition which limited operations."

Lieutenant Commander SUTHERLAND: "We started to put out a spec awhile back asking different manufacturers to submit jet proposals and so we had our design section run through a typical airplane that would meet the spec, and it required about a little over 1,000 gallons of fuel to make a 300-mile flight on a standard Navy combat radius problem. Assuming 100 airplanes on a carrier, which would mean they would be relatively small, a typical CV-9 class carrier would carry 240,000 gallons of fuel. It's a very simple mathematical matter to find out you can only conduct 2.4 missions off one carrier with one load of fuel. We subsequently designed or redesigned our requirements somewhat to make it a little easier on manufacturers in the matter of fuel capacity. That will get to be a very serious item, and I don't think sufficient attention is given to it by anyone here. There is a tremendous amount of fuel required when you start sending out 500 or 600 fighters, as they do now, and how in the hell are you going to keep a carrier filled up unless you attach two or three tankers to each carrier for one short series of raids. That's the main reason somebody brought up composite airplanes – the main and only reason we want that type airplane is to extend cruising radius without getting a gigantic airplane, and if you want to make some quick calculations just look over what kind of fuel capacity you have to have for going out 350 to 400 nautical miles at 15,000 feet and come back at 1,500 feet. I think it will stagger you to know the amount of fuel that is required. For the Navy purposes the composite airplanes offer us the greater advantages which the pure jet airplane doesn't quite show as yet and the prop isn't stubbing its toes at 500 miles per hour which is very little more than that which pure jet airplanes are getting anyway."

Commander RAMSEY: "That follows. Of course, it is not too critical for the tanker-hull carriers but unfortunately the Navy does not have too many tanker hull carriers nor are they fast enough. Of course, I hope I get one when I go to sea – striking for the moon! Any more comments? Does that answer your question, Hugh? The gasoline capacity of carriers space requirements due to the fact that you have enough people to form a small-sized city living on that ship and nothing outside coming in: food, ammunition, gasoline, oil, etc., every little bit of space is quite a critical item and fuel for the aircraft is definitely a critical item. The only carrier that has enough fuel to supply aircraft above own consumption of fuel oil is the *Saratoga*.

"If there is no further comment on jet propulsion, I would like to go to detonation indicators. We have just started running some test work on detonation. Commander Tuttle or Fitz Palmer might be able to tell us how far that has progressed."

Commander PALMER: "I am afraid I can't tell you very much about it, in fact anything. We are just completing the first installation on a PBM-5. That will be our first test on it."

Commander RAMSEY: "Has any contractor done any detonation work at all?"

Lieutenant Colonel MENG: "I will cover what we have done. Maybe Colonel Terhune can tell us what they have in mind. We have run some flights down at Eglin Field. I imagine we have put in the neighborhood of about 50 hours on each one. One of our complaints is that we have had some difficulty keeping it in commission. When it's working it's worked pretty good. The way we've worked it is with 90 octane gas and just over-boosting the engine, and then we get indication of detonation. We're also running some very long-range flights at very low powers and the pilot was supposed to run war emergency when he came back over the field and he noticed right off the bat when he got up to normal military that both engines started to detonate with his indicator which he had not noticed before. There's one thing that we want, to get another model that's a little more functional and works a little bit better than these. The light isn't very bright. The light they have now that's on the tactical you have to stop and stoop and look and if the sun is behind you, you can't tell. We hope to run some more tests down there. Does show promise but don't think there's been any commitment."

Commander RAMSEY: "What is the feeling among the pilots? Do we want a detonation indicator in the cockpit? Mr. Virgin."

Mr. VIRGIN: "We use a spare detonation indicator at North American and in the conduct of some 44-1 fuel tests at high boost. The way we familiarize the pilots or train them in the use of it is to do as it's done at Eglin Field – put 90 octane fuel and overboost on the ground to give them a short course in what to look for. The instrument works satisfactorily during the high boost test, but how practical it would be for a tactical airplane I could not say."

Commander RAMSEY: "Lieutenant Eluhow, would you like to have had a detonation indicator when you were in combat aircraft?"

Lieutenant ELUHOW: "I don't know any too much about that detonation. I have never had too much trouble with it, so I really can't say."

Mr. GUYTON: "We have been playing around with a detonation indicator that you might call one of production design suitable for incorporating in fighter aircraft. We have had quite a bit of flying time on it. However, in my own personal opinion, and it does vary considerably from that of other people in the company, it's not too practical due to the fact that continual readjustment is necessary. However, as a flight test instrument, at the present time, it is a very excellent means of checking out the presence of detonation due to various causes. The pick-up is a very fine pick-up. Incidentally, it's our own design, just putting in a plug for it. It is adequate or sensitive enough so that it can be used to even catch missings at low revolutions per minute, low manifold pressures and that is done by a suitable readjustment of the amplifier unit. It is used both as detonation indicator and as an indicator for spark plugs missing."

Mr. PARKER: "Our company has done quite a bit of research along this line and to me, as a simulated fighter pilot, I think it well pays its weight. We have been able to increase our horsepower over the present limitations of the engine. However, that detonation equipment is the only way that you can actually keep track of the condition of your engine. Under some conditions you will be able to pull 2,300 horsepower, and as the engine grows older and starts pumping oil, you also want to know when it's going to start detonating. With the age of the engine, the power will have to be reduced. I feel certain that we are abusing some of the engines that we have under present conditions by detonation – by trying to keep up with manufacturers' instructions. As far as equipment we have had, I think it's been suitable for flight test, but not practical for combat use. It may be more practicable for bombers or cargo airplanes than fighters because the fighter's attention is usually outside of the airplane, and not on instruments. However, it would be very beneficial for long-range especially air-cooled engines which are trying to compete with liquid-cooled engines on long range. To get long range you have got to pull rpm way down and manifold pressure way up. Now this looks like it's quite a conflict between all these phases, but if some work were put on it to make it practical for fighters, I believe it would be of distinct advantage for war emergency power, and then on returning home, when you do have to conserve fuel on air-cooled engines."

Mr. GOSSELIN: "I think that the detonation indicator has an awful lot of ifs in it as pointed out by some of the people who brought it up previously. You've got the mechanical problem of keeping it operating and so far, while it has worked on the test stand and also in experimental airplanes, I think it's still a laboratory instrument. We have had detonation indicators of one form or another on test for several years – i believe around 7 or 8 years to my knowledge. Insofar as using it in a combat airplane, I feel the same way about that as the controls; that is, for it to be successful in a fighter, it should be tied in with the engine controls in such a way that when you encounter violent detonation, it would pull the power back.

"To comment a little bit on what Mr. Parker said a while ago, it's also a matter of interpretation as to what is dangerous detonation. Very often you may encounter a slight amount of detonation on take-off without perhaps serious damage. Also in the cruising range it may be possible to operate with a slight detonation without serious results, but that's going to take a lot of experience in evaluation, and also bring power back to determine how much detonation you can stand without danger."

Commander RAMSEY: "Is Sherby here? Will you say a few words? Either Sherby or Lawrence."

Lieutenant J.E. LAWRENCE (NAS Patuxent): "Here in flight test we have just begun to use detonation equipment in the testing of engines. We haven't anything to say as to actual tactical use of putting them in fighter aircraft, as they do require a good bit of background in the actual interpretation of their use. I think at the present time the equipment we have would not be satisfactory to carry along."

Commander BOOTH: "During the past several days we have had several remarks which have emphasized the fact that mixture temperature is one of the basic controlling factors on the detonation limits of the engine. In fact, it appears that the allowable manifold pressures which can be drawn are largely governed by the mixture temperature. A thought has occurred to us that it might be possible to pick up the manifold mixture temperature and feed it into a maximum allowable manifold pressure needle on the manifold pressure gauge, thereby giving a pilot at any time an indication of how much manifold pressure he could draw and to tell conditions under which he is operating. This would be only a maximum power feature and probably could not be adapted to cruising conditions."

Commander RAMSEY: "Mr. Winch."

Mr. WINCH: "I agree with most of the other comments on the fact that the detonation indicator in its present stage of development is pretty much a laboratory device. I think at the present time if it were installed generally in combat airplanes or service-type airplanes it would scare the pilot and not do any good. We have been testing and experimenting with detonation indicators for a long time and we helped Sperry develop its present instrument and it is pretty difficult to establish the criterion as to when detonation is serious, or isn't serious. We have been doing quite a bit of work just recently in conjunction with the Army on B-29 tests for use of war emergency ratings and the first laboratory set-up was established, as I recall, of 30 to 50 flashes per minute on the instrument as being the maximum that we should allow. On subsequent tests in our laboratory, we have been running the engine at high powers with water and found we could go to 200-250 flashes per minute and the engine doesn't know the difference between 50 and 250. This was in the early stages and that's the indication at present time. I think I disagree with one of the gentlemen who mentioned previously something about the use of this indicator during cruising conditions. Most of our present-day engines with high-grade fuel in a cruising operation, the engine is not, detonation limited. In other words, you can go beyond best economy and the engine is not in detonation."

Commander PALMER: "It seems to me that assuming we could get an accurate, reliable detonation indicator, you wouldn't want one in a fighter cockpit. You have enough instruments as it is to look at. When you actually need it your throttle and revolutions per minute are full forward and you are not watching for detonation. You lean forward, pushing that job as hard as you can get it to go. If it whistled or sang, you would pay no attention. I recall one pilot's statement returning to Pearl Harbor in an SBD; a Zeke got on his tail, as I recall it, and he exceeded manifold pressure limits about 10 inches for something like 20 minutes and suddenly he had no regard for detonation. I think a fighter pilot's whole business is to assume that his plane will give him the best performance at all times and not have to concentrate on his job or put his neck on a swivel. Keep your eyes out of the office."

Mr. STEPPE: "The only point I wanted to add is some discussion about the use of manifold pressure as a limiting factor for detonation. It might be of interest to know we ran into detonation trouble or the same indication as detonation with low-mixture temperature on the Merlin. We don't know exactly what the answer is yet and as far as we know Packard doesn't either."

Commander RAMSEY: "Is there some comment relative to Commander Palmer's remarks? Colonel Meng."

Lieutenant Colonel MENG: "In the first part of the war I often heard when we were running instead of chasing that they would advance the throttle until it detonated and then pull it back, and their idea of detonation was when the cowl started to jump and black smoke started to come out of the stacks. I think that was done by quite a few pilots. I think that a pilot after a while will realize unless he is in too tight a jam when he is pushing the detonation limit. I think that is also possible in long chase, if one is chasing someone, he might, I believe, watch the detonation indicator pushed up to limit and chase an airplane and get better advantage from it. Also someone being chased, if chased a little bit, after a while he will push it up to the limit and then hold it there and unless he is in dire straits, I think he'll hold back off detonation limit and try to get the maximum power from his airplane."

Lieutenant Colonel RENNER: "I, too, would like to take exception to Commander Palmer's statement. I think he is probably well aware, but let me refresh his memory of the Navy's policy on their engine. Our policy in peacetime was to get an engine that would run 400 or 600 hours without overhauling. And consequently our emphasis was placed on maintenance and reliability. When we first got into the war December 7, 1941, I don't believe we were quite ready for it and I don't think we had adjusted our thinking on engine powers up to the present attitude toward it. So the fact that you could run 10 inches over your manifold pressure did not necessarily mean that you were getting into destructive detonation. Now I don't say this SBD pilot was detonating, but F4F's in the Solomons campaign repeatedly ran over the allowable manifold pressure. When we got down to find out how come the engine could stand up under it, we found out we had limited it in BuAer to allowable power so low that you could afford to run several inches over the red line and not get into any trouble at all. That's just another one of the old dodges to keep you from ruining your engine. The same situation we ran into in peacetime was that if the engine was red-lined at 39 1/2 inches for take-off, the Squadron Commander usually put out an order that no one could use more than 33 inches for take-off, and if you could cruise up to 29 inches with a certain revolution per minute, the Squadron Commander said nobody could cruise above 23 inches or something similar because he wanted the engines to last a long time and didn't want anybody to have to jerk an engine and change it. I for one down at the Eglin Field Fighter Conference told them I would like to have a detonation indicator which would tell me I was getting into or drawing full power, whether it's a hot day, or a cold day and whether at 30,000 feet or sea level, and just telling me the manifold pressure I could run at doesn't tell me what I want to know. I have talked with Captain Spangler and have repeatedly asked for detonation indicators and his answer has always been the same. The detonation indicator at the present time is a large instrument and until we got one that is satisfactory for tactical use, it would be more of a burden to the pilots than anything else to use it. The next step that comes if they develop a detonation indicator that is reliable, that doesn't break down, that you can use out in the field without constant adjustment, they could just as well build a regulator that will restrict your power as soon as you get into detonation so that all the pilot has to do is forget about how many inches manifold pressure he can use. He can shove the throttle up to the stop and when he gets into detonation his power is reduced so that he is not in destructive detonation. From our past experience I can guarantee that when we develop that detonation regulator, shall we call it, that the Bureau of Aeronautics will put a safety factor in there so that we are restricted in our power before we got into destructive detonation."

Lieutenant Colonel TYLER: "I'd like to agree with him entirely on that method of controlling detonation at high power. I'm wondering if there's not a possibility that the detonation warning indicator might be of some value in assisting the pilot to get maximum range out of his aircraft. It appears to me that not all pilots are sure of how far they can go in leaning down their engines in order to get maximum range on long range missions and if this equipment is not too heavy, too great a sacrifice in weight, it might be proven of value in that case. I'd like to comment on this."

Mr. GERDAN: "Again I can repeat one of the comments made previously. Generally, detonation is not a factor. At cruising powers and if you want range, and I think also the ability to pull high power, you want a manifold temperature control. The manifold temperature control would be used at combat ratings to limit maximum manifold pressure that could be pulled and would be used during cruising to extend range by raising the mixture temperature. High manifold pressures would be considerably improved. I think that we engine people can learn one thing from the automotive people – use of heat at very low engine speeds and controlled heat for cruise purposes. As regards to weight of the unit, our particular detonation indicator, as I recall, weighs about 6 1/4 pounds complete. The instrument that the pilot sees in that is a red light and the unit consists of 12 pick-ups, one per cylinder and the red light. The control on the production design is not adjustable in the cockpit. It is a sealed control amplifier, but that can't be reached from the cockpit at all. Of that control, incidentally, I believe there are several hundred being built now as production design that will be tried out at tactical equipment."

Mr. MARTIN: "I would like to back up Commander Palmer's statements. I think that in general the average fighter pilot has a very beautiful disregard for everything in his airplane when he is in combat and is concerned mostly with shooting down the enemy. I object to the use of the detonation indicator in fighter aircraft from another standpoint. In times past when aviation was young neither the engine manufacturers nor the pilot nor airplane manufacturers knew for sure that the engine was going to keep running. For that reason a great many of the instruments used by the engine manufacturer on his test-stand were put in the airplane.

"As the development of engines has continued, we have found that the engine manufacturer and likewise the airplane manufacturer have tended to move practically all instruments off the engine test-stand into the airplane. The tone here of asking us to put in another instrument which belongs only on the engine manufacturer's test stand indicates that the trend has not turned as it should turn towards a simplification of engine instruments in aircraft. I believe this can only be done after the reliability of engines reaches the point where the confidence of both the engine manufacturer and the airplane manufacturer reach the point where they no longer feel it necessary to keep those instruments in the airplane. That has happened in the automotive industry and I think it will eventually happen in airplanes. The first instrument which may be developed along this line is the power indicator: Now, you have the manifold pressure gauge, carburetor air temperature gauge, tachometer and an outside temperature gauge and I think all of those could well be combined into one power indicator. I think the trend toward reducing engine test stand instruments in a plane must get under way real soon. We are crowding our instrument panels tremendously."

Lieutenant ELUHOW: "I'll verify Mr. Martin's talk where he said we don't pay too much attention to instruments but to whether our engine is detonating or not. We learn the best cruising settings for our plane and we try to keep around them unless necessary to either speed up or slow down for our range. If we find out we're low in gas, we don't give a damn about detonation – just to keep enough flying speed to get us back. We want to get back. That instrument in there keeps us doubly worried. I didn't have much trouble with detonation – I probably didn't know – I always got back. So I guess maybe it would be best to eliminate it."

Lieutenant SALLENGER: "The problem of elimination of instruments is well taken in some regards but I think it might be time to learn that when we consider the maintenance angle, the crew chief does have to know a little bit about what went wrong, and I think it is a little too much to expect that no matter how far we go in the development of our planes and our engines, there is going to come the day when nothing is going to go wrong no matter what you do to the engine of the airplane."

Mr. WOOLAMS: "In regards to simplifying the cockpit in relation to engine instruments, for one thing some instruments are put in the cockpit for the benefit of the crew chief for checking out various things on the ground which could be put in some other part of the airplane. There have been some cases in some of our airplanes that there are instruments which are purely for the purpose of checking engine operation on the ground. They could be taken out of the cockpit and put on the nose or some place else where it would be out of the way and you could look at it while the engine is being revved up. There are other things that we have touched on at previous conferences and never had a decision although quite a few people have voiced their approval. Put red lights on the instrument panel to indicate when the instrument has dropped below its required setting. For instance, on the P-59 our oil pressure gauges can indicate anything above a maximum or minimum pressure of 5 to 6 pounds. It doesn't make a damned bit of difference what the instrument indicates, in fact on one engine it might indicate 25 pounds and it might indicate 7 pounds on the other engine and it doesn't make any difference so long as it is above 5 or 6 pounds. So the pilot doesn't care what the actual indication is. It would be much better in that case to put a red light in place of the instrument so that if something did go wrong and our pilot was not paying too great attention to what was going on inside, this light would draw his attention to it. It would draw his attention to it a great deal faster than he would notice by glancing at that gauge. We have tried to get rid of that gauge and have been unable to do so. There is another gauge in there that doesn't mean a thing except when your temperature gets above a certain amount. I would like to hear some discussion on that and what the objections are to taking out these gauges that unnecessarily complicate the cockpit and the installation of red lights in their stead."

Lieutenant Colonel MENG: "The point was brought up a few minutes ago on simplifying the airplane as they do automobiles. However, I don't think it applies to the present-day fighters. If you compare the plane we are flying to the present-day automobile, you would compare it to something you should have on the Indianapolis speed track. We are flying something up to the limit and pushing it farther than they would on the track. The racing automobiles are pretty well instrumented."

Mr. GOSSELIN: "I think Colonel Coates answered Jack Woolam's question with regards to lights versus gauges. For example, if you had a low limit on oil pressure of 70 pounds and if you set the red light to go on at 70 pounds, you wouldn't know when the red light went on whether you had 68 or 10. If you were just below 70 pounds, the chances are that you could come on back in safely and with a red light you wouldn't know where you stood. Whether you should land immediately at the first field, or try to get back home."

Lieutenant Colonel MILNER: "When you start getting lights on the instrument panel you end with something like a Christmas tree, in case you have a red light for wheels, green light for something else – and maybe a red white and blue one for something else. It's just out of the question."

Mr. GUYTON: "After listening to this discussion about lights, I'd sure like to agree with Commander Milner and I just wonder if we are keeping in mind the combat pilot when discussing this subject. Suppose he goes out to an area and he has approximately 1 or 2 hours in enemy territory and over territory he's got to come back. Men coming back from combat have been specially fond of their engines and the care they take of them and they are very serious about when they get away from this rat race where they have been fighting at war emergency power as to whether they land on enemy territory or water or whether they don't. It would seem that – to put a red light in the cockpit would take away some of the means they have of determining the trend of the engine by watching oil pressure go down or fuel pressure go down in which case they might possibly do something about it before it happens. Isn't that a good reason for keeping some instruments there that are of definite value to the pilot, especially to boys in combat, whom I think we're talking about."

Lieutenant Commander SUTHERLAND: "We've heard several rebuttals. I would like to go back to the defense of Jack Woolam's original proposition. They don't have to be red lights, they can be indicators which will flip white when it is on. It is not necessary to have a red light. I think that most of this requirement for instrument reading in fine gradation comes from test pilots who were very skilled in something and they are very sensitive to one-hundredth of a degree some place. I don't believe that modern instruments, modern engines, are inclined to fall off 2 degrees and go that way. If they do, you should be within your allowable limits. Most modern engines fail in an awful hurry. The people who want these instruments can be accused of having a Liberty engine mentality, where oil pressure flutters all over the place, etc. The only danger in putting in positive indicators which read normally within limits is the same danger where the technical people in different sections will put limits so high you will be liable to have your indicator go on when you can get back. I am in favor of having a positive indicator of this sort as long as we can keep the limits so that the indicator does not go on until the engine has very imminent danger of failure. There is no interest in whether it is going at 60-70 pounds pressure until it goes down to 20 or 25 which is below the pressure the engine can run, then it should go on."

Commander RAMSEY: "If I may make a remark – When you have the limits set for these lights on flap valves, or whatever you have, right down to the point where you are going to have a failure. I don't see any point to it, because then you are going to have a failure and there is nothing you can do about it."

Lieutenant SALLENGER: "On my last cruise on a carrier we did considerable night flying and I want to tell you we sweated those instruments out at all times. They were our best friends. And I don't think anybody doing night flying now would think of getting those instruments taken out."

Mr. MARTIN: "I think Boone brought up a very fine point in that people in this room are well aware of pretty near everything that goes on in an airplane and an engine. But I was up at Attu watching the PV-1 operations and I stayed in a tent with an ex-forest ranger. I have nothing against the ex-forest rangers, but I doubt very much if they know as much about the engine as the people in this room do. In fact it was very difficult for me to explain to him why he should do certain things, to get another hundred or two hundred miles out of his engine. It was much more difficult for him to understand it. When I got down to Wallis Island in the Pacific, I bunked with a premed student who was flying a 2800. I think the average operational pilot is not nearly as sensitive about his engine or his airplane as we have been led to believe around here. I know they will exceed their engine limits and airplane limits laid down by the Bureau or manufacturers. I didn't intend that some of the instruments be taken out. I merely intended that some of the many instruments be combined into one. As far as instruments like fuel pressure, oil pressure, and that sort of thing are concerned, if something does go wrong it's generally true that the pilot couldn't do a damn thing about it anyway. If the oil pressure is going down, I don't care whether he has one instrument or 40, there's damned little he can do about it."

Lieutenant ANDREWS: "It seems to me that before the war pilots had much more experience. They seemed to be engineers as well as pilots, and then we reached a period where we had to rush a lot of pilots through and as soon as they learned to work the stick and rudder, they were rushed right out into combat. But it seems to me that the training program is over the hump and the pilots know more. I think that is going to change more and more. They are being given much better training and I think pilots will be more and more trained in engine operation. As Lieutenant Sallenger said, they really watch those instruments and they are being taught what is going on in their engines. I believe that the average pilot now in combat does know something about his engine and is interested in knowing what is taking place."

Lieutenant ELUHOW: "I agree with the lieutenant that just spoke. The average pilot, in fact most pilots, know all about the engines. They do take time to learn about the engine. I had my own plane over there and I wanted it to have a good engine to bring me back and I looked out for it."

Mr. GERDAN: "I believe that Lieutenant Sallenger's remarks about a sort of master group of instruments was directed primarily at having something to take the place of detonation indicator, manifold pressure gauge, etc. There is one difficulty in that and that was the point Mr. Steppe touched on. It is very possible to have a detonation limit – that is a drop in as much as 6-inch manifold pressure when the manifold temperature drops. That is an advancement that's been observed recently on not only the Rolls engines, but also in air-cooled engines as well. The causes for it aren't very clear. We ran into it running some tests in after cooling was used and the knock-limit manifold pressure was actually lower with mixture temperature at 100 degrees F. than it was at mixture temperatures around 160 degrees F. It was lower by as much as 6 inches."

Commander RAMSEY: "I think we could keep this up all night. The next subject is brakes, footbrakes, some good and some bad, some mechanical and some hydraulic; air brakes which the British are very fond of. Let's discuss it for 5 or 10 minutes. I would like to call on the British, Lieutenant Commander Callingham, to tell his reaction to American type brakes versus British type brakes."

Lieutenant Commander CALLINGHAM: "It is a subject very close to my heart. I have had quite a bit of experience with both types of brakes and I am not biased in one direction or the other. We have had hand brakes on all our own aircraft since we started and I think you have had foot brakes on yours since you started. One principal advantage of a hand brake is that you do have a very much finer degree of control with the hand than your foot. Having to push with the feet in a way which stretches the leg out, in my case particularly and with short pilots, reduces his power of foot effectiveness and cuts down his fine control. Some of our modern aircraft are rather touchy on fine control on the runway, and particularly in cross-wind conditions are inclined to grab very quickly. The F4U is very touchy in directional control from time to time and I should be very interested to see results of experiments which, I believe, are going on now to fit that plane with hand brakes. That should be something quite interesting on that plane. I have heard various comments of many pilots around here who have flown the hand brakes and most have been favorable. Those who flew the Seafire and the Firefly have all had their own opinions, and I should be very interested to learn what comments you have written down about them. We are not trying to sell our hand brakes, but I should like to put those points to you."

Commander RAMSEY: "Any other pilots wish to raise their hands to keep this question open?"

Lieutenant Colonel RENNER: "Perhaps I don't have the same attitude towards this brake business that the rest of the pilots do, but my reaction is simply this: Is the braking system satisfactory? If they are both satisfactory and will control the airplane sufficiently to prevent ground loops with the normal pilot and his reactions, then are there any other advantages? If you tell me that the foot brake weighs 300 pounds and the hand brake 25, I'll learn to fly with the hand brake in 5 seconds. I have flown the Mosquito aircraft and I had no trouble at all with the hand brake. However, I have not felt that I was having trouble with the foot brakes on the American planes. With me there has to be some advantage before I'll change."

Commander RAMSEY: "There is a lot in what Renner says, because it is a lot in the way you are brought up."

Mr. BURKE: "Although I have flown the British brake system, I wish this would be construed not so much as an endorsement of their system as a criticism of ours; I think that in a fighter aircraft which will be very likely tricky in landing, we are putting an unnecessary burden on a pilot to demand that he return from the mission in condition to use the full extent of his facilities – both hand and both feet – and I believe, although I don't think it applies so much to carrier activity as it does to land-based activity from rather poor fields, it is likely to lead to injured pilots or tired pilots being over-extended, if you have a system of foot brake where delicate control of both feet as well as delicate control of both hands is required."

Lieutenant Commander MILNER: "I don't know if the British worked the hand brakes out successfully on the carrier or not, but if I were fighting an SB2C and after I had landed, I rolled back a little and had to put on my brakes and bring the hook up and if I had to hold the stick and get the hook up at the same time, I would not be able to hold the plane in place. I couldn't get the hook up and put my brakes on at the same time."

Commander RAMSEY: "Any other comments along that line? Incidentally, parking brakes enter into this thing. Do we want parking brakes or don't we?"

Lieutenant ELUHOW: "I'd say that we definitely want parking brakes because before taking off we rev up our engine and to hold brakes with our feet while revving up the engine puts quite a strain on our legs and we usually set the brakes to the full our legs are relaxed."

Lieutenant Commander CALLINGHAM: "I would like to answer Commander Milner on that question. We use our brakes on deck, but naturally we have so far had snap-up hooks and haven't had to retract them. The case of the SB2C particularly, was probably the worst one because of the massive lever which required a considerable amount of strength to wiggle up the hook in comparison with flip switches which the F6F and some other planes have. I don't think you would have that problem at all to take hand off the throttle for a brief second and flip a switch and put it back and at the same time, in case of hand brakes, you have the brakes on."

Commander RAMSEY: "Can anyone here say what they cost. Are air brakes less costly from the weight factor?"

Lieutenant Commander SUTHERLAND: "I don't think it is a matter of whether they are operated by hand or foot but by hydraulic or air. The Spitfire brakes are among the lightest in the fighter aircraft field in the world, but it's more to do with air than due to the method of operation. There is one neat advantage not yet mentioned, however, of the British hand brakes – that they have parking brakes. All you need is a latch to hold the lever down and you have parking brakes."

Lieutenant ELUHOW: "Another thing I was going to mention is that hand brakes give you another job for your hands, and you have quite a number of things to do with your hands as it is. All you have to do with your feet is steer the rudder and I guess they can do the job on the brakes."

Lieutenant Commander CALLINGHAM: "I don't know if Lieutenant Eluhow has actually seen any of our hand installations, but they fall naturally to your hand on the control column and you don't have to move your hand – just your fingers and thumb."

Commander RAMSEY: "May I suggest that we will either convert the Americans to the British way of thinking on brakes or the British to the American."

Lieutenant Colonel MENG: "This is just a question. We have flown some English planes and most of them are pretty old; I don't know whether it was the fact that the air system didn't pump enough air. But in our Army they do an awful lot of fast taxiing whether they are supposed to or not, and we just can't stop them. They need a brake that you can use often and is dependable."

Lieutenant Commander CALLINGHAM: "These two particular airplanes – the Firefly and the Seafire – took a pretty beating during the past 9 months or so. In the Seafire if you're taxiing in fast and you stuck on both brakes you would stand on your nose. You could also hold the plane to do a good machine check before taking off normally."

Mr. BURROUGHS: "I'd like to ask a question about reliability of the two types of brakes, not comparing the brake that must be energized from an external source with a brake that does not have to be energized from an external source. Our own system is not entirely satisfactory with regard to reliability, but you would think offhand that, when you had to rely on your compressor for energy, there would be more danger of failure."

Commander RAMSEY: "If I may interrupt, we will go on to the next subject – landing gear and flaps. The first question is electric versus hydraulic flaps. Do any of the companies wish to argue one way or another for their particular systems?"

Mr. BURROUGHS: "I am not an expert on the hydraulics, but I think Mr. Schliemann can argue for the hydraulic system."

Mr. SCHLIEMANN (Chance Vought): "The main reason for hydraulic flaps and landing gear on the F4U-1 was because the hydraulic was the lightest when it was brought out. This is not a new airplane and I believe the electric systems have improved. They are now lighter and probably just as dependable as hydraulic. The choice in regard to braking will be whether you are going to have a hydraulically operated plane or an electrically operated one. There have been improvements in both electrical and hydraulic. I don't believe I am in any position to state which is more dependable. However, the hydraulic has been eminently satisfactory."

Mr. STEPPE: "We made a check on a new airplane we were working up and we found that we took a terrific licking on weight on the electrical system. We do it much lighter with hydraulics. The only question I think would decide the use of one or the other would be combat vulnerability. I wonder if possibly some pilots who have flown the P-39 and the P-51 in combat, or some other similar planes that use both systems, could give some dope on that."

Commander RAMSEY: "Anybody had experience on having their system shot up? It looks like hydraulics have the upper hand. Any other comments to make?"

Mr. HAWLEY: "Goodyear Aircraft and Goodyear Tire and Rubber Co., Brake Division, are extremely interested in fighter brakes. I understand that we have down here a new type of brake that has been applied for the first time on heavier aircraft, on F8F. We will have on the F2G a modification of our old brakes, the main point of which we feel is less pilot exertion and about 33 percent longer life than those which were formerly used on the F4U-1 and the FGs. We realize that there's got to be a compromise or else giving in on either one of two points. First, on the present braking system, the pilot has to exert maximum physical pressure on brakes in order to hold when his engines are revving up. Waiting on a field for any length of time is very tiresome to the pilot and he finds his muscles flexing at the time he is ready to take off and really needs control. On the other hand it is possible to give the pilot brakes which are very velvet to touch and very sensitive, with the attendant possibility of fast taxiing and sudden application of the brakes putting him on his nose. We would be very interested in the comparison of the three or four types of brakes which we have here at the conference. The British hand air system, the F4U-1 type or the FG-2 or the F8F which is, I presume, a very easy brake to handle because it is of the new single-disc type, and presumably of high pressure. The one which we have on the XF2G is the multiple disc, but heavier than the old type, but we have a slightly higher hydraulic pressure on there. They have a velvety touch from the pilot's standpoint. We're very interested in the reaction of these four different types of brakes and keeping in mind that we have to sacrifice one thing or another; the brakes we have at the present time are brakes which were designed for around 9,500-pound airplanes, and we are operating them up around 12,000 to 13,000 pounds. That, of course, is the reason for the higher physical exertion by the pilot and short life in operation."

Mr. HALL: "The F8F has the new type of Goodyear brake. It has one or two good advantages and one or two bad ones. The good advantages are it can be used for take-off and landing without trouble of heating up the brake lining and having a loss of effectiveness in brakes. The single-disc seems to cool itself off very nicely. We're operating these brakes on a 8,800-pound plane and they seem to have very light forces for the ability that they have to stop the airplane. They don't seem to have ability to stop the airplane that the internal disc types do. The pilot finds himself putting much more force on the brake than is actually necessary to operate them. That is the only comment we have on these brakes."

Commander RAMSEY: "Getting back to flaps. Do the pilots want a blow-up flap similar to those on the F4F or do we want flaps we can locate in three or four positions as on the F4U, or do we want a flap that we can locate up or down. Any comments relative to that? I assume that everybody is satisfied with the F4U type."

Lieutenant Colonel RENNER: "I know that this is going to raise a storm of controversy but my personal feeling is that if a blow-up feature in the flaps costs 10 or 15 or 20 pounds, I can do without it. I would like to have a flap that I can position 1/3, 2/3, or full down or some such arrangement. But I understand that the complication comes in from putting the blow-up feature in the flap and if that is to be costly from a weight angle, I say take it out. There is no point in putting a premium on stupidity. If you want flaps to go up, put them up."

Mr. STEPPE: "Just for Colonel Renner's information, we have a blow-up system coming out on a new airplane and we get it for exactly nothing. We use the system release valve for the blow-up."

Lieutenant ELUHOW: "The way I understand it from the talk is that when you put them down or pick them up, there is no way to stop them in between. Is that right?"

Commander RAMSEY: "The F4U has a lever with which you can put the flaps up 1/3 in every 10 degrees from full up to full down – but with no blow-up feature. North American has sold me on that blow-up feature that costs nothing. I would like very much to have it."

Lieutenant ANDREWS: "We just finished a combat evaluation project on four planes and it so happens that on the project we used planes that had the different type of flaps. At the completion of the project, the pilots were of unanimous opinion, that they wanted a flap with positions to use in maneuver. We found that the flap such as on F6F, when extended, was just too much. There was too much flap to gain any advantage from maneuvering. We found that the F4U flap was very beneficial in maneuvering. Pilots were of unanimous opinion that they wanted a flap with several positions."

Mr. PARKER: "I have flown both types of flaps and I'm very well pleased with the multiposition flaps. On our particular airplane we have either up or down or stop and neutral to stop at some predetermined position. We have had quite a bit of trouble with pilots trying to estimate when to let up their flaps and how to milk them and this position indicator seems to me to be a positive indication of how many degrees on flaps to use under certain load conditions to assist take-off and so forth. But generally speaking, I'd prefer fast gear action and slow flap action."

Commander RAMSEY: "I agree with you on that fast gear. Lot of planes we find fault with on slow retracting of the gear. As a matter of fact, I believe the F8F could be a little faster."

Mr. BURKE: "Now that the subject of maneuver flaps has come up, I would like to point out that when you design an airplane with maneuver flaps specifically for combat, you get into a completely different set of criteria than you do when you design for landing where you have to have very high-speed operation of the flaps. They should go up or down in 3 seconds or less. They must be designed to be lowered at much higher speeds. It does you no good at all to put a flap down at very high speeds, it does you quite a bit of good to put it down at low speeds, but it does the most good to put it down at just about the speed where you get, say, seven G's with your flaps down and the other fellow can get only about five G's with his flaps down and he will usually stall out. That is usually at about 250 miles per hour on the average present-day airplane. The net effect of all of this is that a good maneuver flap installation is quite a bit heavier than a standard landing flap installation. Flaps for maneuvering have to be considered in light of what it will cost in weight and also of the application of rapid control."

Commander BOOTH: "I just wanted to take exception to Mr. Burke's remarks that a maneuver flap had to be so much heavier than a flap for landing. The flap on the P-51 is designed to be lowered completely down below 170 indicated and that is the critical designed load for it. But I believe you can get around 10 degrees at 410 and something scaled on down, and by no additional beef-up. To further Parker's remarks on instrument adjustment, in the new system we have the flaps move no faster than the pilot moves the flap handle in the cockpit. We have achieved that by actually making a mechanical linkage to the flap and using the hydraulic system merely as a power boost on the flap. That seems to be the simplest way of getting those desired qualities."

Lieutenant ELUHOW: "This man over here, I forget his name, said something about the stronger type of flap for maneuver and the lesser type for landing. When you're in a tight turn, you're in it at very high speed and you can drop the same flaps without any danger, not fully, and I have found in many cases that they do let you turn a bit tighter and you don't need excess speed to maneuver and the flaps are sufficient, especially on the P-47. None of us had any trouble with them and we used them for both maneuver and landing and they worked quite well."

Lieutenant Colonel RENNER: "I hate to show my ignorance in this way, but we in the Pacific often used our flaps in mock dog-fighting when we were up for camera gunnery against each other, but when we got tangled up with the little Nip, nobody screwed around with their flaps and I would just like to know if anybody is actually using their flaps when they get in there hot and heavy."

Lieutenant ANDREWS: "Colonel Renner, I think, was in an F4F. I believe those flaps are very poor for maneuvering. Lieutenant Jorgenson just said negative. I think he was fighting in an F6F whose flaps are inferior for maneuvering but I think I've established for my own satisfaction that the position flap can definitely be used to advantage for maneuvering."

Mr. BURKE: "In further explanation of what I was saying before, I would like to point out that on the airplanes in which the gentlemen say the flaps were not of much use in combat planes the flaps were not specifically designed for combat. The true combat flap should be capable of lowering to its maximum lift position and at the speed at which you stall out if the flaps are not extended at the maximum designed load. For instance if you have a 9G plane and it stalls at 100 miles per hour, it will stall out at 9G at 300 miles per hour and you can hold that down lower to perhaps 220 miles per hour. In that range of a fairly high-speed maneuver, a true combat flap designed specifically for that can give considerably improved turning radius where the landing flap as we now have it is just as satisfactory at the low speed where it will at 1 1/2 or 2 G's. The landing flap is just as satisfactory for combat use at that range."

Lieutenant Colonel RENNER: "I asked that question only because I wanted to know of any combat pilot who when bullets are bouncing off the armor plate, or you see those tracers coming over your shoulder and you twisted the hell out of that thing and maybe got ready to go straight down, which is one saving grace against the Nip – whether he was thoughtful enough or had enough confidence in those flaps that he would reach over and put the flap down when he was in a turn and he might want to go down the next instant. I had enough to do to keep track of the Nip and try to get the proper lead and all the other problems in my mind without groping around for that flap handle."

Mr. SIEGEL (Chance Vought): "I'd like to answer Colonel Renner's question on why a flap blow-up and why pay some weight to get in the airplane as we have done in the F4U1.

Some of them he's already answered for himself. In the use of combat flaps in airplanes with the flap blow-up installed, you can set the flap down at any required degree or any degree that you want for combat. Then when you want to increase the speed of the airplane, the air loads working on the flap will cause it to blow back up to a neutral position where it allows you full maneuverability and saves your flap. Now, you do gain back the weight you add in a flap blow-up system. You save it in the strength you have to build your flaps if you want to use those flaps for combat flaps or in the case of a carrier airplane where you have a wave-off condition. With the blow-up system you can set your flaps to the degree you want to come in on the deck and you don't have to worry about the flaps control at all. The flaps will come up automatically in the case of a wave-off."

Lieutenant Colonel RENNER: "When you roll over with your maneuver flap down and go into a vertical dive and your flap blows up and you get back up to speed, the time that it took me to build up speed before any flap retract may have been just enough to get one in the back that I didn't want to get. I may find that the adversary has already closed in on me before I discovered that I forgot to put my flaps up, plus the fact that when I'm in the zoom climb, if I have been zoom climbing before with my flaps closed, they're coming back down at reduced speed.

"When I'm taking a carrier wave-off, I don't expect to get up to a speed of 150 or 160 to blow my flaps back up without my leveling off some place along the way and realizing that now I'm going around and I have my choice of putting my flaps up or leaving them down."

Commander PALMER: "I think Joe is a little off base there. He was fighting F4F's, mainly defensive. At present it's offensive, with F6F's or F4U's. I think the basic reason that Andrews mentioned the use of maneuver flaps is that against the present Jap plane offensively you can turn with him for a short while only by use of these maneuver flaps. It is a known fact that you can't turn with those jobs indefinitely. It gives you a momentary advantage during an attack. It's also assumed that if you are attacked yourself, you don't want any maneuver flaps. You don't want to get in a dog fight with them. You dive away in a clean condition."

Mr. STEPPE: "I think Colonel Renner's point on that flap motion is very well taken and I am glad to see we arrived at the same answer on them. We analyzed the case the same way and decided the spring system wouldn't work. In our system when flaps blow up they stay up and they don't go back down again. In other words, the way the system works, the pilot puts the handle down – he can hit the handle hard if he wants to – the flaps will go no further down than the design load for them and he can use them for maneuvering. If he decides to peel off and head for the deck, the flaps will come back up and the handle in the cockpit is up in position and the flaps come back up and stay there. If he wants them down again, he's got to put them down."

Mr. ROWLEY: "I would like to point out the fact that our F8F flaps are very similar to Mr.

Steppe's on the P-51, I imagine, with the exception that the flap handle doesn't go back to the "up" position. However, we have pressure relief valves similar to what he explained. If you increase air speed, the flaps will go back and will stay up."

Commander RAMSEY: "We will go on to the next item in order to touch on fuel and fuel systems. Do we want an automatic handle or automatic fuel selection system, similar to the one that's been tested in Florida in an SBD? When you get down to 5 gallons in the tank, electrically the selector valve automatically seeks the next one that has over 5 gallons in it. In other words, if you have a droppable tank, an auxiliary tank, a reserve tank and the main tank and it gets down to 5 gallons, automatically the next tank comes on which has gas in it, always going to the right. It works very successfully and costs only about 4 or 4 1/2 pounds in weight. One PV had 13 different tanks on it and it would have been very successful if it had had that selector on it. Do the pilots want that or would they rather select their own gas? You get in a plane and all 4 tanks are down to 5 gallons, the valve automatically keeps going around till you use every drop out of the system. You start flying the plane and you never have to worry about it until the last drop is gone and then you had better look for a field. It sounds like an excellent idea to me and that weight might be worth sacrificing."

Mr. ROWLEY: "We had the opportunity to fly such an installation at Grumman on an F6F3. It was just about the same as Commander Ramsey explained, but had the addition of a vibrator on the stick. When the tank changed to the next fullest tank, it would set up a vibration in stick grip. Half a dozen different pilots flew this arrangement and the general consensus was that the automatic tank selector was a pretty good thing, but the vibrator on the stick was unnecessary. One feature of the automatic tank selector which I think was a good idea was that if you got down to a minimum of 5 gallons in each tank it would automatically keep running around. If you had three tanks, the tank selector would keep running around and even in the off position you would still be assured of a fuel supply. The designer or inventor of this gadget claims you could drain all your tanks dry. He claims this would not be the case if you had to select manually the tanks and you may run out of gas when in one tank you had less than one gallon."

Commander RAMSEY: "I believe a Mr. Brown, the inventor of this gadget, brought it down here. We tested it and every drop would get out of the tank."

Mr. STEPPE: "The only question I had on this was that when we were studying fuel systems we got into so damned many emergency systems that went along with it that it began looking like a bird cage. That is covering the possibility of one tank getting shot up and the check valves running back and forth. It looked like a rat race to me. I wonder how much emergency systems you want in it."

Lieutenant Commander SUTHERLAND: "We had a lot of simmering in the Bureau of Aeronautics on fuel systems. We looked on the Brown device with interest. It worked most of the time. It is, however, a gadget and as such is subject to failure. We have tried several others. One of the worst in the Navy is in the Grumman F6F. It has three tanks about the same size and pilots are notorious for cracking with two full tanks and one empty tank and the indicator on empty. It runs out very rapidly.

"I noticed another one this morning. I was reading a report on the British Spiteful, which is a supercolossal version of the Spitfire. It has 133 imperial gallons capacity in 6 tanks. I don't know if they all go on or if you have to select them, but there are innumerable 12 3/4-gallon tanks. This doesn't meet my idea of what a combat fighter should have. We approve temporarily at any rate of a system where the pilot turns his fuel on and all the fuel in the plane drains into one master tank. In the case of new planes, we have tried to get designers to build in one large tank. That's frequently impossible, but in the case of medium-sized fighters where you will have perhaps a 150-gallon main tank, and if you have two 150-gallon drop tanks and another 75-gallon auxiliary tank, the fuel should drain from the drop tank into the auxiliary tank into the main tank and then it goes straight from the main tank into the engine without bypassing any selector valve. That is, of course, subject to failure if you have to have vacuum pumps to pump it in or electric pumps to pump it back, but I have never seen anything that wasn't subject to failure unless it was a single tank."

Lieutenant Colonel TERHUNE: "This business of dumping it all into one tank is a good idea, but it has one drawback; if you shoot out any part of the system you are definitely limited in the amount of fuel left. The only reason I can see for using a series system where the fuel dumps into one main tank is in a jet airplane where you should have continuous flow to keep the thing from going out. I don't know how easy it is to start them compared with the reciprocating engines, and I would like to see one large tank more or less like North American had in their first P-51 if it is physically possible."

Commander RAMSEY: "Of course the Navy would have some trouble with the F7F and planes like that. It reminds me of that "dream" airplane sheet – the aerodynamacists want one type, the armament people want another, the power plant people want another and the fuel people came along and want something else. YOU CAN'T HAVE YOUR CAKE AND EAT IT TOO, but this makes for good discussion."

Lieutenant Colonel TYLER: "The Army has an automatic fuel system which may or may not be the same you described. But it works off the fuel pressure. If the fuel pressure drops below 10 pounds, it automatically switches to the next tank. This was tested on the 47 up to almost the last drop. The fuel in the belly tank drains into the main tanks and it is used until the belly tank is dry. I would like to say a word for this automatic fuel system in training activities and also in the combat theaters; I know we have lost planes because the pilot forgot to switch tanks and didn't get started before he hit the water or jumped out. It is also possible that planes are lost because in the heat of combat the pilot forgets to switch and loses his tactical advantage which results in his either hitting the water or being shot down."

Major LAMPHIER: "I would like to endorse Colonel Tyler's remarks. I can tell you, as a matter of fact, that in the last 8 months since the last conference, at which time I made the comment that we were losing a lot of planes due to the indifference of the pilots, particularly in training, who don't give a damn about how much fuel they've got and where, we have lost 44 airplanes and 11 pilots in one air force alone in training in this country due to the fact that these kids got interested in some tactical maneuver or something and forgot to switch their tanks."

Lieutenant Colonel MENG: "Colonel Tyler covered that pretty well. I just wanted to bring up also that another one of the combat hazards is pilots flying along and all of a sudden seeing something, dropping the belly tank and going on merrily on their way. Just about two or three seconds later they find they are running out of gas. I know that has happened several times and some of the boys have been pretty well peppered because the others left them and they were behind with the fan going around in the air."

Mr. MARTIN: "The fuel system we have in the PV-1 and PV-2 is the automatic transfer system and I think that's essentially what we are looking for. It does require that the pilot select a tank, but there is no emergency involved. One front tank is the main sump tank and the fuel from the other tanks is transferred to the front tank automatically. If something happens to that front tank, the emergency bypass system will allow the flow of fuel from every other tank to the engine.

"I would like to be permitted a small chuckle here, if I may. We have heard our friends tell us that pilots have let themselves run out of gas. I wonder what our new high-caliber engineering pilot in the Fleet would do with all of those instruments in the cockpit at that time."

Mr. KUPELION: "We have a system, similar to the one recently described by Commander Sutherland, on the F2G. In this system all the fuel is fed from three auxiliary tanks into a main tank. It also has a stand-by system which, if the transfer system goes to pot, transfers it through the selector valve. The questions about such systems are that in order to transfer from the drop tank say up to the main tank, it is accomplished by pressurizing these tanks or using a transfer pump; transfer pumps are good up to a medium altitude, but above that they are not very effective. We had gone to a transfer pump instead of a pressure system in an effort, in the future perhaps, to use self-sealing drop tanks. I will bring up the point as to whether the Bureau is considering the use of self-sealing tanks or can we go to a pressurization system in an effort to get an adequate transfer at higher altitudes."

Lieutenant Colonel RENNER: "I think I will have to give Mr. Kupelion a double answer on this. I am sure that the Navy does not intend to use self-sealing drop tanks and the Marines do when land-based. If he can figure out how they are going to assign the planes, he will be able to figure out the right answer."

Mr. DEBRUYN (Pratt and Whitney): "I would like to needle Bud Martin a little about his gauge that didn't work when the guy ran out of gas. If we take the gauges out, how does he know what happened?"

Mr. MARTIN: "We don't use a gauge on our transfer system. We have two little red lights. When the little red light comes on, he changes tanks."

Commander RAMSEY: "We will have to stop today's proceedings. We will have night flying tonight with the P-61 only. I would like to caution everyone please to be on time for the scheduled hops tomorrow morning. Also, we have a number of Army pilots scheduled for flights in Navy craft. Let Commander Milner know if you cannot be here. Tomorrow should be a good five-hop day according to the weather experts.

"If you have to leave, please check out with Miss Ryan so that mail, telephone calls, telegrams, etc., can be forwarded as necessary.

"Pictures are available. For requesting prints, please put your name on the back of any print which you desire.

"I wish to reiterate tomorrow's schedule: Five hops; a conference at 1530 tomorrow afternoon, which will be the first time we meet here unless bad weather overtakes us. We will have at that meeting open-house criticism on each plane.

"I'd like to again remind everybody of the cocktail party. All conferees are automatically invited."

ELEVENTH MEETING
22 October 1944
1530-1745

Commander RAMSEY: "We planned to run this afternoon's discussion by compiling a preliminary report on each aircraft, gathering data both from word of mouth and from the cards you filled out after your flights. I will read over those reports for each plane and I know you will all probably wish to make comments on your aircraft. Rather than take one day for each plane which we could very well do, I will read the summary opinion broken down by Navy, Army, British and contractors' comments. Then each airplane will be open to further discussion.

(Hereafter Commander Ramsey read a preliminary consensus on each airplane, which is not reproduced here, but is given in final form under the "Pilots' Flight Comments" section of this report. Discussion from the floor is recorded.)

P-61

Mr. MYERS: "I can't say much but "amen" to that. The ship was never designed as a day fighter so any comments along those lines are very much expected. We are working on most of the items which have been mentioned – visibility, stick position, stick forces, performance, and so on."

• • •

Commander RAMSEY: "The analysis of the results of this conference will be prepared in a report, and so far we have on our list 147 addresses to send copies to. We have tried to leave no one off the list who would be interested in our report. We are going to try with the board of strategists sitting at my left, who are very capable gentlemen and interested in their work, to get out a report from which you can get the meat in four or five pages rather than going through four or five hundred pages which the over-all job will represent. If you will bear with us after the conference is over, you will have a report in your hands. If you have been left off the distribution list be sure to see Lieutenant Commander Wilds and give him your name. Any contractors present, with all branches of the services, will be on the distribution list. Please everybody fill in the questionnaire which was on your desk yesterday. All who did not fill out one or did not receive one, please see Miss Ryan and fill it out. It is very important that we have that to get a good opinion from the conference and in order to incorporate what we want in a fighter for the United States and Allied use."

• • •

F6F5

(No floor comment.)

• • •

P-47

(No floor comment.)

• • •

F7F1

Mr. DEBRUYN: "In connection with the too-cool operation that has been mentioned both for the F6 and F7 I think a comment would be in order. First, in connection with some of the cold weather tests which have been run, I think for the greater part by the Army up in Alaska, it has been found that radial engine operation at which this is particularly applicable is O.K. as low as 25 degrees C. cylinder head temperatures and has been operated considerably at such temperature during those tests. We have also found that the cylinder temperatures do not have an effect on engine operation. In connection with the F7F particularly I believe the comment about it running too rich is really that it is running cold and I would like to point out that the F7 was originally designed for the 2600 and for that reason is possibly somewhat larger and can be cut down quite a bit in opening an exit gap."

Commander RAMSEY: "Does Grumman wish to comment?"

Mr. HALL: "We have known that the elevator was a little deficient in landing for some time and steps have been taken to increase its effectiveness in the later produced airplanes. The elevator throw has been increased so that a great deal more control will be available a little later for landing. We have known that the rudder is a little light with a boost on and steps have also been taken to stiffen the neutral range in the rudder without affecting the force required for full deflection of the rudder in single-engine operation. As far as the cooling of the engine goes, it is true that that nacelle was laid down for a 2600 engine originally but it was laid down for a spinner. A very large spinner which restricted the intake in front of the engine. However, this spinner made the engine accessibility very poor so we ran some extensive tests on how much good the spinner did and we found no effect whatsoever in airplane characteristics with the exception of cooling at low powers. The cooling was not even affected in high powers and slow speeds so that we probably are losing some airplane efficiency by having too-large exits in combination with too-large frontal openings. It is a little bit tough to close down those exits because the only exits are in the way of individual stacks and if we change completely the engine exhausts to siamese the stacks we could cut down the exit opening; but if we use the single stack which we like for a number of reasons I guess we will have to accept the slight loss in efficiency due to some overcooling at low powers."

Commander RAMSEY: "Any other comments?"

Mr. MYERS: "I would just like to make one comment generally and that is that in many cases on these reports you are giving now, we are getting a first flight impression. It looks to me that the men who are most familiar with the airplanes we are discussing now have not flown them at this conference, and consequently have not made any comments and they are the people who are in a position perhaps better than others to report on various characteristics. What we are getting now is in most cases a single flight first impression. That's all."

Commander RAMSEY: "We had quite a meeting of the minds here before this conference was held and I should have made the remark before the assembled throng that these are first impressions, but I believe that in lots of cases first impressions are very valuable. I have talked to some of the contractors' pilots who agree with me. You find out a lot of things on your first hop. I agree with all hands that you cannot possibly fill out one of these cards and do a complete job on it. It would take possibly 20 to 25 hours minimum, but perhaps you have found out a lot of things on a plane which, after repeated test hops on any particular aircraft, you forget and slide over and say "I'll get along with it" because the plane is good. For instance some little thing which is very annoying to you on your first hop but after you repeatedly fly it you forget about. I believe that most of us think this way about it and know that these remarks are first hop remarks and should be treated as such by the contractors and should be used as they see fit."

Lieutenant HALABY: "I would like to emphasize that Commander Ramsey's remarks apply to this plane especially well. It seems to me that there are several of us who have flown it a great deal and have begun to doubt our own perspective a bit, and I am sure that applies to contractor's pilots. I would like to ask pilots from Northrop and Lockheed just what they think the single-engine control speed in that airplane would be. I found a bit of trouble on the fifth or sixth take-off with a minimum of experience in a single-engine type. Could I ask Mr. Martin what he thinks the single-engine speed is in that airplane and then perhaps Lieutenant Twiss after that?"

Mr. MYER: "Yes, I tried the F7F on one engine. I used the right engine which is the worst engine in that condition and found I could hold military power down to approximately 130 knots. That was with full rudder deflection and nearly full aileron deflection and without permitting it to swing. I would say it would take a damned good pilot to get it to 140 knots on the take-off."

Mr. MARTIN: "I hauled back the left throttle on that airplane and I was satisfied to quit at about 135 – between 135 and 140.

"I think with the poor directional stability on that airplane the yawing velocity would be very high and with a sudden engine failure I am sure all the single engine speed would be about 130 which you can hold statically."

Lieutenant TWISS: "I imagine about 145 to 150. At anything below that you could get into trouble flying off a deck and you would not be able to hold it from swinging off the side of the ship or into the island."

Commander RAMSEY: "If there are no further comments on that I will go on with the next plane."

P-38

Commander RAMSEY: "I picked this one out so it would follow behind the F7F. I did not get that aileron snatch myself, but there were several who noted it. All these bail-out comments were obviously based on the person leaving the plane from straight level flight, not some other attitude."

• • •

XF8F

Commander RAMSEY: "Inasmuch as very few pilots got to fly it, we will take the comments of those who did fly it.

"That is one interesting feature. I don't know what the offset is on the vertical fin (about 2 degrees) and you feel no torque at all.

"It is interesting to note that there are only about five or six cards on this plane and everybody voted 10 hours to fly it.

"Grumman is swelling with pride over here and doesn't wish to comment."

• • •

P-51

Commander RAMSEY: "I saw somebody trying to steer 90 degree turns with the tail wheel out here."

Lieutenant Colonel MENG: "One of the nicest things about the 51 was the rudder. Everyone used to complain about the directional stability. I believe they cured it by the addition of a dorsal fin. They had some rudder reversal, I believe, and they put on that boosted torque to reduce your pressure. My opinion is the average man fighting with that plane would be willing to do away with that. They have lost tails but not in combat. I believe it was done by people back here trying to do slow rolls. From a good rudder which had a little change in trim, they have built that in there. Mr. Steppe might be able to explain a little bit about that."

Mr. STEPPE: "That business on the rudder and the change of trim with changes in power and change in speed didn't come about as anyone thinks. It was built up gradually from the P-51A. The A was probably the best airplane in its class for lack of trim changes. We went to the Merlin and put on this four-bladed propeller and got quite a destabilizing effect. We just aggravated that by lack of fuselage area. As Colonel Meng said, we got into some trouble with the tails. We were forced to go to the dorsal fin to get us out of that and along with that, we had to go to reverse boost on the rudder. We had to build up the rudder force to keep from getting reverses at high angles of yaw and we know the stability is marginal. The newest model of the 51 has increased the aspect ratio of the tail and added about 3 square feet on the vertical surfaces and also added on the horizontal surfaces and reduced the CG load."

Captain KESSLER: "I would like to say something about that bail-out. It is all right when you get out, but the way the canopy slides down in front, when it is set all the way up, I believe there have been cases where it has whacked the boys in the head. We have seen them go down and jettison the canopy, but never get out of the plane. We assume that is what happens."

Mr. STEPPE: "I think that point is quite apparent when you climb into the plane. The location of your head is behind the lowest point of the canopy and I think it would be good advice if you have to bail out to drop your head a little. The only thing I can recommend is just duck a little and get your head below the upper level of the windshield."

Lieutenant GAVIN: "Probably somebody here is better qualified to mention this than I am. I was talking to Commander Bottomley at NAMC where they flew a P-51 which has a rather quick and dirty hook on it. They are flying it around 9,000 pounds. The results so far have been surprisingly good. They seem to like it, the way it handles, after the power is cut. The controls, they say, seem to be more effective after the cut. They think it shows some promise of being a carrier airplane."

Mr. STEPPE: "I think that point just emphasizes the destabilizing effect of the propeller."

• • •

Commander RAMSEY: "We will go on to the Mosquito."

Mosquito

Lieutenant Commander SUTHERLAND: "There is one rather petty item, I think. It struck me that the instruments were sort of thrown on there and stuck with glue wherever they happened to fall. Our doctors and technical experts have been telling us that all obstructions in the cockpit should be buried so you won't hurt yourself in case of a crash. I would like some comment from the British on why they put the controls on the outside like the 1921 Stutz I had a few years ago.

"They do have a good feature on the instruments. That is the habit of coloring the instruments with the color of the item which it is measuring."

• • •

Commander RAMSEY: "Lieutenant Commander Wilds, who is in charge of the compilation of the report, asked me to announce the following: On these reports, will all pilots please list on the questionnaire the total number of hours they have had in each type they commented on."

F4U4D

Commander RAMSEY: "One person says one thing and another says something else. I don't know what the contractor is supposed to do. One says too high and the other says too low."

• • •

F4U4C

(No floor comment.)

• • •

FG1A

(No floor comment.)

• • •

Commander RAMSEY: "That completes the compilation of the planes at the conference. Are there any particular comments on any of the planes?"

Mr. PARKER: "Did anybody try the dive-recovery flaps on the P-47?"

Commander RAMSEY: "I tried them. I went just a hair over 400 and they worked very well at low altitude. It felt about 2 1/2 or somewhere along there. I liked it very much. If it does the same at high altitude, I think it will be pretty good."

Commander BOOTH: "I don't think my use of them was more violent than the one you suggested, Paul. The idea appealed to me and I think any modern fighter should have some kind of dive-recovery flap."

Mr. MYER: "I would like to ask a question. I wasn't interested in getting into compressibility to find out for myself. I would rather have you tell me, Joe. I tried it at 400. Is the pull-out similar when the plane is actually in compressibility to the pull-out when you are in a clean dive at low altitude?"

Mr. PARKER: "It is very much similar in compressibility and not in compressibility. The angle of the flaps in the extended position is set at 20 degrees to 22 degrees and at high altitudes, up around 25,000 to 27,000 feet, the recovery is about 4 1/2, after you are in compressibility. Down around 18,000 and 17,000 feet, the recovery is about 6 G's, and then it progressively modifies itself because of the natural tendency of the airplane to come out of compressibility at low altitudes so that it would be about 3 G's at 10,000 feet."

Mr. SCHLIEMANN: "There is one item we are very much interested in and I feel we should get an expression on the FG1A with flat-front windshield and the bubble canopy versus the Vought F4U1 with the flat-front windshield and the formed canopy. I would like to know from those pilots who have flown both of them if they have any comments on that."

Lieutenant Commander CALLINGHAM: "I have flown both of these hoods and I personally feel that the enlarged Vought hood is pretty close to being as good as the bubble canopy on the FG. In the Vought canopy you still maintain the protection of the radio gear. I felt just as happy in the Vought hood as I did in the FG."

Lieutenant Commander OWEN: "I agree with the remarks Mr. Davenport made the other day. One thing we have noticed is that it is necessary to sit higher to see through the gun sight. I think as it is right now, it is necessary to sit too high. I think the removing of the stringers in the new canopy, in consideration of the excellent overhead armor protection, is an improvement over the F4U1."

Mr. KUPELION: "I would like to bring up a point that, in the FG1A with the bubble canopy, the radio is installed in exactly the same location as the FG1 without the bubble canopy so, as far as bullet-tumbling characteristics are concerned, it would be the same in both cases. We think that could be derived from the present raised turtle canopy."

Commander RAMSEY: "Mr. Virgin, do you want to comment on the P-47 dive flaps?"

Mr. VIRGIN: "I used them to recover in a dive at 350 when I trimmed for zero force on the stick, extended the dive flap and although there was an accelerometer in the plane, it appeared to recover between 3 and 4 G's. The operation was smooth, and there seemed to be very little buffeting."

Mr. GUYTON: "I tried them the same as Mr. Virgin did. I checked them out at 400 and I liked them so much I just let the stick go and let it pull me out. It is very nice for maneuvering. If you ever lose your elevators, I think you've got something."

Commander RAMSEY: "I think we had better call a halt to today's meeting due to the time drawing near for the cocktail party.

"Tomorrow morning flying will commence at 0815. There will be three flights before lunch, then lunch, and an afternoon session – the closing session."

TWELFTH MEETING
23 October 1944
1415-1730

Commander RAMSEY: "Gentlemen: This afternoon we will open the final session of the fighter conference. Yesterday we had open discussion on planes, and at that time not all compilations were made, but today we have the remaining ones and we will take them up later in the session.

"I would like to call on contractors' pilots to write up their own planes on one of the comment cards. They will be very valuable to us even though they may be prejudiced somewhat. Each contractor and pilot will please fill out a comment card on his own particular airplane. It will be very beneficial to have you fill those out before you leave the conference. While all the information is fresh in your mind, it is a lot easier than if you get back to the factory and get involved in your red tape and paper work. Fill them out before you leave and we will appreciate it. The knee pads which you used during the meeting, if you do desire, you may have. The cockpit (F7F mockup) in the hangar has been photographed and any contractor desiring a copy of what the Navy considers an ideal cockpit can have it for the asking, by seeing Lieutenant Commander Milner. The data cards which have been posted on the bulletin board have been photographed and anyone desiring a copy of them please let Lieutenant Commander Milner know. The XF2G as most of you saw, is available on the field to look at. Unfortunately, legally we cannot let everyone who might desire to fly it. There will be certain selected people who are concerned with the program of test work on that plane who will be permitted by Goodyear and Navy to fly it. If you have not been notified you are not on the list. A lot of pilots did not get a chance to fly all the planes that they desired during the conference. The Navy will make available on Tuesday and Wednesday the P-59, the Jap Zeke 52, and the British have kindly consented to keep their Mosquito over two more days. Those contractors and Navy pilots who desire to fly it may get on the list for Tuesday and Wednesday for the three planes; see Lieutenant Commander Milner and Lieutenant Sheehan.

"This afternoon, gentlemen, the first subject to take up on the agenda is propellers."

Mr. NEUDECK (Aero Propellers): "As the Aero Propellers representative I cannot have a very large voice; we have only two airplanes equipped on the field here, P-63 and F8F, but I think it can almost go without saying that either or both of these propellers are outstanding in their performance for single engine fighter aircraft. They are quite light in weight and are probably the absolute ultimate in simplicity in both operation and control. The complete operating unit is contained on the propeller. There are only two connections to the airplane, one to the pilot and one on the shaft of the engine. The control unit is composed of an oil reservoir power source and a speed-conscious control valve, and that is about all that is necessary for a smooth-working, speed-conscious propeller. The rate of pitch on that job is more than that initiated by aerobatics requirements. I do not think you can overspeed the propeller by dipping the nose. A good selling point to a procurement bureau would be the case of maintenance. The propeller is composed of several distinct and separate units. It can be replaced as units, and the time required to do such a job has been probably planned so that it will be a minimum. The Navy has never before used this prop to my knowledge. This F8F is perhaps the first installation on that, but as I said before the voice of Aeroproducts cannot be very strong right now, but it is hoped in the future that the voice will grow stronger. Thank you."

Mr. F.C. MACTERNAN (Hamilton Standard): "I would like to comment on that mockup cockpit you have on the lower deck. I notice that the prop feathering switches are on the extreme right of the cockpit and fairly well buried. In fact, they are somewhat blocked by the cowl crank handle. This, to my way of thinking, is a pretty bad arrangement because those switches are used in an emergency and if a man is flying with his stick on the right hand and he wants to feather either one of his props, he has got to switch hands or else lean way out of position to get at those switches. In addition, the switches are located side by side and they are not very well guarded. There have been some embarrassing situations where a fellow flying a twin-engine aircraft has found himself in a spot by feathering the wrong prop. I think an ideal arrangement for a twin-engine fighter type aircraft for feathering switch location is on the F7F. They are right in front of the pilot and he can reach them with either hand and very easily without much movement of his body. In addition to the cockpit, I noticed speed conscious control is arranged lever-wise to be on an even height with the manifold control. In grabbing for the right engine prop constant speed control, I found myself grabbing the left engine manifold control. I would suggest that this prop switch should be raised a little higher and the manifold control switch put a little lower.

"The Navy I believe and most of the contractors are familiar with the hydromatic propeller. There are quite a few of them on the field and I have not heard any disparaging comments on their actions. So I do not think it would be worthwhile to discuss that particular propeller. In regard to newer developments, we have got a new propeller coming out that I think is quite amazing. Those people who have seen it are also of the same opinion. This prop is what they call a superhydromatic and it has been under development for quite a number of years and is now reaching its production stages. The first production propeller has been delivered to the Army. It is an hydraulic propeller and it has the advantage of constant speed, feathering, reverse, lock pitch with electric mechanical heads and synchronizer. It is a self contained hydraulic propeller with hollow steel blades and it has got a rate of pitch change of about 45 degrees per second. This rate of change although it may not be for constant speed operations you can use to some great advantage in reversing and feathering. In fact the constant speed angle will feather in roughly 2 seconds, which is pretty fast. The hollow steel blade has come into its own in that we developed a new type of construction which is very light weight. We have got to a point where we can build a blade that is aerodynamically very good in that thickness in the air foil sections around the blade have been held fairly low and we are still getting some very good strength and vibration characteristics out of the blade. This incidentally will be a step forward in compressibility which was discussed in the earlier meeting, that story being an old one to the propeller manufacturers. And I might add that we don't know any more about it than the airplane people. We are proceeding with the development of new blades and also of dual rotation. I might add that the superhydromatic propeller will be available in both single and dual rotation units, a single rotation unit can be used on a dual. In other words, dual propellers can be chopped in half, and you can have two single rotating props. It is merely a matter of rehooking the control system. Another advantage in the super is that the whole prop and control hangs forward of the engine thrust plate which makes it very nice for a maintenance problem and for accessibility to adjustment for anything you want to do upon the nose. I think that is about all."

Commander RAMSEY: "On your comments relative to the mockup, I do agree with you. Your suggestions as to the prop feathering switches is not new and I made that comment on my card, and I think I am right in saying that most people in the cockpit would suggest changes on that."

Mr. NEUDECK: "This statement by the Hamilton Standard representative on development has made me very much aware that we are also carrying forward on experimental work. We never have had a production installation on a twin-engine airplane and which would require feathering and in some cases negative pitch. But in the case of feathering the propeller, all the advantage of unit construction is still maintained. The only connection to the aircraft is of course on the engine shaft and control in the cockpit. Constant speed control and the feathering control are the same lever. All you must do to feather is kick open the gate and throw the connection on through. There is also some work going ahead on negative pitch. I believe the same thing will hold true on the unit installation. We are using a larger pump and getting the rate of pitch change through that, rather than having any auxiliary drive motor fastened to the adapter plate of the engine. There is an outstanding thing about the propeller pilot combination. I don't believe really a pilot notices performance of his propeller very much unless it is acting up and you can readily see that it can be true. It is really a small part of installation."

Commander RAMSEY: "I think you are right about that. I think any of the pilots whose power plant prop combination are giving no trouble, are perfectly oblivious to the fact that there is a prop out there. You people haven't said how scared you are when you hear jet propulsion and rocket propulsion being spoken of. Have you given any thought to whether you are going to be out of a job or whether you will take over some of that work?"

Mr. MEYER: "That is exactly what I want to ask about if they have any thought on propeller design for speeds above the speed of sound."

Mr. MACTERNAN: "As I mentioned before this superhydromatic propeller has an extremely fast rate of pitch charge and as I also mentioned we are proceeding with blade designs for very high speed airplanes and ones which will give very good propulsive efficiency at high speed. We have been lucky in our high speed propellers in that we have been able to hold very good thicknesses, particularly in the blade tip. I do not know whether you appreciate it or not but our aerodynamicists will take a blade design and will shave it down to a thickness ratio that will make the vibration boys hair curl and after that they analyze the blade for strength and frequency. Then they shove it back up again and the aerodynamicists push it back down again and eventually after that they reach a compromise and as I said before the results on some of the superhydromatic blades have been very fine thickness ratios. Aerodynamically, this is very good for high speed. In addition, certain blades are designed and twisted and have a pitch distribution for high speed, others for low speed. We conducted a very interesting study along that line back a couple of years ago when we put this lightweight hydromatic on DC-3s. The blade design that was originally on the installation was the same blade as is used on the P-51. They found on single engine climb that they lost quite a bit of performance due to the fact that the blade was twisted too much. So we ran some flight tests in connection with American Airlines and discovered this effect and untwisted the blades so to speak, and when we ended up had a prop that was approximately 100 pounds per plane lighter and gave an equal, if not better, climb performance. So in answer to Grumman's question, I will say that every consideration possible is being given to specific blade design whether they are for high-speed fighters or transports."

Commander RAMSEY: "Do you want to say anything about the development of propellers and so forth? Any further comments on props? Let's hear from some of the pilots. What kind of props do you like to ride behind? Johnny Myers, what kind do you like or do you have any preference as long as they are working right?"

Mr. MYERS: "Well, I think we all like props that hold pitch a little better than some of those we are using now. This is particularly amazing on twin-engine aircraft. We also like props that feather pretty quickly and like propellers that keep the blades in the hubs. We have had a little trouble with that recently. Something I would like to ask is whether any of the manufacturers have given much thought to synchronizing the governor controls on two or more propellers in a multiengine airplane so that the propellers will stay in synchronization. I know some early experimental work was done on this several years ago, but I have lost track of that. Would you tell us something on that?"

Mr. MACTERNAN: "We have been working quite a bit on that, and I would say for the past 4 years anyway in conjunction with General Electric, and we have recently service tested what they call a microsynchronizer on a twin-engine American Airline job. This unit holds props together within plus or minus 50 revolutions per minute. In other words, you can set your speed, using one engine as a master unit and the other, a slave, follows whatever the master does. Whatever you set your master at, the props will hold together within plus or minus 50 revolutions per minute. I flew a B-25 where we have an experimental installation, and found the thing to hold revolutions per minute fairly favorable throughout various maneuvers of the airplane. American Airlines piled up quite a few hundred hours of service test on this microsynchronizer and had fairly good results with it. However, in considering microsynchronizer as an application we have talked to both the Army and Navy about it, and we have gotten the general consensus that microsynchronization on a multiengine military job is an unnecessary luxury. What you had was an automatic electric head installation such as you have on a B-24, where you can adjust your speed control, change limiting switches, and you can probably very nearly duplicate actual synchronization. That term, incidentally, is misused quite a bit in that when you have an electric head you actually have not got synchronizers on there but electrical control unit; whereas, a microsynchronizer actually uses one engine as a master and the others as slaves, and the other three engines follow whatever the master does. But, to repeat, that project is still being studied very much."

Mr. NEUDECK: "Well, up to the present time it has been our experience in twin-engine installation with feathered propellers that the governor used in the prop is sensitive enough that with any average care you can bring the controls to within plus or minus 50, and that – in flight – is held rather closely. So in a condition like that it seems that any added electric drive or control would be unnecessary."

Commander RAMSEY: "I don't think plus-minus revolutions per minute is down to the nicety we want if we are going to have that luxury."

Mr. MACTERNAN: "We found that handful of revolutions per minute experience satisfactory in what flying we have done, but as I said, it has been on commercial stuff, and nothing as hot, for example, as the F7F. I am inclined to agree with you if you are going to stick to anything like that in a twin-engine fighter, but in a four-engine transport it has been found to be satisfactory where you are not in the habit of forcing controls."

Commander RAMSEY: "I would like to state that I do think it is a luxury, but I think it is almost a must for a noncombat aircraft. The R60 which Lockheed is working on now for the Navy (that is, the 184,000 pound job) is another point which we should bring up at this time. They are planning on reversing two of those props, I understand, to decelerate the plane on landing. I didn't know that the prop manufacturers had progressed that far to have that in the original design. Have they progressed that far so that they could slip it on a transport?"

Mr. MACTERNAN: "You mean for an actual installation in ground braking? Yes; we have made several, and we are running an experimental-engine aircraft right now. It is a very amazing thing. You can cut down your landing run quite a bit by reversing pitch, and back the damned thing right down. Down at Wright Field a couple of the boys were pretty much surprised at the way they could maneuver this airplane on the ground. Incidentally, however, we have seen some results of reversing of multiengine ships. I think Wright Field did some work on it, and they cut down their landing runs quite a bit. I would like to hear what Navy pilots would think of reversing coming in on a carrier deck, that is, assuming that they have a propeller which would have extremely fast rate of pitch change and can go through reverse pitch so fast you won't stall your engine. I think the super will do that. It has an amazing rate of pitch change on it."

Commander RAMSEY: "I would like to use two words which were in quotes yesterday – "scares me." There is enough to do landing on a carrier without worrying about a prop which you might reverse at the wrong time."

Commander CAMBELL: "It may be of interest to state now that some trials have been carried out in United Kingdom with a Seafire fitted with a reverse roto prop which has been tried at a height of 10 to 15 feet over the ground on approaching. The results so far are fairly satisfactory and they will be issued to the services for proper trial."

Commander RAMSEY: "I will bring up another question on propellers. How much experimentation has the Army done on the use of props for dive brakes on dive bombing?"

Lieutenant Colonel TERHUNE: "We had a P-43 with a reversible prop on it and a 47 and they ran a few tests. Some of the results proved that it was a little bit destabilizing. And also it ran up something like 4,400 revolutions per minute going through the zero pitch, and it isn't good. That is about all I know about it. Sorry I can't give you anything further. Incidentally, I have flown this superhydraulic that the Hamilton Standard man is talking about and the only objection I have is that when it goes through zero it nearly shakes the engine out of the plane. If that were fixed it is probably O.K."

Commander RAMSEY: "Any rebuttal, Mr. Macternan?"

Mr. MACTERNAN: "I wasn't at Wright Field to witness that, in fact, I haven't heard that report before. From what I have seen of slapping it into reverse pitch, it does go through all right but a lot of things are happening that we would still find out about. Pratt Whitney brought up a point the other day. You asked me what happened to the manifold pressure gauge when we slipped through in reverse. That is something I never took particular note of. That's an interesting point, and I am going to look into it when I get back home, and as I said before, there aren't any production installations with reverse on them as yet and certainly no production installations for dive braking. That particular problem is one that we have to approach through an initial start, and delve into blade design. Generally the present blades, the way they are stressed and designed, are for positive thrusts. When you get into using a propeller as a dive brake and impose high negative loads on it we find that the stresses can go up pretty high and we are working along that line on dive brake blades. There is still a big question open though as to what happens stability-wise on the airplane with a dive brake prop. In talking with various manufacturers, they don't exactly know what happens. Perhaps they would elaborate on it if they cared to at this time, but the point is that we are proceeding with blade designs which will be able to be used as dive-brake installations."

COCKPIT
P-38L

COCKPIT
P-38L

COCKPIT
P-38L

COCKPIT
P-47D

COCKPIT
P-47D

COCKPIT
P-47D

COCKPIT
P-47M

COCKPIT
P-47M

COCKPIT
P-47M

COCKPIT
P-51D

COCKPIT
P-51D

COCKPIT
P-51D

COCKPIT
YP-59A

COCKPIT
YP-59A

COCKPIT
YP-59A

COCKPIT
FM-2

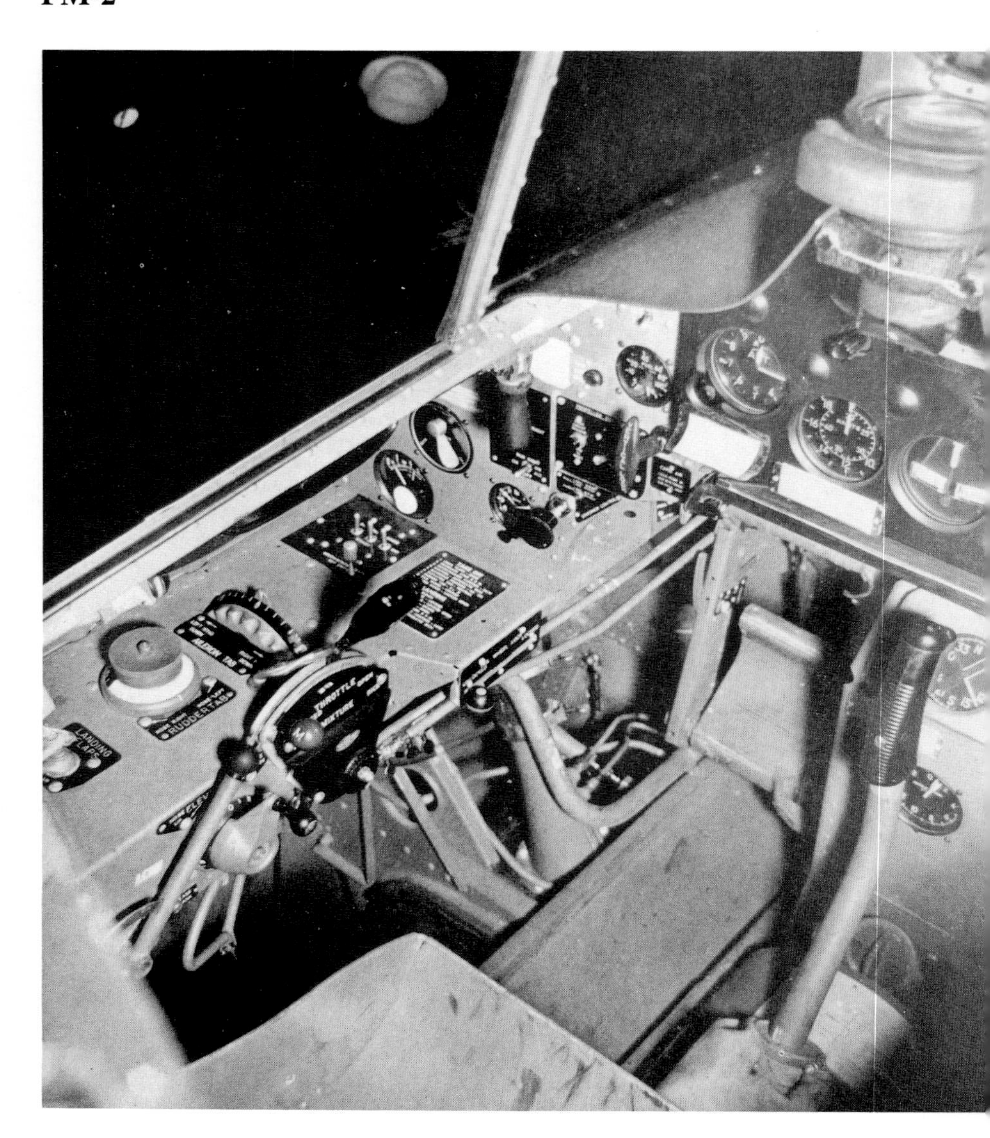

COCKPIT
FM-2

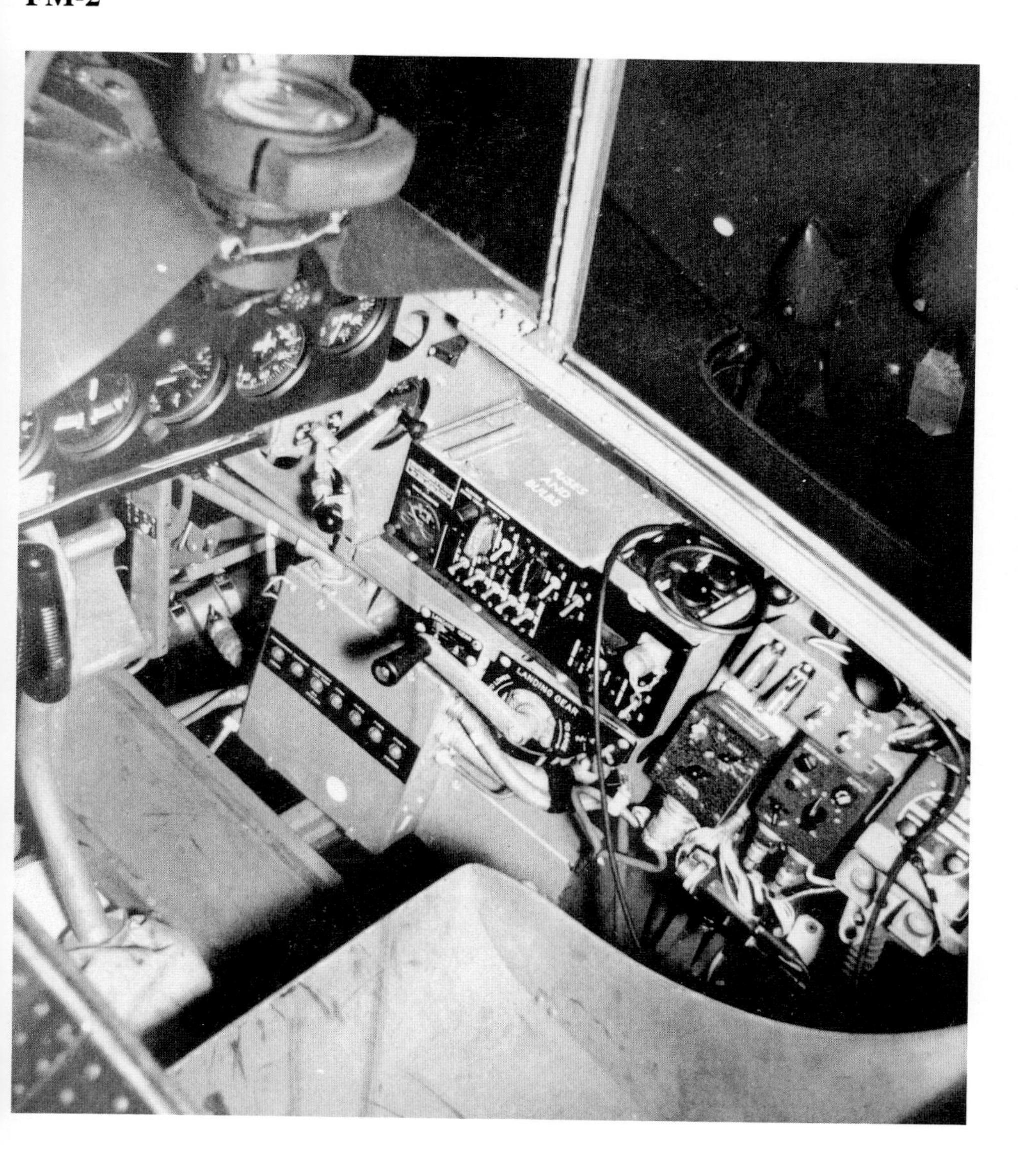

COCKPIT
FM-2

COCKPIT
FG-1

COCKPIT
FG-1

COCKPIT
FG-1

COCKPIT
XF2G-1

COCKPIT
XF2G-1

COCKPIT
XF2G-1

COCKPIT
XF4U-4

COCKPIT
XF4U-4

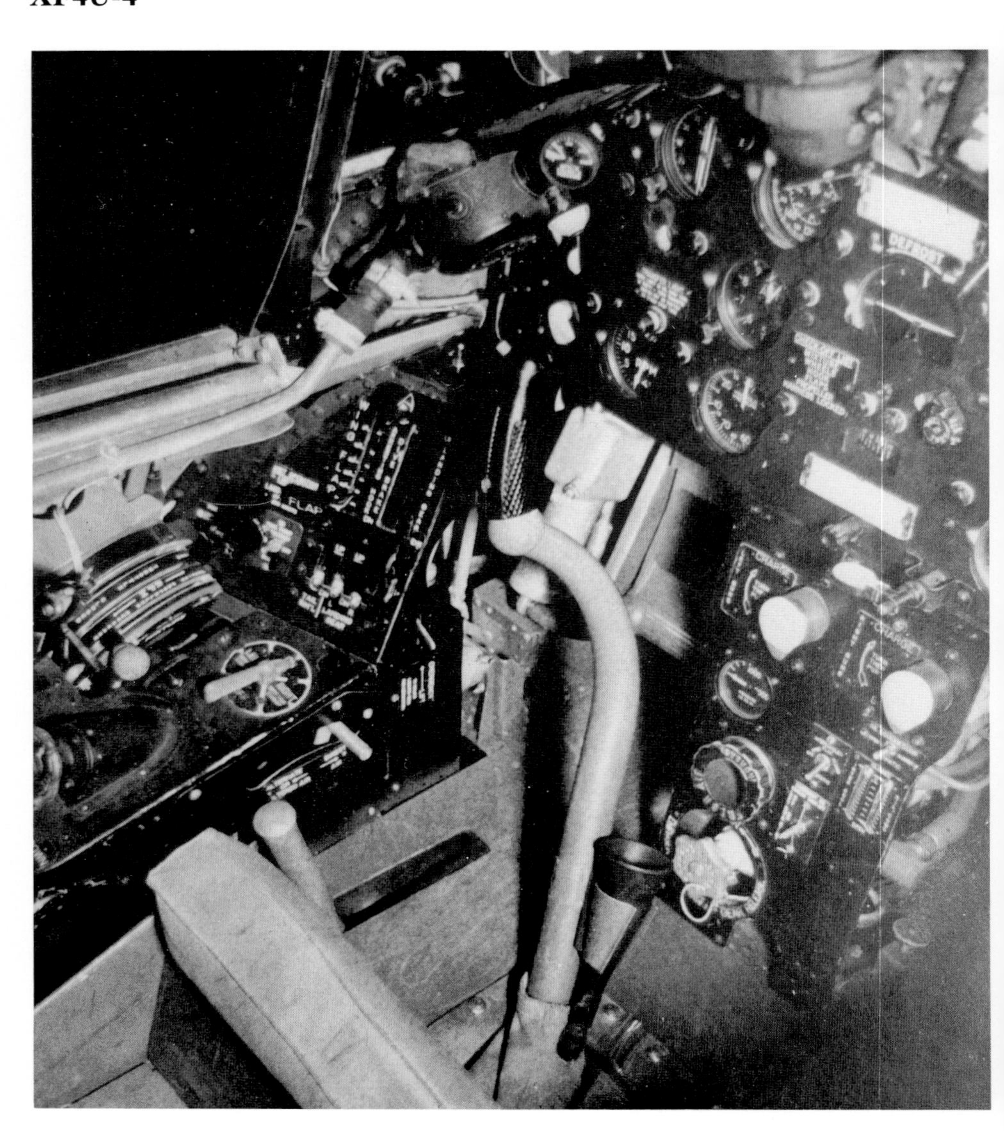

COCKPIT
XF4U-4

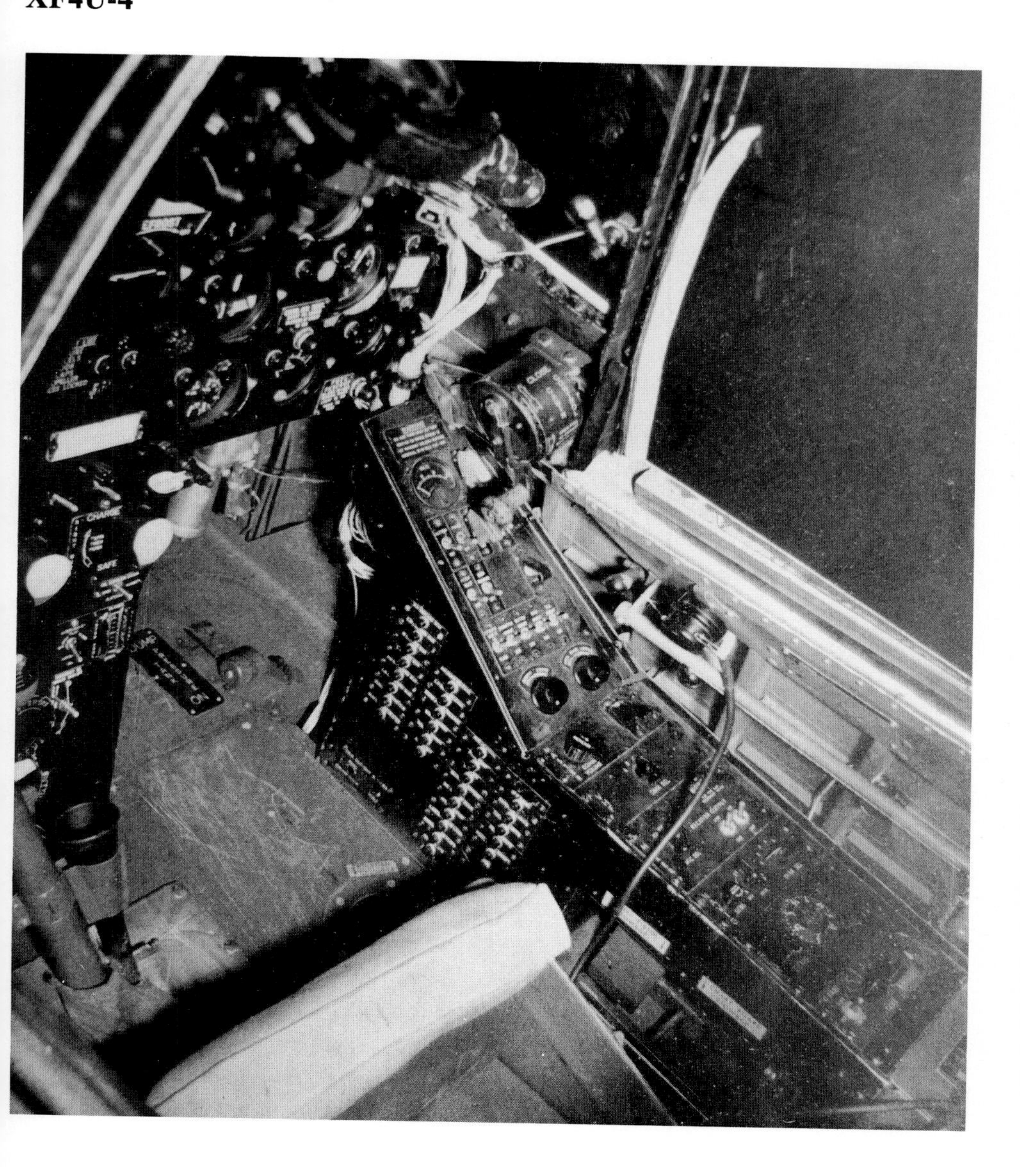

COCKPIT
F6F-5

COCKPIT
F6F-5

COCKPIT
F6F-5

COCKPIT
XF8F-1

COCKPIT
XF8F-1

COCKPIT
XF8F-1

COCKPIT
FG-1A

COCKPIT
FG-1A

COCKPIT
FG-1A

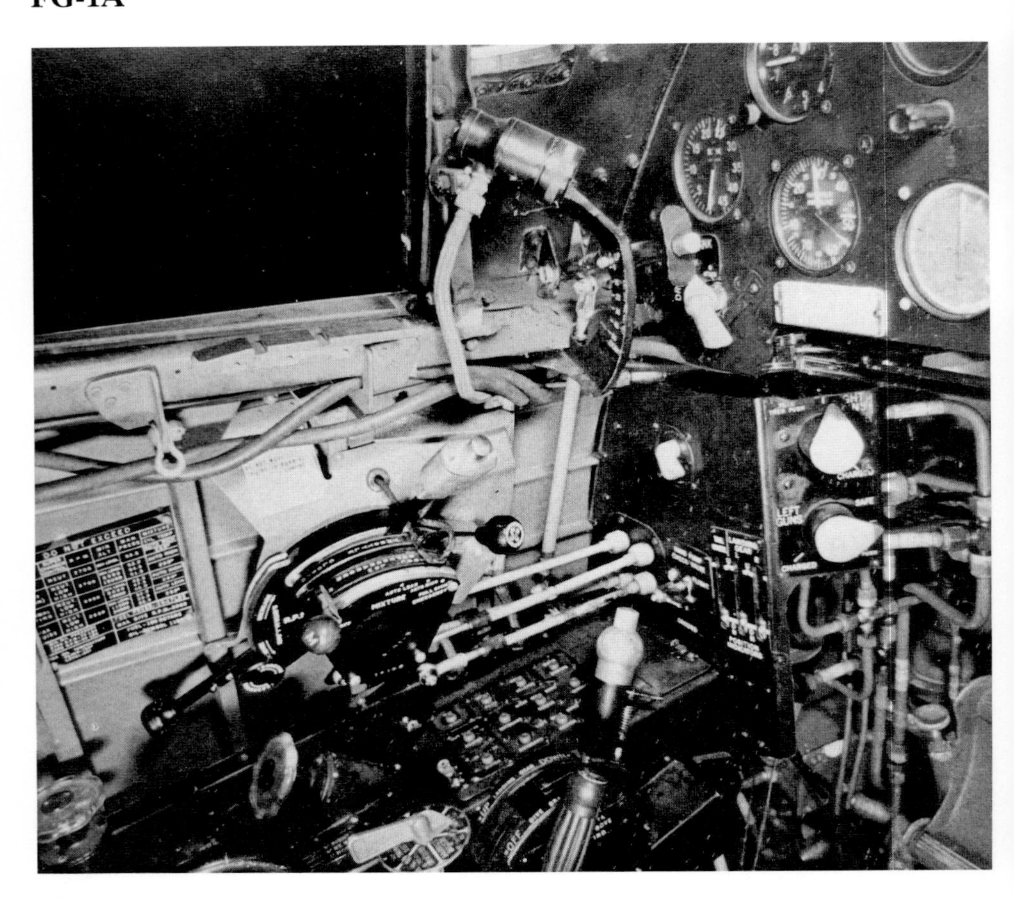

COCKPIT
F7F-1

COCKPIT
F7F-1

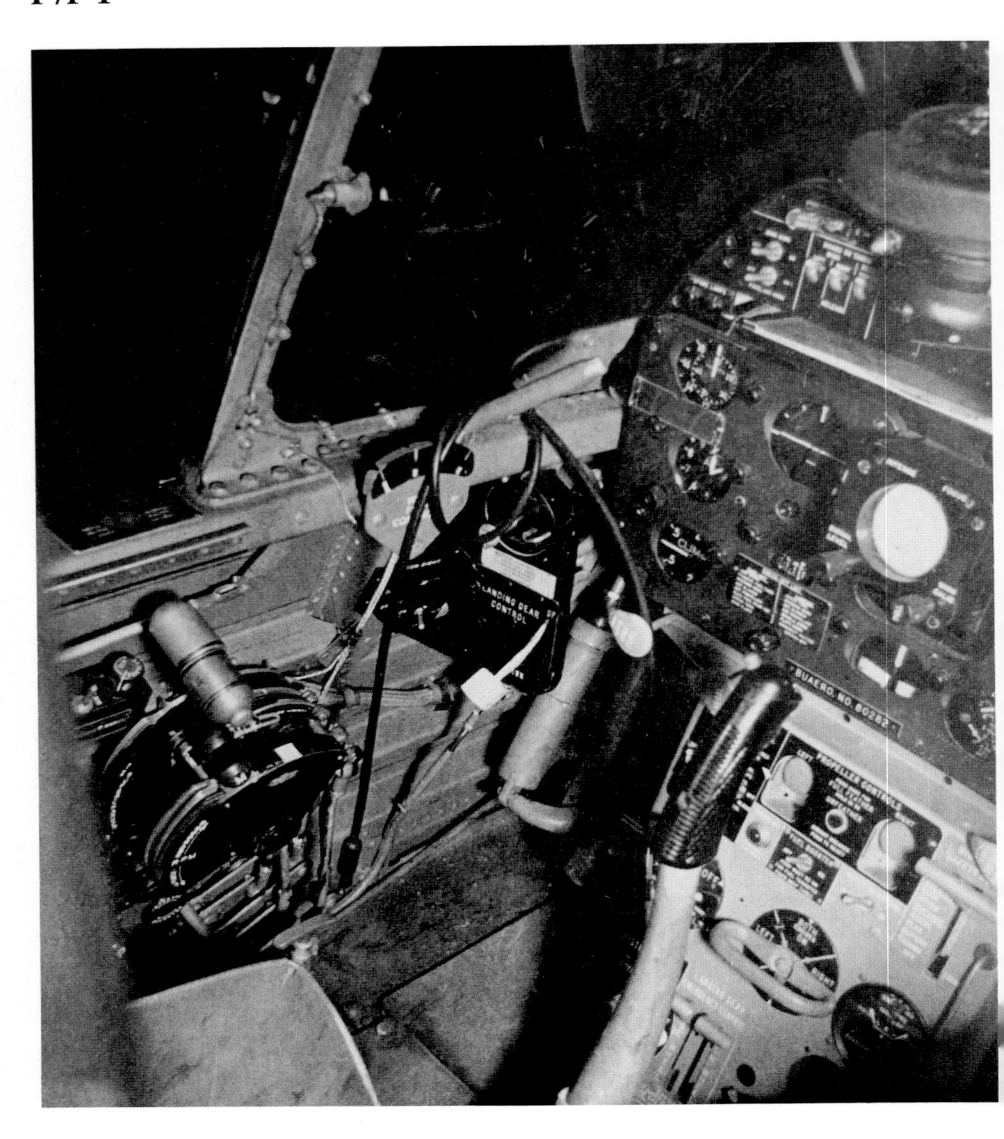

COCKPIT
F7F-1

COCKPIT
MOSQUITO

COCKPIT
MOSQUITO

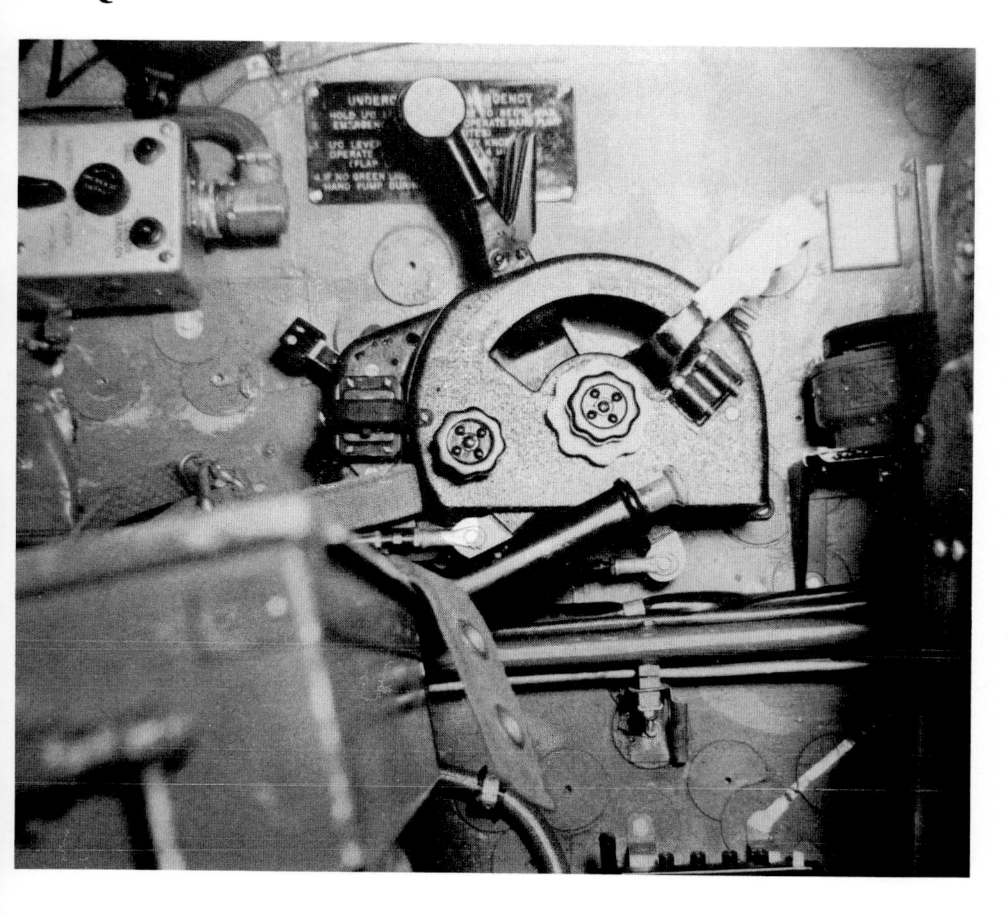

**COCKPIT
MOSQUITO**

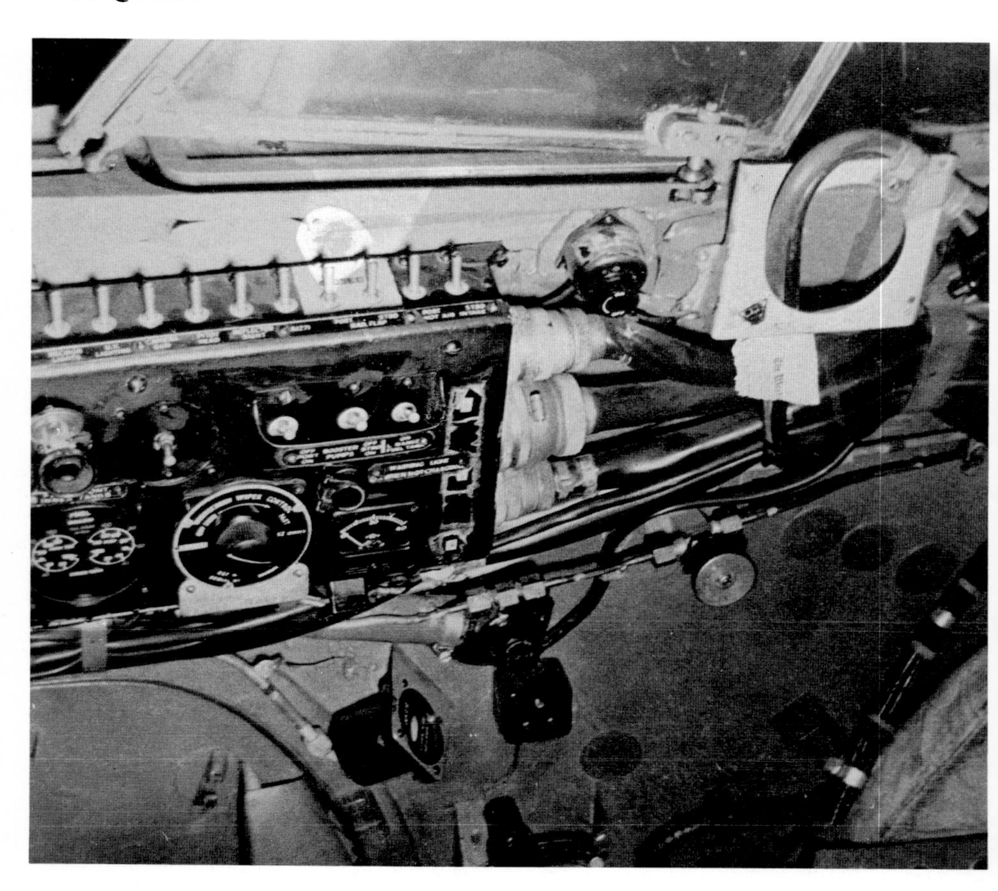

COCKPIT SEAFIRE

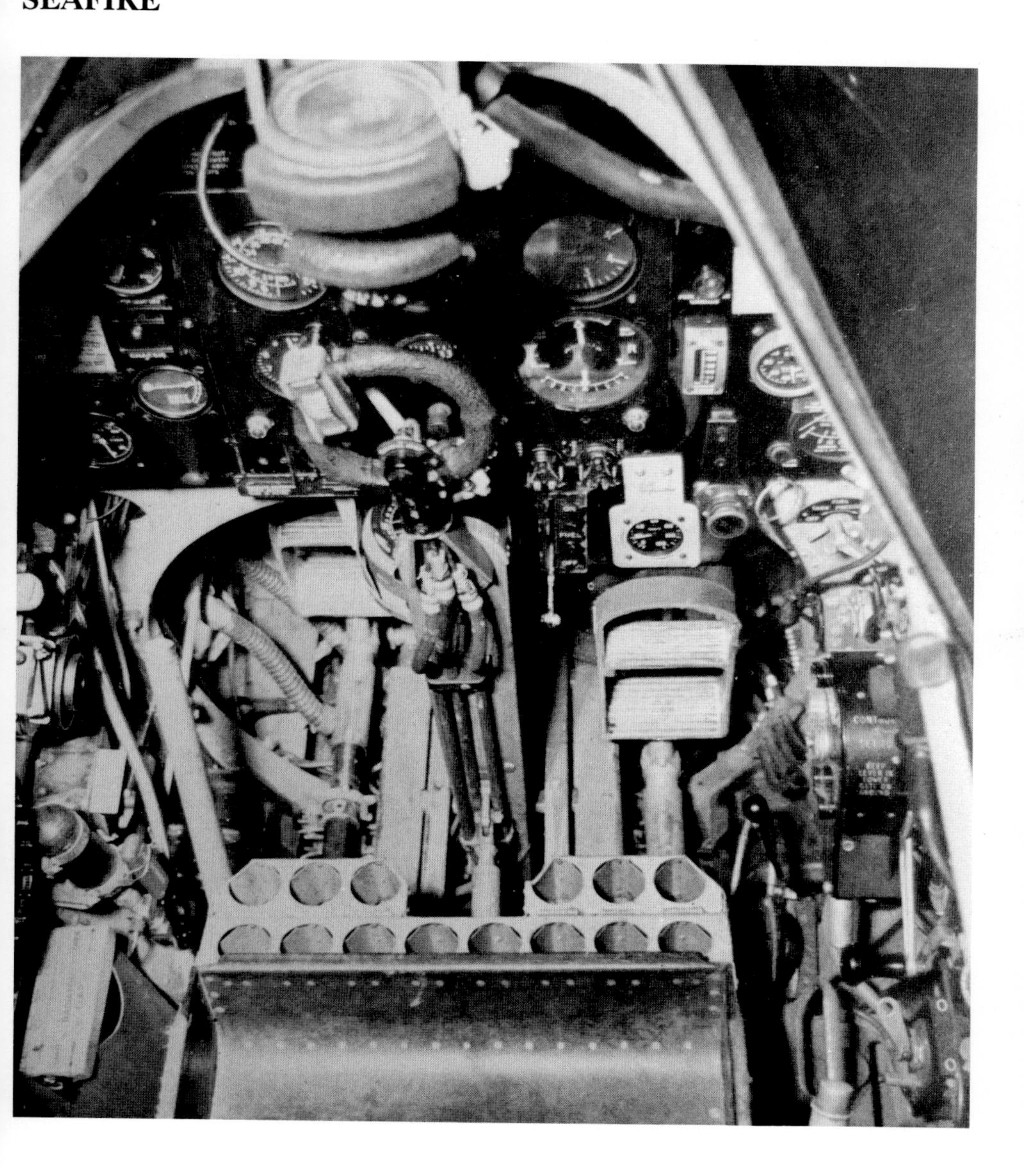

COCKPIT SEAFIRE

COCKPIT
SEAFIRE

COCKPIT
ZEKE

COCKPIT ZEKE

COCKPIT
ZEKE

Lieutenant Colonel TERHUNE: "One thing I forgot to mention is that 4,400 revolutions per minute was in reverse."

Commander RAMSEY: "Did that engine hold together? We have had a lot of engines fly apart at 3,300."

Lieutenant Colonel TERHUNE: "Yes."

Mr. BURROUGHS: "As a matter of interest, Vought some 8 years ago built an airplane, the SB2U1. The Navy wanted a dive brake because it was the first plane that required that. It picked up too much speed in dives. We suggested using the landing gear which didn't go over so well. So the company spent 2 years developing a dive brake prop, burnt several engines in the process and finally got one that worked. They told us to put the gear down so we used the gear as a dive brake instead."

Mr. MEYER: "We went into this project a little bit ourselves, without the horrible experience everyone had with it. We had a Curtiss electric installation on an F6F3 and it was dived to a terminal velocity of 310 miles per hour at 100 degrees dive angle. This was made in a dive from 17,000 feet to 6,000 feet. The reverse pitching of the prop was very simple, and there was never any speed of over 2,700 revolutions per minute encountered in reversing. It had slow speed and the prop was normalized, not according to the instructions which gave overspeed every time, but in a way which we found out ourselves which didn't overspeed above 2,100 revolutions per minute. But the whole idea was excellent except that it didn't work out in practice as it had in theory. We got a reverse stability where one had to stand almost on the left rudder to keep it diving straight, although rolls could be made with the ailerons. The airplane shook fore and aft very rapidly because the plane was loading up air in front and releasing it all at once, which gave one a bit of buffet. The whole idea was more or less junked because the stability was so poor and also accuracy for dive bombing was not very good. The propeller worked very successfully."

Commander RAMSEY: "If there are no further comments on this, we will go to the next item."

Lieutenant Commander MILNER: "I would like to hear something said on counter-rotating propellers. For example, I've heard they are the cause of poor directional control. This would naturally make a fighter equipped with such a prop a poor gun platform."

Mr. MACTERNAN: "I haven't very much to say on dual rotation props. Most of our experimentation work has been in connection with flying a dual to get the initial bugs out of it and we have been using a twin-engine aircraft for that. We have recently installed a prop on a single engine aircraft just to see how it will operate and what the pilots think of it. This work is in the process now and in talking with pilots at the plant, they seem to like it pretty well.

They find in take-off the thing really handles wonderfully and even on the first flight of a dual prop in a single engine airplane, the pilot was comfortable enough to go zipping around the sky a little. He said the prop handled very fine. I think the British might have more experience and might have something to say about that since they have been flying duals for some time. We have a lot to learn about that prop."

Air Commodore BUCKLE: "I mentioned this the other day when the question of prop efficiency and absorption of power in modern engines by conventional types of propellers was discussed. It is true that we are doing a lot of work on counter-rotating propellers. Hamilton Standard and Rotol and all the newer aircraft are being tried out with these props. So far the results are reasonably encouraging. We have had our troubles and we still have plenty more, but the three main reasons why we are carrying out this work and hoping for good results are fairly obvious ones. One is the excessive diameters we have to go to these days and the high powers with a consequent increase in size and weight-high tip speeds and balance problems. Another is that we want to absorb more power. We have more power to absorb and we think there is a definite limit to the maximum diameter you can go to. And the third is that we are in trouble with the small aircraft in torque reaction on take-off and of course for directional control in the air. So far the counter-rotating prop has been found extremely satisfactory from the take-off viewpoint. I would think that this would be of prime importance to carrier operations. We are in some trouble with the wooden blades of the Rotor counter-rotating prop. We are not out of the woods on that. There is some sort of reaction and vibration set up between the two sets of blades. It isn't altogether clear yet. We have had a Griflin Spitfire experimental airplane do a belly landing with a counter-rotating prop on, and it is an interesting point on that incident which may not apply to subsequent incidents that neither set of the blades interfered with the other even after they were bent. Exactly what happened there, I don't know.

"Am I limited to the counter-rotating prop, or can I ask some questions? On the discussion which has preceded this question of counter-rotating propellers, there are one or two things I would like to ask. The mock-up has been referred to and the propeller control criticized. I would like to endorse the statement about the bad location of the feathering switches and while on the subject of feathering switches to say we are in trouble with feathering switches generally, inasmuch as, as soon as you get automatic operation of the engine as far as boost is concerned, when we get engine failure at night it is sometimes quite difficult to determine (A) that you have engine failure and (B) which engine has failed. If you are flying at economic power and an engine fails, the prop goes into fine pitch automatically. That maintains the revolution, the revolution maintains the blowers, the blower maintains engine pressure, and so on. There is no instrument on the panel except the cylinder gauge. That takes much too long to register. We have therefore had to put out a requirement for power failure warning lights so that there shall be no mistake. The ultimate requirement is that the power failure warning lights be located on the feathering button. I have raised this point because I think it might be of interest to you from the night fighter aspect, and there is one other which might be considered at the same time. The F7F is the aircraft I am thinking of, I don't know what the speeds are myself, but I do know that the minimum speed at which that plane can be flown is around 135 indicated. Would it not be an advantage to connect to your prop feathering button an automatic electric trim device to apply approximately the right amount of bias to the rudder for the single engine so that when an engine fails you punch the button and remove the drag and automatically apply sufficient bias to take care of trim.

"The question I wanted to ask was whether the prop manufacturers in this country are tackling the design problems with regard to ducted fans for use with the gas turbines. As gas turbines come forward we shall want something better than the existing type, or prop."

Mr. MACTERNAN: "To answer Commodore Buckle's question, we are not working at this time on fans for gas turbines. The primary reason is that the years of development work on the superhydromatic are now beginning to pay dividends and that we are rolling into production. It has been a tremendous job and we have been short-handed engineering wise and have not been able to carry on as many projects as we would like. For that reason we have not even gotten into the problem of building fans for engine cooling. I don't know what my company's plans are on that at this time, but as it looks at this point, we are planning to proceed in that direction. That policy may change after we get the bugs out of the super and get rolling."

Mr. NEUDECKER: "Before we lose sight of Lieutenant Commander Milner's question as to the suitability of a dual rotation airplane as a gun platform, we might bring to attention here that really no airplane that has been supplied with dual rotation propeller and it was also brought out earlier last week that even in props of larger diameters and more horsepower absorbing area in single rotation you still get that same type of instability. I am sorry, I don't know whether I should say this or not, but it would have been nice if there had been a P-75 down here – they had an accident I think; and I don't know whether or not that was the reason why none was here. I think that would have been a marvelous opportunity for all the pilots to get a crack at dual rotation. I can't go into that very deeply because I am ignorant of just exactly what happened. But I think everybody, including our company, is looking forward to the performance reports on the new Boeing F8B. That will be an interesting installation."

Commander RAMSEY: "Unfortunately the F8B is going to be delivered three planes total, I think. I hope I haven't given away any Navy trade secrets."

Mr. FINLEY: "We did not put the dual rotation prop on the F8B in any attempt to greatly improve the performance of the airplane. About the only thing we were after was finer handling qualities and elimination of torque particularly for the wave off condition. We believe we can approximately equal the performance of the dual rotation airplane with single rotation propellers and probably in production would plan to use the single rotation prop. We have listened with considerable interest to the stories of planes becoming unstable with dual rotation propellers. For the reason that we haven't flown, I have refrained making any comments on the subject. We have recognized on our larger four-engine airplanes the destabilizing effect on single rotation props, particularly on the B-29. The very large prop used there and the long distance forward of the aerodynamic center of the airplane make the difference between power on and power off stability very noticeable. We think it is just a matter of propeller area projected onto either the vertical center plane of the airplane or projected on to the horizontal plane, that the area of the prop might be considered as horizontal or vertical tail area put on the wrong end of the airplane. It requires that you put some additional tail area on the other end of the plane to achieve stability. When planes such as the F4U, which is one of those modified by the addition of dual rotating propellers for experiment work, when those planes are initially marginally stable or weakly positively stable and have this additional fin area put on the wrong end of them, it is bound to result in some undesirable characteristics. We have in wind tunnel tests on the F8B simulated the effect of the dual rotation prop and are waiting to see in the next few weeks whether our calculations are correct or not. In the next Fighter Conference, we might be able to tell you more about it."

Commander RAMSEY: "Was the F14C designed for the counter-rotating prop? It was. Mr. Heller, can you tell us something about propellers?"

Mr. HELLER: "Commander, after just flying the squirt job, I am out of business and I don't think there is any reason why we should talk about dual rotation. If anybody needs a good tug boat captain or something, that is about all I can talk about. I just finished the test in the squirt job and I have about two pages of comments. First I am going to hand in my resignation.

"As regards dual rotation, we have temporarily shelved it until such time as higher powers are available to absorb that prop area. However, we are going into a series of surveys in vibration stresses on it within the next month, on the P-62. Possibly most of the boys here know about the other work we are doing and I would like to go over it briefly to give a short picture. We are stressing right now, and it has had universal acceptance, the reverse pitch feature we are using on the P-47 and also on a B-17F. Everyone who has flown it with me is impressed very favorably in its qualities and though we have a great direction to go in research, we are in the right direction. The next thing I think you would like to hear about are thermo heated propellers in solving the icing problem. Most of you are familiar with the Ice Bureau at Minneapolis and I think heated wings are becoming standard equipment on the B-32s, but the icing situation for propellers has never been completely solved and I believe, I hope, we have an answer for it. I believe that is about all I can add to the discussion."

Commodore BUCKLE: "Could I ask what results you got with your reversing propeller for braking purposes? Our results have been the equivalent retardation that we would normally get with wheel brakes."

Mr. HELLER: "Commodore, in the B-17 on a runway at Caldwell, at which the entire length of the runway is only 3,100 feet, the approach to that runway in a northwest direction is exceptionally poor with hills and large trees. I have been able to land the B-17 using the reverse pitch thrust only and no brakes in 900 feet with 1,400 gallons of gas aboard and 11 people. The gross of the airplane the way we have it rigged is about 35,000. The figures that I'd quote for the P-47 on reverse pitch dive testing. I don't have the figures available – but I remember in one dive from 20,000 to 6,000 feet the terminal velocity of the plane was about 310 miles per hour with reverse pitch. We intend to continue this reverse pitch dive testing and also incorporate four-engine reverse pitch; we are using only the two inboards on the B-17, but I believe British Overseas Airways have equipped a Liberator with four engine reversibles and Captain Percy, who brought the plane down, was very impressed flying it as grosses up to about 52,000. At a future date we expect to do some reverse pitch dive testing with an SB2C, which I believe the Commander would know about. I think that's all."

Commander RAMSEY: "If there are no further comments on propellers, I would like to c on Lieutenant Selby to speak on the question of radio interference."

Lieutenant SELBY: "The question of radio interference is something that possibly most of the men present aren't too interested in, although it is a subject that constantly gives the pilots a lot of grief and possibly causes them to gripe more than almost any other item on the aircraft. I propose to deal with this subject briefly and cover it as quickly as I can in the short time we have available here; so I will start out and give you an indication of how this interference gets into the receiver, then cover the types of interference encountered in fighters; then go on to tell you what the Navy is doing about reducing the interference; finally design practices which manufacturers should incorporate in their installations and responsibility which we are now placing on the manufacturer to keep this interference down to as low a level as is possible in the present state of the art. First of all, I don't know if you people realize that interference can come into a receiver in five different ways. Normally we think of it just appearing on the antenna only, but that isn't the only way it can come in. It can come in by the internal antenna lead in, besides the external antenna. It also can come in on the power and control wire, the receiver case, or even through output telephone leads. An example of that is if you have an electrically heated flying suit, the telephone cords run over your suit and interference in the wiring can couple into the telephone outlet cords and appear again as interference in the output in your ear phones.

"The types of interference that we are concerned with are of two types. One is interference caused by external sources. In this discussion, I won't be too concerned about that. It is the interference that is produced by the equipment installed in the plane that I am going to discuss. However, there are other types of static caused from external sources, as you know. We have atmospheric static, precipitation static, and static that is caused by enemy jamming equipment. I think that one of the things that pilots are interested in is precipitation static and what has been done about it. Very little information has been available on this subject.

Recently a great deal of research work is taking place up in Minneapolis. They have made some very interesting discoveries and at present are able to state that they have the problem about 80 percent solved. That is, they can reduce the interference by 80 percent of what it normally would be. The way this is done is by cleaning out the airplane externally; that is, removing paint from the plane, eliminating sharp protruding surfaces that would cause the discharge to be precipitated in the atmosphere, and they also are putting a special type of antenna wire which prevents the antenna from going into what we call corona. We have installed some of these static discharges and cleaned up one airplane here at this base, that was the RY and from tests that were conducted through precipitation static conditions flying to the West coast, we were able to indicate that our results were fairly promising. Flying to the West coast in the precipitation conditions, they were so bad that there was what we call St. Elmo's fire dancing up and down the wings. I don't know if any of you boys have ever seen that condition exist, but it will scare the life out of you. The airplane is charged to a very high degree. In fact the charge can be in the vicinity of many hundreds of thousands of volts. The potential difference from one wing tip to the other may be many thousands of volts and that is why bonding the various parts of an airplane is so very important. On this flight the static discharges were emitting corona right out in front of the pilot. They had two on the pilot tubes out in front and several on the wing tips and trailing edge of the stabilizers and these were in corona and shining forth like a Christmas tree. They were able to maintain communication through the worst conditions going across the country. That is all I will say as to external interference.

"Now there are many types of interference produced by the equipment installed in the airplane and I will now deal with these. First of all, there is interference produced from rotating equipment, such as generators, alternators, and electric motors. This interference is conducted through the receivers usually along the power lines and can very easily be eliminated by filtering at the receiver. Of course, these remarks pertain particularly to the contractors and not so much to the pilots as they are not so interested in how the interference is eliminated so long as it is eliminated as much as possible. By filtering at the receiver we can reduce many kinds of interference that are accompanied by conduction on the wires to the receiver by one filter. If we filtered everything at the source, there might be as many as 25 to 30 different sources of interference, and you can see that it would involve an enormous amount of filtering to attack each unit in turn. So the contractors might remember that in eliminating power line interference, a filter at the receiver will do it. If the lines radiate, the radiation can be reduced considerably by by-passing the interference by a small condenser at the source and that is what we do in the case of turret interference which cannot be entirely reduced by a filter at the receiver.

"The next is ignition interference. On fighters that is probably the most intense source that we have and probably the most troublesome. Now there is no reason or excuse that any pilot should tolerate ignition interference any more, from what we have learned about it. We can cure it and I would like to tell the pilots that if they experience ignition interference they should immediately report it to their radio officer and have it cleared. We have what we call fixes for all the various types of engines that are used on Navy fighters and if these fixes are applied to the ignition systems you should have no more interference from the ignition, so bear that in mind when you fellows go back to the Fleet. There are fixes for ignition interference, and you do not have to tolerate it. The way that the interference is reduced to a satisfactory level is by arranging the shielding assembly and we have contacted the various engine manufacturers and ignition manufacturers in the past 2 years and have really been jumping on them and gotten excellent results. We have gotten very favorable cooperation from Pratt and Whitney and Wright on this type of work, although much is still to be done.

"Now there are other forms of interference besides your ignition and that is from radar modulators and transmitters. On night fighters you will experience this interference and it is very difficult to cure, in the present state of the art, as very little has been done by the manufacturers of the radar equipment to eliminate the interference. However, you have field men – air-borne coordinating field men – who are now available to most units in the Fleet and you can refer this type of interference to these men and they have the latest dope available to help you in the solution of it. That type of interference that I was referring to is interference emanating from the radar equipment and appearing in the communications equipment at the pulse recurrence frequency of the radar. It is fairly easily recognized. There is another type of radar interference caused by ignition. The ignition can throw spurious traces on the radar scope to foul you up just about the time when you are interested in making a strike at night and what is very important and rather sad is the fact that ignition interference can foul up your IFF equipment so it would be advisable to check with the man responsible for the correct operation of your radar equipment to make sure the ignition interference isn't interfering with the IFF. What happens there is that the ignition triggers off the IFF and gives you the wrong indications on the shipboard interrogating devices. So you can see that it is very important to have ignition interference to as low a level as practicable. Another thing, VHF transmitters we have found will cause triggering of the IFF gear giving you also spurious responses.

"Now, as far as the Navy is concerned, about 2 years ago we got together at Anacostia with the manufacturers and thrashed this interference problem out and after about 2 weeks of discussion we formed a noise spec called SR 125 and in the spec the interference limits were laid down and the method of measuring the interference was also stipulated. We found out in that that it was far better to use the receivers installed in the plane as the noise measuring instruments rather than trust the result of some external instrument that was brought to the plane and used to measure the interference. The Army previous to this had used a certain type of instrument which they had taken from one plane to another to check the interference on the antenna. It didn't give a complete picture. Using the actual receiver that you are going to use in the plane is the best instrument you can have and the only way you can covert it into a noise meter is merely to calibrate it with what we call a signal generator. So you can determine the output of interference in terms of some value which is reproducible. In this case it is called micro-volts input. Also in this spec it was decided that if we were to hold the manufacturers to a limit, we were to tell them how to do it or at least assist them in maintaining this noise limit and so therefore in the spec we have an appendix attached which gives the suggested practices that, if followed, will produce the desired results. Unfortunately many of our installations don't follow those practices and we would like to see the manufacturers pay more attention, if they can, to these practices which we have found in the past to be very effective in reducing interference. These practices stipulate placement of units, wire record, placement of antenna, etc., and as I said before, if followed, will greatly assist in reducing interference. Manufacturers have a great responsibility in keeping their noise levels down because right now the Bureau of Aeronautics is requiring manufacturers, rather some of the manufacturers, to install radar equipment besides the normal radio communicating equipment and they have been able to reduce interference satisfactorily on communicating equipment, but when it comes to radar it is a horse of a different color. So we would strongly recommend that they follow the practice of SR 125.

"There is a responsibility on the engine and ignition manufacturers to clean up their ignition systems so that they will operate satisfactorily from the low frequencies we are now using, up to frequencies in the centimeters where radar equipment in the future will be operating. There is a responsibility for receiver designers to clean up the design of their receivers and incorporate in the design some of the desirable features in automobile radio sets. It may seem silly for me to point out that we have not taken advantage of what the automotive industry has learned in interference reduction in aircraft receivers, but we have not. And the Navy would like to see receiver manufacturers take cognizance of these items. In radar units, the manufacturers of radar equipment should include case shielding and filtering of their units to eliminate the interference.

"I would like to say in conclusion that we have come a long way in cleaning up some of our troubles that we have had in the past, but there is room for considerable improvement."

Commander RAMSEY: "Anyone else wish to talk further on radio or radio interference?"

Mr. VonEss: "I should like to ask if there is anything at the moment available for incorporation on production airplanes for the elimination of precipitation static. We have had a great deal of trouble, particularly in the Aleutians, with our PV-1; and PV-2, just arrived here for test, still carries nothing in the way of a device for the elimination of precipitation static. Also, in investigating the P2V we have been unable definitely to lay our hands on any device such as a trailing wire, wire brush on the extremities of the plane, anything of that sort actually to incorporate on the original drawings."

Lieutenant SELBY: "I can't speak with authority on that subject, as I haven't had much work to do with it. But I have read many reports produced by Dr. Dunn at NRL, who was in charge of the project at Minneapolis. The reports indicate that they have developed a static discharger which can be placed on the trailing edges of the wings and on the trailing edges of tail surfaces and these enormously reduce the point at which the plane goes into discharge. In other words, when you are flying in an area containing these charged particles there is a certain potential that the plane is charged to and above which it will go into discharge. By installing these discharging devices, the potential is enormously increased before the plane will discharge and it is the discharge that creates the interference. So I would say that they have available at this time equipment in BuAer that will greatly assist you in operating in precipitating static conditions. Besides the static discharger they have also a special type of antenna wire which is the normal antenna wire with a coating of insulation to make the wire three-sixteenths of an inch thick. This greatly improves the situation, although it increases the drag on the antenna and may create a problem in icing. The other item is to clean up the plane. If it is painted, it will give you more trouble in precipitation static than if it is a clean plane without paint, and if there are any sharp edges projecting from the plane, they have to be cleaned up or covered with insulating tape to prevent the sharp edges from discharging."

Commander RAMSEY: "Now we will take up the planes which were not ready yesterday for discussion and read off our preliminary data on those. I also want to cover the Hoover horizon and the anti-G suit. That horizon instrument is under development under high priority, and the Bureau hopes to have it down to 30 percent less space requirement. I think that is an excellent instrument myself and most pilots found it fairly easy to handle."

F6F-5N

Commander RAMSEY: "This plane was not built as a night fighter, as you know. I am not going to defend any contractor, but it should be said in passing."

P-63

Commander RAMSEY: "Commodore, could you get into that plane and shut all the windows? Just as a matter of academic interest, could you get into that plane? The commodore said he could get into the 63 with a chute, but he had to leave out the chute in getting into the 39. I might make a comment here. I don't know who the contractor was who said "very good for waveoff," but I personally would hate to have a wave-off in that plane."

Seafire

Commander RAMSEY: "I'll add my comment that the elevators are very touchy in the landing condition."

FM-2

Lieutenant ANDREWS: "All through the conference we have been wanting to ask a question on carbon monoxide and we would like to know what the manufacturers are doing to prevent it entering the cockpit. This seems to be a good time for that question, since it was mentioned here. I have been on two projects to determine the concentration of monoxide which enters the cabin, not only on the FM but also the F4F. The results were quite alarming to the doctors and they might have been to me also, had I known enough about it. I think it would be interesting to know what not only Vought and Eastern have been doing, but also what the other manufacturers have done to get away from it."

MR. SCHLIEMANN: "We ran into that problem and we ran a series of tests that have taken quite some time. For the entire past summer it has had highest priority. Our greatest trouble was determining what the exact CO concentration was. We got very conflicting results. We have determined to our satisfaction and the Navy's that the CO entering the cockpit was through the bulkhead aft of the pilot. The CO comes in around the tail wheel or the arresting hook door and then built up pressure inside the cockpit and leaked through control cable holes in that bulkhead. We have now developed the sealing method which has effectively cut down the CO concentration in the cockpit to below .005 in the majority of cases. We are checking every tenth airplane and any airplane above that limit is resealed. Obviously any plane that shows below that limit on a spot check of every 10 indicates that you have to check the production run. So far for the past month or two, we have been very successful in spot checking, and have indicated a CO concentration of below .005."

Commander RAMSEY: "Any other comments on CO?"

Lieutenant SALLENGER: "I would like to ask if the sealing is permanent. I know on the TBF's and fighters in our squadron, the planes were O.K. for about 10 hours and then the sealing strips came off."

Mr. SCHLIEMANN: "The new sealing material that we are using has not been in use for a long enough time to determine that. It does appear to be a more satisfactory sealing compound. It is applied in a much more intelligent way, in that we are setting up a procedure for pressurizing the bulkhead and I believe a great deal of the difficulty was because it was not applied properly before. To apply it properly aids to keep it in the proper place. We definitely believe that if it is applied properly and hardens properly, it will not chip out but provide effective sealing for a fair length of time. We will continue our duration tests, however, to determine just how long this sealing will be effective."

FG-1

Mr. KUPELION: "Just one comment – that the comments on this airplane should be tempered with the comments on the F4U-1 since they are the identical airplane. I notice that there is quite a variation between this and the F4U-1. That probably bears out the Commander's comment that they hit left and right indiscriminately."

Commander RAMSEY: "Any other comments on the planes? We are going back to our list of topics to see if we have missed any. Does anyone want to bring up one while I shuffle my papers?"

Group Captain CLARKSON: "I don't know whether I missed it, but while I have been in the conferences I have not heard any real discussion or expressions of opinion on the question of automatic engine controls – automatic cooling flaps, cowl flaps, radiator flaps, automatic mixture, and entirely automatic blower changes. I don't know if anybody has any ideas on that."

Commander RAMSEY: "I have an idea on it. I should like to have as many things along that category as automatic as possible. However I like the automatic override features and many of the Army planes are well ahead of the Navy on that. I like the automatic cooler flaps. I'd like even to have automatic cowl flaps. You can't get those automatic flaps to think as well as the other controls, but I do think that the automatic controls should be there; supercharger, oil cooling, etc. I think that point was raised during the power-plant discussion when Captain Spangler was here and the opinion on the comment cards we have perused is that most pilots like automatic controls.

"Does anyone want to speak further on night fighters, or was that covered to the satisfaction of everyone?

"Are there any other comments anyone wishes to make on the Hoover horizon?"

Mr. BURROUGHS: "I don't want to make a comment on the Hoover, but I wondered if anyone had flown the F4U4 so that we would get some discussion on the cockpit and power plant arrangements or is that a secret?"

Commander RAMSEY: "We had kept that out of the conference purposely as a sort of state secret, but I don't think there is any reason why we can't tell them if people are interested in hearing about it.

"Tom, would you like to comment on that, or shall we read it off the cards?"

Commander TYRA: "I would rather have some of the flight test pilots who have flown it comment on it."

Lieutenant Commander OWEN: "I think the new cockpit arrangement in the F4U4 is a decided improvement. I like the position in the cockpit. I like the arrangement of the controls. The airplane feels 50 percent more comfortable to me than it did before, but I think Lieutenant Davenport can probably give us a comment on that, too."

Lieutenant DAVENPORT: "The duration of my hop this morning wasn't very long, but the few things that I did notice were the lack of real up and down adjustment in the seat. I find that many pilots have many different ideas about where they want their heads to be. Some of them want it behind the armor and some want it in front so they can see everything and the inability to adjust the seat up as high as they want it is one thing. I don't believe cushions will be enough. Another thing is the hydraulic difficulty. They are still having hydraulic difficulty with the airplane. The bubble, I understand, will be different and they should have at least a bulge canopy on it. The stick is too long for the cockpit. You have to push upward and forward on it as you are raising the tail for take-off. The gun sight isn't much of an improvement. The automatic controls for cowl intercooler and oil cooler are fine. The landing characteristic of course is much improved by the new oleos and high tail wheel."

Lieutenant Commander CALLINGHAM: "The cockpit in the F4U4 is undoubtedly a vast improvement on the F4U1, but I think there are still one or two points which in general cockpit and instrument arrangement still require a little bit more thought. I think it is axiomatic that the fighter pilot should keep his eyes outside the cockpit as much as possible while he is in flight and he should be able to drop his hand on any instrument without looking at it. Some of the levers in the F4U4 are still too close together, on the carriage and wing flaps and in a hurry without looking down or at night you can quite easily make a mistake. The same applies to instruments. I think your eye should be caught by the instrument for which you are looking. It gets easier as you become more familiar with your cockpit, of course. I think that is one point which has not been brought out before. Separation of your vital controls so you can place your hand on them without having to look for them."

Mr. WOOLAMS: "Along that same line we have had automatic shutters for both our prestone and oil coolers in the P-63 ever since it came out and I have had experience with it and think the best place to locate the switches for these shutters is immediately adjacent to the gauge that the shutters work for. When the shutters are not working properly, it is a heck of a lot easier to look at the gauge and you automatically see what position the switch was in and if you have to make an adjustment, the switch is right next to the gauge and you automatically see the switch and don't have to go looking around the cockpit for it. I believe that is the best place to put the switches – right next to the gauge which the switches work for."

Lieutenant ANDREWS: "The F4U is one plane that is rather critical to change the trim for changes of speed and therefore I have noticed the plane is rather hard to trim in a dive-bombing run or just a normal gunnery run. You have to change your hand back and forth from the rudder trim to the aileron trim. Some of these planes here have the trim controls so that you can work them both with your hand in one position without jumping back and forth. I believe in the F4U that could be done if they were just a little closer together. They are in the right position, but just a little too far apart."

Commander RAMSEY: "I think Jack Woolams brought up a good point which I will endorse – having switches next to the gauges that tell you something is wrong."

Mr. MARTIN: "I agree with Jack Woolams, but it does offer a very difficult situation for the designer. There has been a trend in a lot of new airplanes to put all the electrical wiring into one panel if at all possible so that if there is a fair amount of maintenance on that equipment you don't have to stand on your head to get under the instrument panel to find out what's wrong. If that applied to only two switches it would be fine. But the particular philosophy applies to many things in the cockpit. It will save weight and maintenance time to have units on one panel that would be very easily removable."

Commander RAMSEY: "Of course with a lot of the instruments you could not possibly have the handle next to the landing gear indicator, for instance. It would be very awkward to have the stick up in the cockpit. But I agree with you Bud that it would save weight and be valuable out in the field."

Mr. KUPELION: "We have heard comments either from the rostrum or the floor on most planes except the F2G. I appreciate that no one here except our pilots have flown it, but we believe most of you have seen it parked out there during the past 2 or 3 days. We would like to get some comments on power-plant arrangement, on the general arrangement. If you wish, we can tear out a panel later. Until then, if there are any comments, we would like to hear about it now."

Commander RAMSEY: "Anyone want to comment on the F2G? Has anybody flown that? Is Don Armstrong here? We can come back to it. Maybe Don can discuss it. Does anyone want to go into stability and control further."

Commander CAMBELL: "I would like to touch on an aspect of stability which has not been dealt with before. The very special aspect of stability for landing on carriers. I notice on the questionnaire we had to fill out two rather misleading questions to me. If you bear with me I would like to read: "To what degree should a land-based fighter be dynamically stable?" Naturally, dynamically, longitudinally. And "To what degree should a shore-based fighter exhibit dynamic stability along all three axes?" I am not quite sure what the second question is meant to differentiate between as far as the first question is concerned. I can only suppose that it was meant to mean what degree of stability should a carrier-based fighter have for the landing condition. Or else the assumption was whatever degree of stability it would have for fighting it would also have for landing. Everyone will agree, I think, that it isn't necessarily so. What I want to say on the stability for the landing condition is my personal opinion and not based on anything else. Landing on a carrier is a special condition of landing, not like landing on a runway even though the runway may be the same width as a carrier deck. Old carrier pilots will agree with me, I think, that direction in landing is everything on the deck. The line is all important whereas it matters little whether you are in the middle of the runway or to one side. The correct arrangement of the plane on the deck is the difference between success and failure. It seems to me that to do that you acquire a high degree of response in control and not necessarily a high degree of stability and I suggest that the two things are paradoxical – that you can't have a high degree of response if you have excessive stability. I bring this forward because among the Navy airplanes we have here, I think there are two very definite differences. If I can particularize a bit, I would say that all the Grumman fighters here are completely within what I believe is the Navy's requirement for a high degree of stability in all three axes in the landing condition. I submit that this requires you to use your controls more than is convenient to direct it from each of the axes to pick up the correct groove. On the other hand, these fighters which have not got the excessive stability are much easier to pick up the groove in a plane which is just weak positively stable in the landing condition, and very stable ones I find require considerable effort and a conscious effort at the same time."

Lieutenant Commander OWEN: "What I really want to do is ask Commander Cambell if he will give us the same ideas regarding lateral stability. As he expressed his desire for a lesser degree of directional stability, I would like to know his ideas of lateral stability in the same respect."

Commander CAMBELL: "My remarks were intended to apply to all three axes – lateral, directional, longitudinal."

Lieutenant Commander OWEN: "I agree with Commander Cambell in certain respects. I think the question of stability and control becomes confused and I think that the apparent effectiveness of the control is probably more important. It is my opinion that a plane that is, in the landing condition, very weak positively stable laterally, but which has very responsive aileron control (that is very much effect from the aileron control) is actually more comfortable than one that is less positively stable laterally. That is probably true of rudder control, although I hadn't thought of it quite that way. In combat we have seen planes come back in several cases completely without aileron control and in numerous cases with reduced aileron control – one aileron being shot away. You have only to do that once to appreciate inherent lateral stability. Of course you can't have everything. But assuming that all controls are operating, I wonder if we might have someone else's view on the idea of a carrier airplane with positive, although weak lateral stability, the same directional stability, but good response from each control. That of course is dependent upon the effectiveness of the ailerons and the rudder at the slow speed."

Commander BOOTH: "In listening to the remarks that have come from the pilots who have flown during this conference, I have been quite surprised at the apparent confusion over the use of the terms control, controllability, maneuverability and stability. I believe it is possible to have both stability and controllability. If you have those two things, it seems to me you should have maneuverability. When it comes to landing conditions, if you have an airplane which is directionally unstable or which is directionally weak you have one which will produce side-slip easily and side-slip is something you don't want when you land aboard. Yet you can have good stability – an airplane which desires to stay in its path, but control – a good rudder. I don't believe we have seen an airplane here that has the proper directional stability for a fighter unless it is the P-61 which comes as close to it as I have seen. If you have weak control, or negative longitudinal stability, you have, by definition, erratic speed control. That certainly is undesirable for a carrier plane. In my opinion, the way we have been working here in flight test, is toward positive static stability to a degree and then we want good control forces, and from those good control forces, good response. We don't always get it, but that's what we want."

Mr. MARTIN: "I want to agree with the Commander and offer you the proposition that fundamentally pilots don't want to fly. They want the airplane to do its own share of the flying, which means, first of all, that the plane must be stable around all three axes; the second requirement of anything moving, as your automobile, is that if it desires to change its attitude that the airplane tell the pilot about it which means a positive slope to the stick force curve and low enough friction in the controls so the signal will get to either the pilot's feet or the pilot's hands. We realize the rather great difficulty of obtaining an airplane which is both stable and also satisfactorily controllable under some conditions. Aerodynamicists tell us that the two problems are not two problems at all, but only one and yet I think that in the actual flying of an airplane from a pilot's viewpoint, the two items are quite separate.

"I would like also to point out that if our plane is stable under normal flight conditions in the carrier landing condition, the stability will be reduced appreciably what with the great increase in power and in the friction gear being down. I think the goal certainly is attainable and I think the problem rests squarely on the shoulders of the aerodynamics boys who can with an intelligent wind tunnel program come a lot closer to giving you what you want.

"I have one general comment in regard to most of the planes I have flown around here. I think everyone of them, except the Black Widow are very poor directionally. The tail was built behind the plane and that is where it should stay. The pilot should not be required to keep it there. I think a high degree of directional stability is certainly indicated in all airplanes."

Mr. MYERS: "I just wanted to toss something out for discussion that has puzzled me. Naturally I am keenly interested in the F7F. I do consider that it has a very serious problem of the relation of its directional stability to its lateral stability. I am astounded that the Navy has apparently made the same requirements for lateral stability for a twin engine plane as they do for a single engine plane for this same carrier landing condition that Commander Owen has mentioned.

"The P-61 is very weak laterally and yet because of its excellent directional stability, I have made a half dozen landings with it, and with no difficulty, with no aileron control whatsoever. We had a few trips when we started out on the spoilers and I submit that the same thing can be done with any airplane that has good directional stability. I'd like some discussion on that."

Mr. BURROUGHS: "I wanted to raise a separate question while we are on the subject of positive static stability in the carrier approach condition which Group Captain Clarkson brought up. The issue is confused, I believe, by the fact that most, in fact all contemporary airplanes have very high changes in trim with changes in power in the carrier approach condition. These changes in trim due to changes in power mask any weak static stability the plane may or may not have. Until such time when we can build planes that don't have these changes in trim with changes in power, it seems to me that the question of static stability for the carrier approach condition is unimportant."

Mr. WOOLAMS: "I wish to disagree with Bud Martin that none of the planes down here are stable enough directionally. The F4U is too stable directionally. It is so stable it does not have proper dampening. Less stability, less directional stability, on that plane would make it a better gun platform. It has so much directional stability it keeps over-shooting and jumping back and forth and takes too long to dampen out, which is especially objectionable in rough air. As regards lateral stability, I believe that you can design a good airplane which is very positively laterally stable and will not have the maneuverability characteristics suffer too much provided you have powerful ailerons to overcome the stability in high speeds. As regards longitudinal static stability, it is necessary for you to overcome the stable characteristics by use of the elevators, which means that if you have an airplane which is very positively statically stable longitudinally, that you will have a great deal of change of elevator trim with changes in speed and changes in power, which is very undesirable from a gunnery viewpoint and from the pilot's viewpoint – attention to trim tabs, etc. I believe a good fighter should have just a very slight amount of positive static longitudinal stability, just enough that the plane has the tendency to return and not so much that there is any necessity to use the elevator trim tabs."

Commander RAMSEY: "Getting back to Johnny Myers' question, do we apply the same requirements for single engine as we do for twin engine aircraft. I will turn the mike over to Commander Tyra, who can tell whether they are wavering in one direction or the other."

Commander TYRA: "Well, from a design standpoint, we would like to have the airplane able to come back aboard the carrier whether it is a single or a twin engine plane. I think that has been the thought behind flight test work to get good controllability and good stability in bringing the airplane back. It is one thing that we have not been able to change on the F7F. We tried a set of dropped wing panels making the upper surface of the wing flat in an attempt to reduce some of the lateral stability and it apparently didn't have very much effect. I think that a further program on that particular plane will include more directional stability and a bigger fin and probably a de-boosted rudder so that there won't be quite the same tendency to yaw the airplane."

Commander RAMSEY: "There seems to be a difference of opinion of the directional stability of the F4U."

Lieutenant HALABY: "I think it is rather interesting that no one has mentioned here the fact that both the Army and the Navy have on paper specifications regarding stability and control. And to me it's indicative of the whole trend that the pressure of military requirements have put on the airplane. That the P-38, for example, for the P-51, the original Mustang and in either case built for the British, the Lightning and the Mustang, from the time they were first built until the latest version, the P-38L and the P-51D, there has been no improvement in stability or control. The P-38 has resorted to an artificial device – the boosted aileron – to get light forces and high rates of roll. It seems to me that we can get good rolling velocity without that device. Moreover, the weight of the P-38 has increased 3,000 pounds and the cockpit has gotten lousier and lousier, and more and more uncomfortable and you can see fewer and fewer gauges. The P-51, by admission of the British and Army, and North American itself, has gotten to be a less nice flying airplane and yet those two airplanes are doing the workhorse work and are winning the war. I feel that they are doing it by quantity rather than by quality. It seems to me if we can go along now designing the airplanes for the post-war Security Patrol and the next war, the competition between manufacturers is going to require us to put out a good flying airplane as well as a high performance airplane. Every one of the features that have made the P-51 a worse airplane to fly have added to its performance and to maintain that, the manufacturers are going to have to concentrate on C-1815, the Army specification, or SR-119, the Navy specification, if they are going to sell airplanes after the war. Even Bell and North American have gone to instrumentation and quantitative analysis. Generally the NACA, supported by the taxpayers and certain funds, is available to manufacturers to measure these things that the Army and Navy have agreed on as being the specified qualities and right now they are trying to get together on the Army-Navy-C.A.A. specification.

"Another thing that seems to me has been brought up by these airplanes at this conference is that the manufacturers have been forced to get more performance and at the same time not required to make the airplanes nicer to fly. Nice flying is not just some theoretical concept, it is something that protects lives and saves airplanes for greater operational efficiency. Just my own personal opinion, a lieutenant junior grade in the Navy, is that we might well require more stability and control compliance, that is more compliance with these specifications we have set up and also more demonstration of the compliance with those specifications. Every time we put on a four-bladed prop and a new bubble canopy, etc., the airplane should be redemonstrated and it should be shown that it has not deteriorated or become unsafe. I think that we have the brains and the ability in this country to put out a nice flying, high performance airplane instead of one that we can build in great number and will go fast or far, but flown at great risk and a lot of energy and wasted ability."

Mr. MARTIN: "I would like to agree with Lieutenant Halaby to some great extent, and also to point out that a great many of these airplanes were designed some time ago when power plants, for one thing if you please, were not what they are today. I guess most of the airplanes out on the line today have been increased from 10 to 30 percent in horsepower with practically no change in the tail surfaces. I believe there is a very good reason for that. If the manufacturer finds out that he can get 10 percent more horsepower in the same power plant, he can make that engine change without too much difficulty, but to put a tail section in the wind tunnel and retool for a completely new tail is quite a different story, and I think the pressure of time in this production war of ours has actually prevented the increased flying characteristics with the increased power. I have no doubt at all but that a good flying, fast airplane can be designed and built, but as in the case of the ailerons on the P-38, it is obviously a short cut. I think it is a rather poor substitute for a solution of a very difficult aerodynamic problem, but in many cases we have to take short cuts today. When that will change, I can't guarantee. In general, I agree with Lieutenant Halaby."

Commander RAMSEY: "I think a lot could be said in defense of contractors and I also think that a lot of blame is on the Bureau of Aeronautics and the Army, and they in turn are pushed by flight demands or service demands. In other words, some Joe wants better gunpower, so they put in a lot more guns. The contractor hasn't got the time, as Bud remarked, to retool, etc. I think Lieutenant Halaby struck a very keen note that after the war in order to keep their doors open, contractors are going to have to give us better flying machines for safety, etc. Don Campbell brought up a good point on what stability we want for carrier approach and carrier landings and so forth, and I would like to go on record as saying that we have to have definitely positive stability around the three axes or we aren't going to have the planes always come back and be able to fly another day, or the pilots come back and do it. I think that those differences in these planes have been brought out by this conference.

"To digress a moment: In one instance I went out to a factory, I won't mention any names, to try out a plane before it was delivered back here for flight test. Remarkably good pilots were available to that company, some of the best in the field, test pilots, and it surprised them a whole lot that after my first hop, I said, "This plane is longitudinally unstable to all conditions," which it was. They said, "By George, we forgot all about that." That is one of the biggest companies in the United States and its chief test pilot told me frankly that he had forgotten to check the longitudinal stability."

Mr. MARTIN: "To add just another comment to this stability and control problem, I do not think that too many people flying this class of airplane have been concerned quite as much as the boys flying the big airplanes where the actual physical work expended in controlling the unstable airplane adds greatly to pilot fatigue and crew fatigue in general, and as you will realize, changing a large airplane is much more difficult than changing a small one, but I think the basic assumption that all airplanes must be stable and that they must all tell the pilot when they are doing something applies equally well to the fighter class, especially since the fighter class is reaching out for longer ranges. The instability of some of the larger airplanes is quite startling and damned irritating to fly them for long periods. A fighter is essentially a maneuvering airplane and it has only been recently that they have been asked to go on longer flights and I think the trend toward stability and a finer control will develop rapidly very shortly."

Commander RAMSEY: "Armament was covered fairly well by Commander Monroe and those people who spoke during discussion of that topic. Are there any further comments relative to armament or armor?"

Commander TYRA: "There was one point on armament which wasn't touched and where the Army and the Navy are quite divergent. That is the number of switches you should have to control the gun. On some of the Army models, you have only a single master switch and once that is thrown the trigger is fired for all guns, whereas on the Navy we have a master armament switch and selector switches for the various guns and finally the trigger. We would like to have some comment from the Army as to their thought on that."

Lieutenant Colonel TERHUNE: "The Army has been trying to reduce the number of switches and confusion in the cockpits. The consensus is that if you are going to shoot at someone, shoot with all the guns you have. It was felt that the addition of selector switches for each pair of guns, which I believe you have in the Navy fighters, was not needed from a practical standpoint.

"Just recently, within the last 2 or 3 weeks, we have rigged together what we consider the nearly perfect armament panel containing all of our gun switches and switches for rocket, chemical tank, bomb release, etc. We are in the process of trying to get that standardized for all airplanes, and if it is so standardized or if it isn't, we should be glad to show you what we have if you are interested."

Commander RAMSEY: "What is the British thought on that subject? Are they going to separate gun switches, or one master, or what?"

Commander BUCKLE: "I believe I can speak for the RAF and the Royal Navy, and Commander Campbell will correct me if I am wrong. We want to get rid of as many switches and controls as we can. At the present time, I should say that the situation changes from time to time, but at present we are firing everything we've got off one button. But we have the permanent safeguard against a change in policy which enables us to fire either pair of guns or together as we wish without the need to have selector switches. That is done by means of a rocking bar form of switch on the control column. If you press one end you get .50 cal. and if you press the other, you get the 20 mm. The middle fires with everything you have – the works. You can change your mind in an attack without the need for reaching out and selecting another set of armament, and it obviates the possibility of going into combat having forgotten to select your gun switch. I think it is a simpler and far more reliable system."

Commander RAMSEY: "Tom, have we given thought to such a device on our sticks?"

Commander TYRA: "The only way that sort of device can be used practically is to use a circular grip or some similar arrangement, and the Navy itself has never looked with too much favor on the circular grip. Practically, I think it would be a good idea. If you have to hold your hand on it for long periods, it is easier. It should be most useful for long-range flying. But the reason we haven't used a system like the British is because of our pistol-grip arrangement. We have tried splitting the guns – some on the trigger and some on the bomb button. But that gets to be awkward, particularly if you have any force on the control stick."

Mr. SIEGEL: "Along with the idea of simplifying armament switches, I would like to ask Commander Tyra why the Navy insists on a separate switch and a rheostat for the gunsight. Why not combine the two on one?"

Commander TYRA: "The Navy policy on armament circuits changes from week to week almost, and they also get more complicated as weeks go by. Perhaps we can start all over again soon. The thought behind having the separate switch for the sight is that if you want to make dummy runs, you can turn on the sight without turning on the guns, so there won't be any danger of shooting anybody up."

Commodore BUCKLE: "I should have mentioned also in connection with this simplification that there is a safety aspect that you have with your selector switches, and we have taken care of with this push-button idea of ours by having a flat cover which, when it is over the switch bar, prevents you pressing it, and also turns off the gun circuit. It also makes possible the gun camera circuit so that you can do camera run training with guns loaded without the slightest danger of shooting a friend. Altogether in one small item on equipment we have combined what might run into three push buttons or three or four tumbler switches."

Commander RAMSEY: "What is the Army opinion on the stick versus the wheel or spade-grip handle?"

Colonel CALDWELL: "I don't believe we know enough about it or have given enough thought to it to comment."

Mr. DONALD ARMSTRONG (Goodyear): "I personally prefer the British type of control column throughout and in addition I greatly prefer the type of gun selector or the gun trigger mechanism as pointed out by the Air Commodore. I have used both in air firing and in general flight and I like that circular grip very much."

Commander RAMSEY: "Don, will you also speak a few words on the F2G."

Mr. ARMSTRONG: "The F2G aircraft is very similar to the FG1 and the F4U with the exception of the bubble canopy and some changes in the wing section and the after turtle deck for the installation of the bubble canopy and 4360 Wasp Major engine. Our bubble canopy has exceptionally fine visibility. The night vision and the reflection of that canopy are particularly good. The amount of distortion around the edges of the canopy has been reduced so that it only covers an area of about 1/8 inch. I think this is particularly fine at the forward edges of the canopy in comparison to other canopies I have flown. The cockpit arrangement is similar to the F4U4 and it is a very simple transition naturally from either the FG or F4U aircraft. The cockpit size is very comfortable and the seat is big enough to permit the jungle pack and the large life raft which is another feature. Several people have asked me about the visibility with that induction air scoop on the forward part directly in front of the wind screen. From eye level to the forward part of the cowl places that air scoop about 1/2 inch below and out of range of vision and it does not interfere any at all. The ground handling of this aircraft is about the same as the standard FG or F4U and I believe that most of you saw its take-off characteristics, they are very short with an excellent initial climb and I would like to point out that the torque effect is not adverse in any way. And those of you that saw the take-off here and the other day might be interested to know that I had 50 inches at the time of releasing the brakes and yet no difficulty in controlling it and I haven't had any difficulty using full 3,000 horsepower on take-off's here and back at the plant. I think the initial climb spoke for itself. The aircraft gets right up in the air retaining nice control throughout. As you might have noticed, at the end of the initial climb the aircraft rolled very easily. We also have incorporated the automatic cowl flap and automatic oil cooler flaps that have been so popular on the P-51 aircraft: I think those are a nice addition. The cruising characteristics of this aircraft are exceptionally fine; with very low manifold pressure and rpm settings the indicated speeds are well up with a nice, quiet, smooth-running engine and somewhat similar, although a greater horsepower than the old Samson engine, which ran like a sewing machine. I don't know that Pratt and Whitney will like the comparison. The V Max level as can be imagined is well up due to the horsepower and of interest are the engine cooling characteristics, it is pretty well known this engine cools exceptionally well, we actually have less cooling drag on this aircraft than on the 2800 installation. The diving acceleration of this aircraft is very fine. It picks up speed very amazingly. From level flight cruise, you are up over 400 in no time at all. The maneuverability is about the same as the standard Corsair at the present time. The rudder control however, for some reason has lightened down and although we anticipated a little trouble, we haven't gotten into it as yet and I am quite pleased with the rudder. The general engine operation, I might add, has been satisfactory on that aircraft out here. We have run into a really characteristic trouble on the other engine but fortunately we are running very smoothly on this aircraft.

"The approach on this aircraft has been rather interesting to me. The idling horsepower of the 4,360 is above that of the 2,800, which reduces the stalling speed of the tail surfaces and we have been able to fly this thing around very nicely and comfortably at speeds slightly below the FG1. The closed throttle landings are also very nice. I don't have any trouble with them, and the carrier approach should prove very satisfactory. I have only made a couple of them myself due to the limited time, but I was very pleased with it. The wave-off characteristics are very excellent. With everything down, the aircraft will still pull up and get away. You can leave everything down and still get away. The bailing out characteristics of this aircraft are satisfactory, I would say. It is a little higher cockpit than the standard FG and bubbles should jettison very nicely. One interesting thing which the short-legged Navy pilot will note: we have made the ease of boarding this thing a little better by installation of a new step system, which is a step on the flap and a step on the side of the fuselage. I believe I mentioned we have the automatic cowl and oil cooler flaps. As a whole I can summarize the airplane very easily in that it is a real fighter pilot's airplane. I don't think anybody will fly it and not like it. Regarding stability and controllability questions, I would like a little time. I haven't flown it very much myself."

Commander RAMSEY: "On the fuel and fuel system discussion we did not touch on what we wanted internally and what we wanted to carry externally. We have a fighter that carries 1,000 gallons of fuel internally and externally. What do we want internally? What does the Navy require on this?"

Commander TYRA: "The Navy thought on that is to require sufficient external fuel just to match the internal fuel when working on a standard combat problem. That problem requires warm up on deck, minute take-off, 20 minutes rendezvous and climb to 15,000 feet, out as far as you can go and 20 minutes combat, and then return with an hour's reserve after you get back. Of that the droppable fuel is used for the rendezvous after take-off and for the climb and trip out and protective fuel is used for the take-off and the trip back."

Lieutenant Colonel TERHUNE: "We have a policy somewhat to what Commander Tyra mentioned and we require 45 percent external, although this is deviated from in the case of the P-47 if I am not mistaken. However, that is a particular case. We have also a way of figuring our range similar to what Commander Tyra mentioned except that we haven't the requirement to last about 1 hour. We have one-half hour because it is not as hard to find the airport as a ship."

Commander RAMSEY: "What do the British require?"

Commodore BUCKLE: "We wish to limit our external fuel to not more than 40 percent of the built in fuel. That takes care of it. We should have it not more than one-third. We don't want to go beyond 40 percent of the fuel to be carried externally."

Commander TYRA: "One development that is just beginning to come into effect in the Navy now is the use of self-sealing drop tanks. During recent phases of the war we found that we would carry non-sealing drop tanks and have them shot full of holes, and the tanks even though they caught fire could be dropped and the plane could come back. It is customary now unless the fighter pilot gets into a pretty tight spot to bring the drop tank back after carrying it through the mission. We are now going into self-sealing drop tanks with planes such as the Corsair and Hellcat where there are two points which will carry one self-sealing drop tank on one side and a non-self-sealing drop tank on the other and we use that on special missions where we particularly want long ranges."

Commander RAMSEY: "Gentlemen, regrettably the conference is coming to a finish. However, we will keep the facilities available for those of you who are remaining over for the next day or so who wish to make clean-up flights and carry on discussions informally. Therefore the same facilities as far as the headquarters office and this meeting room, the locker room and the facilities below and on the station will be, for your convenience, available to you.

"I want to express appreciation for your grand cooperation during the conference. It has been a pleasure to be chairman of the meeting and I hope we get as much out of it as we planned. I wish to bid you Godspeed, and with the hope that we will get together next year and carry on throughout the years to come. Gentlemen, it is now my pleasure to present Captain Storrs."

Captain STORRS: "I know that some of you are very anxious to get going and there isn't much to be said, except that you all know where this place is now and we expect to have you come back. Good luck to you all."

Commander RAMSEY: "Please call on any members of the committee, if there is anything that we can do for you. The meeting is over."

Appendix

For each flight of conference aircraft, a comment card was filled out by the pilot. This section summarizes the data and opinions so reported.

Comments for each airplane are compiled separately, with a paragraph devoted to the data tabulated under each heading.

The total number of cards turned in by each of the services is listed at the beginning of the summary for each aircraft. In general, the British representation in pilot comments is small; furthermore, few pilots flew airplanes of their own service. The comments are therefore about half from contractors' pilots, a fifth or so from British pilots, and the remainder from the service not sponsoring the aircraft in question.

Separate test break-downs for each service and the contractors showed no significant differences in the results; the comments have therefore not been distinguished as to the auspices of the pilot making them.

The comments were of two general types; Simple, general evaluation of the item – such as "Good," "Fair," "Poor," "Liked it very much," "NG," etc.: and detailed comments on specific features – such as "Stick too short," "Brakes spongy," etc.

The column at the left of each paragraph lists the number of comments of each type, the number of cards with specific comments only, the number of blank cards, and the total of cards turned in. For example:

Good 8	Means 8 cards said "Good" "Liked it" "O.K." etc.
Fair 3	Means 3 cards said "Fair."
Poor 3	Means 3 cards said "Poor" "NG" etc.
Other 12	Means 12 cards had detailed comment only.
Blank 2	Means 2 cards had no comment at all.
Total 28	Means 28 cards were turned in.

Detailed comments are listed under "Remarks" in the paragraphs at the right. Each comment is followed by a figure indicating the number of cards giving it.

The summary of comments on stability and control is self-explanatory, as is that for dive characteristics and stalling.

Opinion on combat qualities has in general been quoted in full.

FM–2

Army—6 British — 0
Navy —5 Contractors—12

Remarks

Cockpit		Remarks
Good	10	Simple—3. Small and compact—2. Small, but comfortable—1. Comfortable—1. Roomy—1. Good arrangement—2. Slightly scattered and mixed—1. Messy like F4U–1, only more manual labor—1. Simple instrument panel—2. Controls well placed, accessible—2. Some controls are too far away—1. Radio controls awkward to operate—1. Oxygen control not accessible. Electric control panel should be rotated 90°—1. Rear vision bad—1. Feet should be higher—1. Monoxide while taxiing—1. Much more noise and vibration than in F4F–4—1.
Fair	1	
Poor	1	
Other	11	
Blank	0	
Total	23	

Comfort		Remarks
Good	10	Too hot in cockpit—2. Vibration and noise—1. Raise rudders and tilt seat. Would be cramped after two hours—1.
Fair	1	
Poor	2	
Other	4	
Blank	6	
Total	23	

Ground Handling		Remarks
Good	8	Good visibility—4. Vision blocked by cowl flaps—1. Poor forward visibility 1.—Gear too narrow—2. Difficult in crosswind—2. Directionally unstable—1. Bad, due to excessive "Leaning"—1. Hard to taxi without excessive use of brakes—1. Better than old F4F–1. Oleos drop alternately—1.
Fair	4	
Poor	4	
Other	6	
Blank	1	
Total	23	

Power Plant Operation		Remarks
Good	13	Simple—3. Smooth—2. Slightly rough—1. Rough—1. Noisy—2. Good acceleration—1. Prop control should be on throttle quadrant—4. Prop pitch control and supercharger control should be closer to throttle—1. Prop control awkward—1. Prop overspeeds applying power—1. Pitch change with speed slow—1. Throttle creeps—1. Carbon monoxide fumes very noticeable—1. Manual cowl flaps no good—1. Cylinder temperature gauge not in front of pilot—1.
Fair	0	
Poor	0	
Other	9	
Blank	1	
Total	23	

Take-Off		Remarks
Good	11	Simple, normal, easy—3. Tendency to swing—3. Leaning tendency—2. Lots of torque, almost full aileron—1. Straight with tendency of wing to dip—1. No difficulty in slight cross wind—1. Good directional control—1. Rudder effective at start of run—1. Left rudder—1. Necessary to use full right rudder tab until half of take-off run—1.
Fair	0	
Poor	0	
Other	7	
Blank	5	
Total	23	

Approach and Landing		Remarks
Good	11	Landing uncomfortable due to "leaning"—2. Rocking and dipping of wings in landing roll—1. Oleos drop unevenly—1. Tends to swerve on ground. Good visibility—2. Easy—1. Best landing fighter flown by this pilot—1. Very good control—1. Slowly-runs true—1. Flaps very effective—1. Slight tendency for elevator force reversal—1.
Fair	0	
Poor	0	
Other	8	
Blank	4	
Total	23	

FM-2—Continued

Wave-Off

Good	10
Fair	1
Poor	0
Other	2
Blank	10
Total	23

Best ever experienced—1. No difficulty due to stall—1. Climbs away—1. Job cranking wheels up—1.

Bail-Out

Good	7
Fair	1
Poor	0
Other	1
Blank	14
Total	23

Tail too low—1.

Combat Qualities

Good	0
Fair	0
Poor	2
Other	13
Blank	8
Total	23

Needs lighter controls, better diving ability, more speed (lots), better climb, more stable landing platform—1. Very useful as defensive fighter, carrier based (airplane vibrates more than Army types). Slow, but should be able to use effectively—very good scrapper if top speed is disregarded—1. Lacks high speed, dive and performance about 15,000 ft.—1. Needs flat front windshield. Mike in bad spot. Considering size it is one of the best—it is shy speed—1. Very good for low level although it lacks speed—1. Insufficient level speed and dive speed. Vibration in engine-prop combination. All other qualities excellent—1. Noisy. Power control needs pilot attention. Fumes and CO in cockpit during high power climbs. Mirror not available when canopy open. Slow flaps very good. Not enough available thrust horsepower. O. K. at low altitude. Ailerons too heavy. Climb too slow. O. K. for small fields. Probably marginal against competition. Excessive vibration—particularly control stick. Wind shield distorted. Speed poor. Not good enough any more. Too slow and rate of roll too slow. Prefer F6F—1. If the ailerons and rudder were lightened, this would be a very good fighter for small carrier operation—1. Nice maneuvering plane—good rate of climb in comparison to level flight speed; short radius—1.

Stability

Longitudinal	*Good*	*Fair*	*Poor*	*Positive*	*Week Positive*	*Neutral*	*Unstable*
Climb	5	0	0	0	0	2	2
Level	7	0	0	0	0	2	1
L/C	3	0	0	0	0	0	0
General	5	0	1	0	0	2	0

Remarks.—Becomes slightly unstable at high angles with power—1. Marginal stability—1.

Lateral	*Good*	*Fair*	*Poor*	*Positive*	*Weak Positive*	*Neutral*	*Unstable*
Climb	7	0	0	0	0	1	0
Level	9	0	0	0	0	1	0
L/C	2	0	0	0	0	0	0
General	7	1	0	0	0	0	0

Remarks.—Slight tendency to roll—1.

Directional	*Good*	*Fair*	*Poor*	*Positive*	*Weak Positive*	*Neutral*	*Unstable*
Climb	8	0	0	0	0	0	0
Level	10	0	0	0	0	0	0
L/C	2	0	0	0	0	0	0
General	8	0	0	0	0	0	0

Remarks.—Rudder load reverses—1. Damping weaker than F4U; positive but not strong statically—1.

Controllability

Elevators	*Good*	*Fair*	*Poor*	*High*	*Moderate*	*Light*
Force	8	1	0	0	2	1
Effectiveness	11	0	0	0	0	0
General	1	0	0	0	0	0

Remarks.—Slightly heavy—1. Stiff at high speed—1. Lighter with G—1. Reasonable—1. Plane doesn't "squash" in pull out as heavier planes do—1.

FM-2—Continued

Ailerons	*Good*	*Fair*	*Poor*	*High*	*Moderate*	*Light*
Force	2	0	0	6	0	1
Effectiveness	5	3	0	0	0	0
General	1	0	0	0	0	0

Remarks.—Too heavy—7. Stiff—2. Slow—2. Rate of roll slow—3.

Rudder	*Good*	*Fair*	*Poor*	*High*	*Moderate*	*Light*
Force	4	0	0	4	1	2

Remarks.—Too heavy—2. Stiff—2. Slightly heavy—1. Heavy in dive—1. Lighter than ailerons (which are too heavy)—1. Reasonable relative to other controls but heavy at 300 knots for jinking—1.

Ailerons	*Good*	*Fair*	*Poor*	*High*	*Moderate*	*Light*
Effectiveness	7	3	0	0	0	0
General	1	0	0	0	0	0

TRIMMABILITY

		Remarks
Good	16	Elevator trim too sensitive—3. Elevator trim especially good—2. Aileron trim changes with speed—1. Aileron tab no good due to spring in system—1. Rudder requires too much trim—1. Requires trim too often—2. Requires moderate use of tabs—1.
Fair	0	
Poor	0	
Other	5	
Blank	2	
Total	23	

	A	*B*	*C*	*D*	*Other*	*Blank*
Maneuverability	8	8	4	0	2	1

Remarks.—A—turning, D—Aileron—1. Aileron forces are limiting—1. Vibrate—1. Control loads are too heavy and rate of roll low—1.

	A	*B*	*C*	*D*	*Other*	*Blank*
Night and Instrument Flying Rating	0	9	2	1	0	11

Remarks.—B—generally, C—landing night—1. Accepting it as a single-engine ship, B—1.

	A	*B*	*C*	*D*	*Other*	*Blank*
Gun Platform Rating	3	10	2	0	0	8

Remarks.—Heavy rudder and aileron in turn—1. Heavy controls—1. Excessive rudder trimming—1.

DIVING CHARACTERISTICS

	Good	*Fair*	*Poor*	*High*	*Moderate*	*Light*
Acceleration	5	1	3	0	0	0
Stick Force	4	2	0	1	0	1

	Good	*Fair*	*Poor*	*Fast*	*Moderate*	*Slow*
Recovery	7	0	0	0	0	0
General	1	0	0	0	0	0

Remarks.—Acceleration: Slow—1. Normal—1.
Stick Force: Too heavy—1. Ailerons very heavy—4.
Recovery: Elevator satisfactory—2.
General: Excessive trim changes—1.
Rudder trim excessive—1.

STALL CHARACTERISTICS

IAS (KTS):	*Clean*		*L/C*		*Accelerated*
	Power Off	*Power On*	*Power Off*	*Power On*	*3G*
Average	68	62	62	56	103
Range	60–72	55–75	58–72	50–62	103

Warning.—Good—9. Adequate—1. Fair—1. Not Much—1. Buffet—2.
Recovery.—Good—12.

F6F-5

Army—11 Navy — 4 British — 2 Contractors—10

Remarks

Cockpit

Good	17
Fair	0
Poor	0
Other	9
Blank	1
Total	27

Clean or neat—4. Good instrument lay-out—7. Radio controls inaccessible—2. Stick position bad—2. Prop control poor—2. Comfortable—4. Fuel selector unhandy—1. Fuel pump, battery-ignition might be grouped—1. Fuel pump control should be on right panel—1. Suggest neutral position on flap control—1.

Comfort

Good	15
Fair	1
Poor	1
Other	4
Blank	6
Total	27

Too much vibration—1. Rudder vibration—1. Too noisy, slightly rough—1. Seat can't be adjusted high enough—1.

Ground Handling

Good	21
Fair	2
Poor	0
Other	4
Blank	0
Total	27

Brakes good—4. Brakes spongy—2. Brakes worn—1. Brake action heavy—1. Visibility good—3. Visibility fair—1. Visibility poor—2. Cowl flaps block vision—1.

Power-Plant Operation

Good	17
Fair	0
Poor	0
Other	8
Blank	2
Total	27

Vibration—2. Rough between 1,700 and 1,950 revolutions per minute—1. Very smooth except in restricted region—1. Noisy—1. Noise not excessive—1. Do not like prop control—1. Fuel pressure fluctuates—1. Water injection switch in awkward position—1.

Take-Off

Good	21
Fair	0
Poor	0
Other	4
Blank	2
Total	27

Good rudder control—2. Slight torque—1. No torque—1. Very stable—1.

Approach and Landing

Good	20
Fair	0
Poor	0
Other	1
Blank	6
Total	27

Good control—3. Lacks directional trim—1. Rudder least effective; aileron and elevator good—1.

F6F-5—Continued

WAVE-OFF

Good	11	Rudder weak, aileron good—1. Requires strong forward pressure or change in trim tabs—1.
Fair	0	
Poor	0	
Other	2	
Blank	14	
Total	27	

BAIL-OUT

Good	10	Complicated—1. Average for Navy ships—1.
Fair	1	
Poor	0	
Other	2	
Blank	14.	
Total	27	

COMBAT QUALITIES

Good	8	Like F4U better—1. More suited for Pacific than European theater—1. Ideal for close work. Rate of climb most surprising feature—1. Excellent for carrier, useless for land—1. A good rugged work-horse. A little lacking in performance. Ideal for operation from small fields—1. Excellent up to 20,000 feet—1. Has sufficient combat qualities for fighting present Jap planes. Probably not fast enough in near future—1. Maneuverability limited by longitudinal and lateral stability—1. Performance and control forces probably slight to marginal advantage against first line enemy opposition, but operational accidents should be practically nil—1. Not up to present competition—1. With more performance this ship would be one of the best—1. Should be good for the Pacific until speed becomes inadequate—1.
Fair	0	
Poor	0	
Other	11	
Blank	8	
Total	27	

STABILITY

Longitudinal	*Good*	*Fair*	*Poor*	*Positive*	*Weak, Positive*	*Neutral*	*Unstable*
Climb	18	1	0	1	0	1	0
Level	18	0	0	2	0	0	0
L/C	10	0	0	0	0	0	0
General	1	0	0	0	0	0	0

Remarks.—Oscillations good—1. Too strong—1. Too strong—1.

Lateral	*Good*	*Fair*	*Poor*	*Positive*	*Weak, Positive*	*Neutral*	*Unstable*
Climb	18	3	0	1	0	0	0
Level	16	0	0	2	0	0	0
L/C	8	0	0	1	0	0	0
General	1	0	0	0	0	0	0

Remarks.—Too strong—1. Too strong—1. Roll due to yaw proper sign and normal slope—1.

Directional	*Good*	*Fair*	*Poor*	*Positive*	*Positive*	*Neutral*	*Unstable*
Climb	18	0	0	1	0	0	0
Level	15	0	0	1	0	0	0
L/C	9	0	0	1	0	0	0
General	1	0	0	0	0	0	0

Remarks.—None—1. Dead Beat—1. Strong—1. Dead Beat—1. Yaw due to yaw weaker than F4U but seems O. K

CONTROLLABILITY

Elevators	*Good*	*Fair*	*Poor*	*High*	*Moderate*	*Light*
Force	12	0	0	1	2	4
Effectiveness	17	0	0	0	1	0
General	2	0	0	0	0	0

Remarks.—Heavy at high speed—1. Sensitive—1. None.

F6F-5—Continued

Ailerons	Good	Fair	Poor	High	Moderate	Light
Force	9	1	0	2	3	6
Effectiveness	11	4	1	0	1	0
General	2	0	0	0	0	0

Remarks.—Heavy at high speed—1. Average—1. Very responsive to controlling action—1. None.

Rudder	Good	Fair	Poor	High	Moderate	Light
Force	4	1	2	10	3	0
Effectiveness	18	0	0	0	0	0
General	2	0	0	0	0	0

Remarks.—Heavy at high speed—1. Sensitive—1. None.

Trimmability

Good	14
Fair	2
Poor	0
Other	8
Blank	3
Total	27

Remarks

Too much trim change for change in speed—4. Trim tabs not sensitive enough—5.

	A	B	C	D	Other	Blank
Maneuverability	12	12	2	0	0	1

Remarks.—Rate of roll should be faster—1.

	A	B	C	D	Other	Blank
Night and Instrument Flying Rating	9	5	2	0	0	11

Remarks.—None.

	A	B	C	D	Other	Blank
Gun Platform Rating	9	8	2	0	0	8

Remarks.—None.

Diving Characteristics

	Good	Fair	Poor	High	Moderate	Light
Acceleration	8	2	1	0	0	0
Stick Force	10	0	0	2	0	0
	Good	Fair	Poor	Fast	Moderate	Slow
Recovery	13	0	0	0	0	0
General	1	0	0	0	0	0

Remarks.—Acceleration: Normal—1. Fast—2. Slow—2. Slower then F4U—1.
Stick Force: Normal—1. Could be lighter—1. Ailerons tighten somewhat—1. Ailerons good, elevator slightly heavy—1.
Recovery: None.
General: None.

Stall Characteristics

	Clean		L/C		Accelerated
IAS (Kts):	Power Off	Power On	Power Off	Power On	3g
Average	77	69	66	65	121
Range	65-81	60-79	55-75	60-70	105-150
Warning:					
Good	12	10	10	9	9
Fair	0	0	0	0	0
Poor	0	0	0	0	0
Type	Slight Buffet—3.	Slight Buffet—2.	Slight Buffet—4.	Slight Buffet—2.	Strong Tail Buffet—2.

F6F-5—Continued

Recovery:					
Good	15	10	11	8	7
Fair	0	0	0	0	0
Poor	0	0	0	0	0
Remarks	Drops nose slightly—1.	Drops nose slightly—1.	Drops nose slightly and right aileron—1.	None.	Relax back pressure—1.

F6F-5N

Army—1 British —0
Navy —1 Contractors—2

		Remarks
Cockpit		
Good	3	Want engine instruments on panel—1. Throttle and prop control should be together—1. O_2 gauge and regulator should be in sight—1.
Fair	0	
Poor	0	
Other	1	
Blank	0	
Total	4	
Comfort		
Good	3	
Fair	0	
Poor	0	
Other	0	
Blank	1	
Total	4	
Ground Handling		
Good	4	
Fair	0	
Poor	0	
Other	0	
Blank	0	
Total	4	
Power Plant Operation		
Good	3	Throttle quadrant too low—1. Hard to start. Throws some oil—1. Rough, 1,900-2,200—1.
Fair	0	
Poor	0	
Other	1	
Blank	0	
Total	4	
Take-Off		
Good	2	Short—1. Appeared to yaw to left—1. Torque high at high power—1.
Fair	0	
Poor	0	
Other	2	
Blank	0	
Total	4	

F6F-5N—Continued

APPROACH AND LANDING

Good	4
Fair	0
Poor	0
Other	0
Blank	0
Total	4

WAVE-OFF

Good	3
Fair	0
Poor	0
Other	0
Blank	1
Total	4

BAIL-OUT

Good	2
Fair	0
Poor	0
Other	0
Blank	2
Total	4

COMBAT QUALITIES

Good	3
Fair	0
Poor	0
Other	1
Blank	0
Total	4

Airplane definitely underpowered and performance will not permit a great enough advantage over Japanese aircraft. Visibility poor to rear. Airplane critical altitude too low for work in later Pacific operations—1. Should be a fairly good night fighter. However, due to lack of trimmability, pilot will have to fight the controls all the time—1. An excellent carrier fighter. Lack of speed and climb is only serious disadvantage. Needs a bubble canopy for better rearward visibility. Excessive rudder trim change and hard-to-get-at rudder trim tab control are bad features for gunnery—1.

STABILITY

Longitudinal	*Good*	*Fair*	*Poor*	*Positive*	*Weak Positive*	*Neutral*	*Unstable*
Climb	1	1	0	0	0	0	0
Level	2	0	0	0	0	0	0
L/C	0	0	0	0	0	0	0
General	1	0	0	0	0	0	0

Lateral	*Good*	*Fair*	*Poor*	*Positive*	*Weak Positive*	*Neutral*	*Unstable*
Climb	1	0	0	0	0	0	0
Level	1	0	0	0	0	0	0
L/C	0	0	0	0	0	0	0
General	2	0	0	0	0	0	0

Remarks.—Marginal—1.

Directional	*Good*	*Fair*	*Poor*	*Positive*	*Weak Positive*	*Neutral*	*Unstable*
Climb	1	0	0	0	0	0	0
Level	1	0	0	0	0	0	0
L/C	0	0	0	0	0	0	0
General	2	0	0	0	0	0	0

F6F-5N—Continued

Controllability

Elevators	Good	Fair	Poor	High	Moderate	Light
Force	0	1	0	1	0	0
Effectiveness	1	0	0	0	.0	0
General	2	0	0	0	0	0

Remarks.—Heavy at high speeds—1.

Ailerons	Good	Fair	Poor	High	Moderate	Light
Force	0	1	0	1	0	0
Effectiveness	0	1	0	0	0	0
General	2	0	0	0	0	0

Remarks.—Poor at high speeds—2. Slow response—1.

Rudder	Good	Fair	Poor	High	Moderate	Light
Force	1	1	0	1	0	0
Effectiveness	1	0	0	0	0	0
General	1	0	0	0	0	0

Trimmability

		Remarks
Good	2	
Fair	0	Requires excessive trim—3.
Poor	1	
Other	1	
Blank	0	
Total	4	

	A	B	C	D	Other	Blank
Maneuverability	2	2	0	0	0	0
Night and Instrument Flying Rating	0	2	0	0	0	2
Gun Platform Rating	1	2	0	0	4	1

Diving Characteristics

	Good	Fair	Poor	High	Moderate	Light
Acceleration	0	0	0	0	0	0
Stick Force	0	1	0	0	0	0

	Good	Fair	Poor	Fast	Moderate	Slow
Recovery	1	0	0	0	0	0
General	0	0	0	0	0	0

Remarks.—Noisy and vibrates. Rudder trembles—1.

Stall Characteristics

IAS (kts):	Clean		L/C		Accelerated
	Power Off	Power On	Power Off	Power On	3g
Average	60	50	50	20	

Remarks.—Excellent stall characteristics both clean, and wheels and flaps down. Cannot get a definite break when stalls occur—1. Pitot static error is very bad at low speeds. Indicated 25 kts when stalled—1.

F7F-1

Army—12 British — 4
Navy — 3 Contractors—20

Remarks

Cockpit

Good	18
Fair	0
Poor	0
Other	21
Blank	0
Total	39

Well laid out—5. Comfortable—1. Cramped, too narrow—6. Good forward vision—2. Poor rear vision—3. Should have bubble—2. Cabin hard to open and close—4. Throttles awkward—1. Throttle creeps—1. Don't like prop pitch control—1. Oil cooler shutters should be auto—1. Artificial horizon knob difficult to pull out—1. Engine speed, boost instruments very good—1. Radio gear is too far back—5. Hit radio controls with right elbow—2. Hard to reach side controls aft—2. Case aft of throttle hit left elbow—1. Trim tabs not accessible—2. Rudder boost not sufficiently accessible in emergency—1. Throttle quadrant too low—1. Gyro compass too low—1. Would like oxygen regulator higher and more visible—1. Pitch control should be workable with same hand as throttle—1. Too many instruments (apart from blind flying instruments) on dash board—1. Relocate switches above instrument panel to avoid distraction of attention—1.

Comfort

Good	20
Fair	1
Poor	1
Other	8
Blank	9
Total	29

Not suitable for long mission—2. Too small for long mission—1. Cramped—2. Plenty of room—1. Cockpit too cold without heater, too hot with heater; suggest finer adjustment—1. Cold cockpit—1. Noise level fairly high—1. Visibility to rear closed for 20°—1. Seat not too comfortable—1. Most switches can be reached with straps locked—1.

Ground Handling

Good	37
Fair	0
Poor	0
Other	2
Blank	0
Total	39

Good brakes—4. Brakes tend to grab a little—1. Brakes a little too sensitive—1. Visibility good—3. Easy directional control with rudder or engines—1.

Power Plant Operation

Good	29
Fair	0
Poor	0
Other	9
Blank	1
Total	39

Smooth—3. Runs cool—2. Cylinders overcool in autorich or autolean—1. Prop controls should be placed near throttles—2. Prop controls awkward—2. Throttle and prop change quadrants could be improved—1. Throttle grips too big—1. Ratchet on mixture controls makes operations awkward—1. Nice power plant controls—1. Ran quite rich in autorich—2. Required lean settings for very low altitudes—1. Attention is required in synchronizing manifold pressures when changing quickly—1. RPM synchronizer is fine—1. Supercharger operation should be separated—1. Auto boost regulator inadequate—1. Oil shutters should be automatic—1. Automatic subsidiary controls would help—1. Eliminate mixture control and employ separate cut-out control near mag. switches—1.

Take-Off

Good	30
Fair	0
Poor	0
Other	8
Blank	1
Total	39

Acceleration very good—3. Good climb—2. Excellent vision—1. Rudder too light—2. More directional control needed—1. If done quickly, right swing noticeable—1. No swing—1. Handles well—3 Stable—1. Controls felt sloppy at first—1. Very small trim changes during initial climb—2.

Approach and Landing

Good	20
Fair	0
Poor	1
Other	10
Blank	8
Total	39

Normal—2. Good visibility—2. Comfortable—1. Control in roll good—1. Rather hot for short field night work—1. Elevators insufficient—4. Nose heavy in landing—4. Lack of elevator control makes high rate of descent—1. Full flaps blanket elevators; 30° flaps result in normal characteristics—1. Excessive stability makes large back stick force necessary—1. High aileron forces at low speed—1 Ailerons weak near stall—1. Flaps reduce effectiveness of ailerons—1. Plenty of trim required—1 Heavy control forces at stall—1. Cannot hold at carrier speed on one engine—1.

F7F-1—Continued

Wave-Off

Good	11
Fair	0
Poor	0
Other	6
Blank	22
Total	39

Bad for single engine—2. Don't know about single engine—1. Control satisfactory—1. Doubt elevator effectiveness—1. Requires time getting used to changes in trim—1.

Bail-Out

Good	9
Fair	1
Poor	4
Other	2
Blank	23
Total	39

Wing would kick you—1. Maybe O. K. over L. E. wing—1.

Combat Qualities

Good	14
Fair	0
Poor	2
Other	11
Blank	12
Total	39

Excellent with exception of bad rear view—1. Climbs better, dives better, flies faster than P-38. Could turn with it. Except for range, therefore, is better than P-38 against Nip, from 20,000 feet down—1. Would be a very difficult ship for night instrument work during turbulence—1. That rear visibility just "ain't there." Dog fought with an F6F, I didn't have a chance—1. Excellent. Nothing can catch it. A superb interceptor (with minor control changes). Potentially the best night fighter if it could be slowed down—1.

Excellent as either day or night fighter, if rear visibility were improved and ailerons were boosted to equality with rudder—1. Fast cruise for low altitude, good fire power. Could be developed as good attack plane with fighter potentialities—1. Excellent as reconnaissance, target support airplane, patrol, or night fighter. Visibility and handling characteristics good for strafing and bombing. Probably the finest twin engine medium and low altitude fighter in the world today. Needs more fuel—1. The plane has wonderful possibilities. Internal gas limits range however—1. Should be best night fighter in the area and should rank with the best day fighter bombers. If the Navy ever gets any I want to be in them—1. A high performance, pleasant airplane, but owing to poor turning circle should not attempt to mix it with current Jap types—1. Should make a popular, effective, and valuable fighter airplane. Maneuverability is good. Firepower well up to modern standards. Rear view requires improvement for day use. Should be easy to fly at night but may require more positive stability for night fighter duty. Is a "friendly" airplane and one is immediately at home in it—1. A very fine and delightful airplane to fly and would certainly feel confident to fight in it. As a night fighter the view is excellent, though the side panels distort downward vision. Reflections on windshield from instruments might prove undesirable—1.

Very good. Rear view should be improved. Coordination of controls could be improved with advantage—1.

Excellent plane. Possibly short of fuel range for a twin-engine fighter—1. Rear visibility very poor—inadequate for combat. Directional and longitudinal stability also inadequate. Will be a good fighter *after* these items have been remedied—1. Has great possibilities and is delightful to fly. Performance and maneuverability (considering its size) are very good. Extremely dangerous to operate at night. Minimum speed at which pilot could *catch* the ship on take-off if an engine failed is around 140 knots although it can be held to a somewhat lower speed. Take-off acceleration is so rapid that the problem is not too serious on take-off, but the airplane is so clean, even with flaps and gear down, that a single-engine approach at a speed sufficient to go around makes landing difficult—1.

Excellent. Erratic buffet at high speeds on tail. Elevator control questionable. Rudder boost too sensitive—1. Control and stability needs correction—1. Airplane could be improved by: (1) Reducing dihedral effect. (2) Increasing directional stability. (3) Increasing force per g in accelerated flight—it is probably intolerably unstable at high altitudes—certainly with more aft c. g.'s. (4) Improve lateral control effectiveness at low and medium speeds (up to 250). Forces are O. K. item (1) above would contribute to this. (5) Increase rudder forces booster or decrease forces with booster off. Item (2) above would alleviate the snaking condition and considerably improve control with booster on. (6) Airplane is uncontrollable at take-off with power on left, gear down, right throttled. Item (2) above would improve force per g for stable conditions if mechanical advantage is reduced—1. Excellent. Only doubt would be slightly high speed for carrier operations—1. Looks very good. Can beat F6F in dog fight at low altitude. Rudder needs improvement—too heavy without boost and too light with boost. A little unsteady in aiming—suggest more fin area and more powerful ailerons—1. Visibility aft very poor.

F7F-1—Continued

Needs bubble or bulge canopy—1. Prefer single-engine type. Don't enjoy landing characteristics—1. Sweet ship to fly. Good climb and V max at low altitude and excellent maneuverability for a large twin-engine fighter. Should be an excellent night ship—1. Should be fairly good—1

STABILITY

Longitudinal	*Good*	*Fair*	*Poor*	*Positive*	*Weak Positive*	*Neutral*	*Unstable*
Climb	7	1	3	0	1	2	1
Level	7	1	2	0	1	2	1
L/C	1	0	2	0	1	1	0
General	5	1	0	0	2	0	1

Remarks.—Climb: Marginal—2.
Level: Marginal—1. Weak—1. Sensitive—1. Slightly sluggish—1.
L/C: Too strong—1. Very weak—1.
General: Marginal—1. Dynamically inadequate—statically marginal—1. Due to elevator friction—tends to be neutrally stable to unstable statically—1.

Lateral	*Good*	*Fair*	*Poor*	*Positive*	*Weak Positive*	*Neutral*	*Unstable*
Climb	9	1	0	0	0	0	0
Level	9	1	1	1	0	0	0
L/C	3	1	0	1	0	0	1
General	7	0	0	1	0	1	1

Remarks:—Climb: Slight wobble—1.
Level: Bad with present directional stability—almost dutch roll—1. Roll due to excessive yaw—1. Could be better with heavier rudder—1.
L/C: Very weak—1.
General: Excessive stability—2. Too much dihedral effect—2. Tends to climb or dive too much—1. Has tendency to fall to right—1.

Directional	*Good*	*Fair*	*Poor*	*Positive*	*Weak Positive*	*Neutral*	*Unstable*
Climb	3	0	2	0	0	1	1
Level	3	1	2	1	0	1	1
L/C	2	0	1	1	0	1	1
General	6	1	2	0	0	2	0

Remarks.—Climb: Weak—2. Weak with boost on—1. Marginal—1.
Level: Weak—3. Sensitive—1. Marginal—1. Yaw due to yaw—1.
L/C: Very weak—1.
General: Too weak—3. Marginal—1. Bad with power boost—1. Unstable with boost—1. Bad dutch roll with bad wing roll—1. Has rolling characteristics—1. Lateral tendency to fall to right causes plane to drop wing and turn—1.

CONTROLLABILITY

Elevators	*Good*	*Fair*	*Poor*	*High*	*Moderate*	*Light*
Force	8	0	0	0	1	10
Effectiveness	19	0	0	0	0	0
General	0	1	0	0	0	0

Remarks.—Force: Too light—6. Little high—1. Normal—1.
Effectiveness: Insufficient for landing—2. Slight at slow speeds (under 120 knots)—1. High—1. Inadequate—1.
General: Poor at slow speeds—2. Has initial tendency to keep going in direction pressure applied—1. Mushy—too light in accelerated turns—1.

Ailerons	*Good*	*Fair*	*Poor*	*High*	*Moderate*	*Light*
Force	6	1	0	9	0	2
Effectiveness	11	4	0	0	0	0
General	5	1	0	0	0	0

Remarks.—Force: Slightly heavy—2. Too much—1. Stiff—should be boosted to equality with rudder—1. Stiff at 350 knots—1. Slight force but requires abnormal amount of travel—1.
Effectiveness: Poor below 150 knots—1. Ineffective in landing condition—1. Good rolling velocity for span involved—1.
General: Good ailerons at high speed—1.

F7F-1—Continued

Rudder	*Good*	*Fair*	*Poor*	*High*	*Moderate*	*Light*
Force	5	1	0	1	0	7
Effectiveness	14	0	0	0	0	0
General	3	1	1	0	0	0

Remarks.—Force: Too light—4. Normal—1. Boost too touchy—1. With boost—too light; without boost—too heavy—2. O. K. with boost; too heavy without boost—8. Rudder forces are much too light for the ailerons—1. Could not hold airplane level without boost, if not in operation—1.

General: Will not return to neutral. Lag in operation. Not enough for slow speed single engine—1.

Trimmability

		Remarks
Good	23	Requires little trim—4. Needs plenty of trim in landing condition—2. Does not trim low enough speed condition—1. Would like trim to be more effective—1. Rudder tab should be more sensitive and require less travel—1. Not readily trimmed—1. Difficult longitudinally—1. Difficult to trim plane for level flight—1.
Fair	3	
Poor	2	
Other	4	
Blank	7	
Total	39	

	A	*B*	*C*	*D*	*Other*	*Blank*
Maneuverability	5	20	6	0	6	2

Remarks.—A in comparison with twin engine aircraft—3. Excellent for plane of this size—2.

C in comparison with F6F—1. Rate of roll poor at high speed—2. Aileron combined with rudder gave rapid response—1. Would be better if ailerons were a little more easy—1. Must watch elevator forces due to lightness of control—1. B with rudder boost modified—1.

	A	*B*	*C*	*D*	*Other*	*Blank*
Night and Instrument Flying Rating	3	10	3	3	0	20

Remarks.—Longitudinally unstable—1. Extremely dangerous to operate at night. With the present very weak lateral and directional stability (almost dutch roll), it would be impossible to handle on instruments in turbulence—1.

	A	*B*	*C*	*D*	*Other*	*Blank*
Gun Platform Rating	7	9	8	0	1	14

Remarks.—Poor directional stability—1. Poor directional control—1. Too much roll due to yaw—too easy to yaw—1. Rudder too light—2. Lateral oscillation in dive—1.

Diving characteristics

	Good	*Fair*	*Poor*	*High*	*Moderate*	*Light*
Acceleration	14	1	0	0	0	0
Stick Force	12	0	0	0	0	1
Recovery	15	0	0	0	0	0
General	2	0	0	0	0	0

Remarks.—Acceleration: Rapid—3. Normal—1. Steady—1. Lateral oscillation—1. Could be more stable—1.

Stick Force: Too light—4. Too heavy on ailerons—6. Elevator light—1. Rudder and elevator too light—1. Elevator and rudder lack force building with speed increase—1.

Recovery: Mushes—2. Normal—1.

Stall Characteristics

	Clean		*L/C*		*Accelerated*
	Power Off	*Power On*	*Power Off*	*Power On*	*3g*
IAS(kts):					
Average	86	79	76	66	131
Range	78–95	63–95	60–90	50–85	120–140

Warning:
- Good—15.
- Poor—1.
- Shudder, buffet—9.
- Very little warning in L/C—1.
- No warning in condition of stall power off—1.

Recovery:
- Good—15.
- Fair—1.
- Normal, simple, easy—9.

XF8F-1

Army—1 British —1
Navy —3 Contractors—0

Remarks

Cockpit

Good	4
Fair	0
Poor	0
Other	1
Blank	0
Total	5

Nicely laid out—1. Transition from F6F would be easy—1. Might be tiring in long hops due to inability to change position—1. Very good—simple—access fair (needs extra step)—1. Very good indeed. Very comfortable, good stick position, good high position. No criticism of lay-out—1.

Comfort

Good	4
Fair	0
Poor	0
Other	1
Blank	0
Total	5

A little cramped. Probably be hot in South Pacific—1.

Ground Handling

Good	4
Fair	0
Poor	0
Other	1
Blank	0
Total	5

Good visibility—3. Brakes fairly good, but won't hold for a trim-up of engine—1.

Power Plant Operation

Good	2
Fair	0
Poor	0
Other	3
Blank	0
Total	5

Simple—exceptionally smooth and quiet—1. Would like propeller control moved to top of quadrant with throttle—1. Inspires confidence. Runs rough at high powers but not objectionable. Will probably run hot in South Pacific in CV approach—1. Air-screw lever too inaccurate, as usual in United States cockpits—1.

Take-Off

Good	3
Fair	0
Poor	0
Other	2
Blank	0
Total	5

Very little swing—2. Short run—2. Very little torque noted—1. Port wing dips slightly—1. Brakes don't hold take-off power—1.

Approach and Landing

Good	2
Fair	0
Poor	0
Other	3
Blank	0
Total	5

View over nose excellent. Controls effective right down to landing, maintaining a feeling of control at low speeds—1. Don't like (1) lack of longitudinal trim, (2) negative lateral stability, (3) erratic air-speed system. Do like good ailerons. Airplane tends to bounce on landing—1. Ailerons not very effective. Large stick movements required to hold off bank when rudder is applied. Nose down load on landing typical of Grumman—1.

XF8F-1—Continued

WAVE-OFF

Good	3
Fair	0
Poor	0
Other	0
Blank	2
Total	5

BAIL-OUT

Good	2
Fair	0
Poor	0
Other	1
Blank	2
Total	5

A little tight—1.

COMBAT QUALITIES

Good	4
Fair	0
Poor	0
Other	1
Blank	0
Total	5

Outstanding rate of climb—3. If the aircraft has the performance it appears to have, it should be rated as the top interceptor—1. Excellent if firepower is sufficient—1. This is a first-class fighter which has very high maneuverability and a well-knit feeling. I rate it easily tops in Navy fighters from the combat point of view—1.

STABILITY

Longitudinal	*Good*	*Fair*	*Poor*	*Positive*	*Weak Positive*	*Neutral*	*Unstable*
Climb	0	0	0	0	3	0	0
Level	0	0	0	1	1	0	0
L/C	0	0	0	1	0	1	1
General	1	0	0	0	0	0	0

Remarks.—Level: Just on right side of positive—1.

Lateral	*Good*	*Fair*	*Poor*	*Positive*	*Weak Positive*	*Neutral*	*Unstable*
Climb	0	0	0	0	3	0	0
Level	0	0	0	1	1	0	0
L/C	0	0	0	1	1	0	1
General	1	0	0	0	0	0	0

Directional	*Good*	*Fair*	*Poor*	*Positive*	*Weak Positive*	*Neutral*	*Unstable*
Climb	1	0	0	1	1	0	0
Level	0	0	0	1	1	0	0
L/C	0	0	0	0	1	1	0
General	2	0	0	0	0	0	0

Remarks.—Climb: Damps in two to three swings—1.

CONTROLLABILITY

Elevators	*Good*	*Fair*	*Poor*	*High*	*Moderate*	*Light*
Force	2	0	0	0	0	3
Effectiveness	4	0	0	0	0	0
General	2	0	0	0	0	0

Remarks.—Force: 3 to 4 #/g—1.

Ailerons	*Good*	*Fair*	*Poor*	*High*	*Moderate*	*Light*
Force	3	0	0	0	0	2
Effectiveness	4	0	0	0	0	0
General	1	0	0	0	0	0

XF8F-1—Continued

Rudder	Good	Fair	Poor	High	Moderate	Light
Force	2	0	0	0	1	0
Effectiveness	3	0	0	0	0	0
General	2	0	0	0	0	0

Remarks.—Effectiveness: Too much—1. Rudder was quite sensitive, but believe pilots would become accustomed—1.

TRIMMABILITY

Good	2
Fair	0
Poor	0
Other	2
Blank	1
Total	5

Trim used only for landing—effective—1. Ship was wing-heavy; could not be trimmed—1. Heavy nose down change with wheels and flaps—1. Needs lateral trim. Needs more longitudinal trim in L/C—1. Ideal for a fighter—1.

	A	B	C	D	Other	Blank
Maneuverability	5	0	0	0	0	0
Night and Instrument Flying Rating	0	1	1	0	0	3
Gun Platform Rating	2	3	0	0	0	0

DIVING CHARACTERISTICS

	Good	Fair	Poor	High	Moderate	Light
Acceleration	2	0	0	0	0	0
Stick Force	1	0	0	0	0	1

	Good	Fair	Poor	Fast	Moderate	Slow
Recovery	1	0	0	0	0	0
General	2	0	0	0	0	0

Remarks.—Stick Force: Loads build up very slowly—1. High above 375 A/S—1.
Recovery: Force/g O. K.—1.

General: Rudder very light, requires little or no trim up to 400—1. Very slight change of trim—1. Cosiderable change in rudder trim—1. Ailerons deform at high speed—1.

STALL CHARACTERISTICS

	Clean		L/C		Accelerated 3g
	Power Off	Power On	Power Off	Power On	
IAS (kts): Average	98	--	--	80	170 to left 150 to right

Warning:	Clean Power Off	Clean Power on	L/C Power off	L/C Power on	Accelerated 3g
Good	--	--	1	1	--
Fair	1	1	--	--	--
Poor	--	--	--	--	--
Type	Aileron snatch to left—1.	Almost impossible to stall without climbing vertically—1.	O. K., unstable—1.	Straight forward—1. O. K., unstable—1.	O. K., no tendency to whip—1.
Recovery:					
Good	2	1	1	1	--

Remarks.—General: About 10 miles per hour air speed indicator was very erratic. Only stalled in clean condition straight forward and if held tight, falling off on one wing. Recovery normal and immediate with very little loss of altitude—1. A/S indicator unreliable at low speeds—1.

F4U-1C, D

Army—13 British —3
Navy — 4 Contractors—8

Remarks

Cockpit

Good	8
Fair	3
Poor	3
Other	12
Blank	2
Total	28

Untidy, messed up—4. Too many gadgets—2. Excellent arrangement—2. Comfortable—1. Convenient—2. Instruments difficult to see—1. Stick a little low and far forward—1. Throttle quadrant too low—1. Gunsight is bulky and seems badly placed—1. Armament arming switches badly placed for night vision—1. No landing gear locked down indicator—1. Sliding hood difficult to operate—2. Flat plate windshield is an improvement—1. Bulged canopy is great improvement over standard canopy. Hood shows tendency to crack open—1. Rear view is distorted | 3.

Comfort

Good	12
Fair	2
Poor	0
Other	7
Blank	7
Total	28

Roomy enough—2. A bit cramped—1. Stick not high enough—2. Seat too low—1. Seat too high—1. Develop headache trying to read instruments—1. Too full of bits and pieces—1. Vibration at cruise—1.

Ground Handling

Good	9
Fair	5
Poor	3
Other	8
Blank	3
Total	28

Forward visibility poor—5. Visibility good—1. Excessive "S" turns—2. Directionally unstable—1. Tendency to yaw—1. Unsatisfactory in cross wind—2. Have to operate brakes too much—3. Brakes good—2. Brakes very sensitive—1. Brakes harsh—1. Tail wheel lock unsatisfactory—1. Recommend allowing more tail swing when tail wheel is in locked position—could then taxi with tail wheel locked—1.

Power Plant Operation

Good	11
Fair	0
Poor	0
Other	14
Blank	3
Total	28

Rough—4. Rough below 17-20" MP—2. Smooth—3. Engine vibration—1. Vibration excessive at cruise and above—1. Vibration from 1,850 to 2,100 revolutions per minute—1. Vibration too much with gear down and flaps open and at idling—1. Propeller rough in 1,800 revolutions per minute range—1. Cough on blower switch (neutral to low)—1. Excessive smoke in cockpit at take-off—1. Propeller control preferred adjacent to throttle lever. Ignition switch inaccessible. Blower shift, intercooler, oil cooler, and cowl flaps should be automatic—1. Pitch lever badly positioned—1.

Take-Off

Good	22
Fair	0
Poor	0
Other	4
Blank	2
Total	28

Smooth, rapid—2. Good visibility—1. Forward visibility limited—1. Torque—2. Inclined to swing—1. Tends to yaw—1. Good control—1. Peculiar snatch when wheels retracted—1.

Approach and Landing

Good	8
Fair	1
Poor	0
Other	14
Blank	5
Total	28

Good approach characteristics—3. Comfortable, good control—7. Good visibility—2. Forward vision limited—1. Rudder ineffective at semistalling speed—1. Rudder will not bring up wing—1. Left wing drops on full stall—1. Inclined to swing after landing—1. Bad on ground even with raised tail wheel—1. Brakes necessary in roll—1. Like the way the flaps come down quickly—1.

F4U-1C, D—Continued

WAVE-OFF		
Good	7	Requires retrim—1. Aileron O. K.—1. Rudder weak, aileron necessary—1.
Fair	0	
Poor	0	
Other	4	
Blank	17	
Total	28	

BAIL-OUT		
Good	5	Emergency exits well placed—1. Cabin stiff to open—prefer cabin crank—1.
Fair	1	
Poor	0	
Other	4	
Blank	18	
Total	28	

COMBAT QUALITIES	
Good	12
Fair	0
Poor	0
Other	7
Blank	9
Total	28

Want more visibility and more view over nose. Clean up cockpit. Good ailerons, but lighten up rudder. Oxygen system not in best position. Want ignition switch more accessible. Need bubble canopy—1. Looks and feels O. K.—2. Would prefer it against Jap—third to F7F and P51, in that order—1. Rather blind especially below—1. I think it's a fine A/C. Flies and handles easily. The arrangement of the instruments in the cockpit is considered extremely bad. You can't see them without standing on your head—and even then the cockpit is so dark you have trouble reading them—1. Should outclass Japs with fair margin. I'd like it—1. Speed of this model is its main attribute, it's only fair—1. Ship fairly rugged for diving maneuvers—1. Generally excellent but poor visibility and cockpit comfort leaves much to be desired—1. Marines have proven its worth. I heartily concur. Plane might be made more maneuverable to great advantage. Trim tabs extremely well placed and designed—best I've flown. Flap handle too prominent. Recommend "instrument" power switch (gas gauge, oil temperature, etc.) be removed and instruments hooked through battery switch. Instrument panel 6–8 inches too close to pilot. Move radio equipment farther forward; have not yet seen any VF satisfactory in this respect. Recommend small auxiliary panel for "engine" and "armament" instruments (like F6F and F7F)—1. One of the best—1. Has already proven to be a good combat airplane—1. A very fine fighting airplane with great possibilities—1. Excellent, all around—2. Airplane requires too much changing of lateral and directional trim. I prefer the F6F in all respects—1. Very good vibration engine prop combination. Power control needs pilot attention—1. Speed is good—rate of climb not what I expected—firing platform definitely poor—raised canopy helps—rear vision mirrors good—1.

STABILITY

Longitudinal	*Good*	*Fair*	*Poor*	*Positive*	*Weak Positive*	*Neutral*	*Unstable*
Climb	6	0	0	0	1	1	0
Level	9	1	0	2	1	0	0
L/C	5	1	1	1	0	0	1
General	9	0	0	0	0	2	0

Remarks.—Marginal—2. Neutral (seems to be friction)—1.

Lateral	*Good*	*Fair*	*Poor*	*Positive*	*Weak Positive*	*Neutral*	*Unstable*
Climb	9	2	0	1	0	0	0
Level	10	1	0	2	0	0	0
L/C	6	1	0	1	0	2	0
General	9	1	0	0	0	0	0

Remarks.—Slow but good—1. Sloppy—1.

Directional	*Good*	*Fair*	*Poor*	*Positive*	*Weak Positive*	*Neutral*	*Unstable*
Climb	9	0	1	1	0	0	0
Level	10	0	0	2	0	0	0
L/C	6	0	0	3	0	0	0
General	11	0	0	0	0	0	0

Remarks.—Soft rudder—1.

F4U-1C, D—Continued

Controllability

Elevators

	Good	Fair	Poor	High	Moderate	Light
Force	9	0	0	1	0	4
Effectiveness	14	0	0	0	0	0
General	6	0	0	0	0	0

Remarks.—Force: Normal—1. Force high in dive if trimmed. Comes out of dive on release—1.
Effectiveness: Sensitive—1.
General: It is believed that this plane should be the criterion for U. S. Navy in regard to control forces—1. Nicest coordination of controls of any airplane at this conference I have flown—1.

Ailerons

	Good	Fair	Poor	High	Moderate	Light
Force	8	0	0	0	0	8
Effectiveness	15	0	0	0	0	0
General	6	0	0	0	0	0

Remarks.—Force: Very sensitive—1. Unusually light at high speed—1. Necessary to move stick too far—2.
Effectiveness: Sloppy—2. Overbalance and lock at high speeds (400 kts.)—1. Ailerons overbalanced, when full deflection was tried at 300 kts. IAS—1. Four second aileron rolls at 250 kts.—1.

Rudder

	Good	Fair	Poor	High	Moderate	Light
Force	8	2	0	2	0	3
Effectiveness	12	2	0	0	0	0
General	4	0	0	0	0	0

Remarks.—Force: Fair at cruising—stiff above 280 knots—2. Forces build up particularly if trimmed for dive—1. Loads build up excessively with speed—1.
Effectiveness: Effective except carrier landing conditions—1.

Trimmability

Remarks

Good	18
Fair	1
Poor	0
Other	7
Blank	2
Total	28

Requires too much trim (particularly rudder and elevator—2), with changes in speed—6. Rapid—3. Sensitive—2. Acute on change in attitude—1. Slightly out of reach—1. Trim wheels accessible—1.

	A	B	C	D	Other	Blank
Maneuverability	6	17	3	0	1	1

Remarks.—Couldn't stay on the tail of FM-2.—1. Free control stability in pitch causes roll and yaw—1.

	A	B	C	D	Other	Blank
Night and Instrument Flying Rating	1	9	2	1	0	15
Gun Platform Rating	3	12	5	1	0	7

Diving Characteristics

	Good	Fair	Poor	High	Moderate	Light
Acceleration	6	2	0	0	0	0
Stick Force	4	0	0	1	1	1
Recovery	10	0	1	0	0	0
General	5	1	0	0	0	0

Remarks.—Acceleration: Rapid—2. Slow—1.
Stick Force: Rudder heavy—4. Ailerons good—2. Elevator good—1. Elevator average—1. Elevator slightly heavy—1. Lateral heavy—1.
Recovery: Easy—1. Shudders on recovery—feels like engine vibration—1. Shudders in full auto—1.
General: Like the easy way it picks up speed in dive. Gives time to get squared away. Good plane for rockets—1. Dive brakes functioned OK but caused vibration—1.

F4U–1C, D—Continued

STALL CHARACTERISTICS

	Clean		*L/C*		*Accelerated*
IAS (kts):	*Power Off*	*Power On*	*Power Off*	*Power On*	*3g*
Average	82	76	74	70	150
Range	65–88	60–83	63–90	63–84	130–190

Warning:
- Good—8.
- Poor—1.
- Buffet, shake, shudder—8.

Recovery:
- Good—12.

XF4U–4

Army—0	**British —1**
Navy —2	**Contractors—0**

Remarks

COCKPIT

Cockpit excellent, well arranged.

Cowl flaps will not remain closed. Lower mirrors—double reflection of sight at top of windshield. Instrument arrangement excellent. Prop control an excellent improvement. Do not like chart-board arrangement; frequent use will be very awkward—1.

Rudder length indicators hard to see. Just possible to confuse wing folding with tail wheel lock although unlikely with wing hinge pin lock. Radio lay-out very good. Chart board is difficult to grasp in stowed position. Harness release awkward to find without looking for it. Prop and throttle control excellent. Instrument dials are too similar in size and appearance, making quick reference to any particular one difficult. Landing gear lever may inadvertently be put to dive brake instead of gear—1.

GROUND HANDLING

Good—1.

POWER PLANT OPERATION

Good—engine is smooth—1.

The above are the only comments on the XF4U–4 received.

FG–1, FG–1A

Army—3 British —4
Navy —2 Contractors—1

Cockpit—FG-1		*Remarks*
Good	1	Poor arrangement—1.
Fair	0	
Poor	0	
Other	4	
Blank	0	
Total	5	
Cockpit—FG-1A		(Bubble Canopy.)
Good	3	Bubble head gives improved vision—4. Bubble canopy no better than flat front windshield—1. Hood handle awkward due to high forces involved—3. Locking mechanism failed and hood came open while yawing—1.
Fair	0	
Poor	2	
Other	4	
Blank	0	
Total	9	
Comfort		
Good	6	Seat too upright—2. Seat too high—1. Stick position bad—1.
Fair	1	
Poor	1	
Other	2	
Blank	0	
Total	10	
Ground Handling		
Good	4	Poor forward visibility—3. Directionally unstable—3. Brakes strong and sensitive—1. Brakes awkward—1.
Fair	0	
Poor	1	
Other	4	
Blank	0	
Total	9	
Power Plant Operation		
Good	6	Like having electric primer next to starter—1. Prefer automatic supercharger control—1. Prefer pitch control beside throttle—1. Airscrew lever inaccessible—1. Smooth running above 2000, rather rough below—1. Rough—1.
Fair	0	
Poor	0	
Other	3	
Blank	1	
Total	10	
Take-Off		
Good	7	Torque fairly heavy—1. Stability on take-off neutral—1.
Fair	0	
Poor	0	
Other	3	
Blank	0	
Total	10	
Approach and Landing		
Good	3	Poor visibility over nose—1. Very good vision—1. Comfortable—1. Good approach—1. Stable, slow—1. Harder to land than most fighters—1. Rudder useless last two-thirds of landing run—1.
Fair	1	
Poor	0	
Other	5	
Blank	1	
Total	10	

FG–1, FG–1A—Continued

Wave-Off		
Good	4	Power OK—1.
Fair	0	
Poor	0	
Other	1	
Blank	5	
Total	10	

Bail-Out	
Good	4
Fair	0
Poor	0
Other	0
Blank	6
Total	10

Combat Qualities		
Good	4	Aircraft of proven ability—3. Poor climb, otherwise good—1. Good performance range and maneuverability, poor gun platform, poor rear visibility—1.
Fair	0	
Poor	0	
Other	3	
Blank	3	
Total	10	

Stability

	Good	*Fair*	*Poor*	*Positive*	*Weak Positive*	*Neutral*	*Unstable*
Longitudinal							
Climb	8	0	0	0	0	0	0
Level	5	0	0	0	0	0	0
L/C	4	0	0	0	0	0	0
General	2	0	0	0	0	0	0
Lateral							
Climb	6	0	0	1	0	0	0
Level	4	0	0	1	0	0	0
L/C	3	0	0	1	0	0	0
Directional							
General	2	0	0	0	0	0	0
Directional							
Climb	7	0	0	0	0	1	0
Level	4	0	0	1	0	0	0
L/C	3	0	0	0	0	1	0
General	1	0	0	0	0	0	0

Remarks.—Poor damping in dive—1. Directional stiffness and directional damping both less than for conventional F4U—1

Controllability

	Good	*Fair*	*Poor*	*High*	*Moderate*	*Light*
Elevators						
Force	1	0	0	3	1	2
Effectiveness	4	0	0	0	0	0
General	4	0	0	0	0	0
Ailerons						
Force	1	0	0	0	2	2
Effectiveness	4	0	0	0	0	0
General	4	0	0	0	0	0

Remarks.—Effectiveness: Reaction slow—2.

Rudder	*Good*	*Fair*	*Poor*	*High*	*Moderate*	*Light*
Force	0	0	0	2	2	0
Effectiveness	3	1	0	0	0	0
General	3	0	0	0	0	0

Remarks.—Force: 30#—1.
Effectiveness: Small rudder buffet in climb—1. No rudder overbalance at 105 kts clean with full rudder—1.

FG-1, FG-1A—Continued

Trimmability

Remarks

Good	9	Excessive trim with change of speed—3. Excessive trim change when flaps and U/C are lowered—1.
Fair	1	
Poor	0	
Other	1	
Blank	0	
Total	11	

	A	*B*	*C*	*D*	*Other*	*Blank*
Maneuverability	1	10	0	0	0	0
Night & Instrument Flying Rating	0	4	2	0	0	4
Gun Platform Rating	2	3	2	1	0	2

Diving Characteristics

	Good	*Fair*	*Poor*	*High*	*Moderate*	*Light*
Acceleration	2	0	0	0	0	0
Stick Force	2	0	0	1	0	0
	Good	*Fair*	*Poor*	*Fast*	*Moderate*	*Slow*
Recovery	3	0	0	0	0	0
General	2	0	0	0	0	0

Remarks.—Acceleration: Slow—1. Normal—1.
General: Nose light in dive; requires trim—1.

Stall Characteristics

FG-1 (Clipped Wing).

	Clean		L/C		*Accelerated*
IAS (Kts):	*Power Off*	*Power On*	*Power Off*	*Power On*	*3g*
Average	82	82	74	70	130
Range	73–91	80–83	72–78	69–75	----
Warning:					
Good	1	0	0	0	0

Remarks.—Clean Power Off: Not much—1.
Clean Power On: Not much—1.
L/C Power Off: Airplane does not know which way to go, rolls right and left, pitches and finally breaks violently—1.
L/C Power On: Gets sloppy—1. Stall break is to the left with the same snap as (or less than) the conventional F4U—1.
Accelerated 3g: Stall right or left is nice with slight (controllable) left wing dropping tendency—1. Fairly sharp—1.
General: Not much improvement noticed at the four conditions on the card—1.

Recovery:					
Good	2	1	1	1	0

Remarks.—Clean Power Off: Break left wing down 15°—1.
Clean Power On: Breaks left—1.
L/C Power Off: Stick fixed, stability increases at stall—1.
L/C Power On: Stick free, stability goes sour—1.
Accelerated 3g: Nice either to right or left—1.

FG-1A.

	Clean		L/C		*Accelerated*
IAS (Kts):	*Power Off*	*Power On*	*Power Off*	*Power On*	*3g*
Average	89	84	75	69	----
Range	82–98	78–92	70–85	61–80	----
Warning:					
Good	2	2	2	2	1
Type	Buffet—1.	Buffet—1.	Buffet—1.	Buffet—1.	
Recovery:					
Good	3	3	3	3	0

Remarks.—Normal, but control forces very heavy and left wing drops violently—stall is too vicious for naval aircraft, —1.

XF2G-1

Remarks

No Pilots' Flight Reports were submitted for the XF2G-1.

P-38L

Army—1	British — 5
Navy —9	Contractors—13

Remarks

Cockpit

Good	2
Fair	1
Poor	11
Other	13
Blank	1
Total	28

Yoke hides instruments—10. Complicated—3. Strong gasoline smell after rolls—1. Controls inaccessible—1. Crowded—1. Instrument panel and windshield too far away—1. Many switches could not be reached with harness locked—including auto override switches—1. Position of tabs poor—1. No landing gear position indicator—1. Comfortable and quiet—1. Visibility not too good—.1

Comfort

Good	9
Fair	4
Poor	5
Other	7
Blank	3
Total	28

Cabin roof too low—4. Crowded—3. Noise level good—1. Fair except for visibility; all I could see were engines and boom—1. O. K. except have to duck to see up—1.

Ground Handling

Good	21
Fair	2
Poor	0
Other	4
Blank	1
Total	28

Good like most tricycles—1. Taxis nicely—good control and vision—1. Throttles too long and far apart for good directional control—1. Brakes very soft—1. Lot of brakes required to overcome nose wheel—1.

Power Plant Operation

Good	16
Fair	0
Poor	1
Other	10
Blank	1
Total	28

Smooth and easy—6. Controls well placed—operation simple—3. Tendency to oil up under 800 revolutions per minute—1. Auto controls worked perfectly—1. Elimination of mixture and pitch controls would make power plant operation extremely good—1. Noticed raw gas fumes while maneuvering—1.

Take-off

Good	10
Fair	0
Poor	0
Other	14
Blank	4
Total	28

Long—5. Aileron snatch in air with wheels down—3. Very easy—7. Straightforward—6.

Approach and Landing

Good	8
Fair	4
Poor	0
Other	9
Blank	7
Total	28

Straightforward, very pleasant—4. Excellent landing and ground roll—2. Ailerons too sensitive—2—Heavy—1. Trim necessary and trimmer rather inaccessible—1. Weak lateral effectiveness—2.

Wave-off

Good	6
Fair	2
Poor	0
Other	3
Blank	17
Total	28

Good with two engines—1. Couldn't be used on carrier—1. Elevator trim tab too awkward to relieve trim change—1.

P–38L—Continued

Bail-out	
Good	2
Fair	1
Poor	12
Other	3
Blank	10
Total	28

Easy emergency release—no awkward projections except yoke over knees—1. O. K. as far as actual egress from cockpit—1. O. K. upside down—1. NG—would hit stabilizer—1. Dangerous—1.

Combat Qualities	
Good	3
Fair	0
Poor	1
Other	15
Blank	9
Total	28

Bad visibility to sides and down. Would rather have F4U or F6F for Pacific—1. I would not consider this a modern fighting aircraft. Poor coordination of control forces and effectiveness, combined with very weak directional stability make it a poor gun platform, and its maneuverability rating is so low as to preclude its use in modern combat—1. As a fighter-bomber—good; for fighter sweep—just fair; as escort—poor—1.

Good due to: (1) Twin-engine reliability; (2) altitude performance; (3) good accelerated stall; (4) versatility; (5) dive recovery flaps, which make prolonged zero-lift dives possible—1.

Apart from very queer ailerons, the aircraft is quite pleasant to fly, and would probably make a very good strike fighter. There is, however, an objectionable wobble in bumpy air—1. An excellent escort fighter. Speed should be sufficient for most present-day Jap fighters. View is poor—too many struts in the way. Rudder makes aircraft very hard to maneuver on first flight—1. Too complicated and full of gadgets—would make unserviceability rate very high—1. Query on maintenance and operational problems with liquid cooled engines in hot climates—1. Too much mechanical equipment for one man to operate in combat—1. Record speaks for itself—1.

Stability

Longitudinal	*Good*	*Fair*	*Poor*	*Positive*	*Weak, Positive*	*Neutral*	*Unstable*
Climb	4	0	1	5	3	4	0
Level	5	1	1	5	0	5	0
L/C	2	1	0	2	1	2	0
General	0	0	0	0	0	0	0

Remarks.—Positive, but slow—1.

Lateral	*Good*	*Fair*	*Poor*	*Positive*	*Weak, Positive*	*Neutral*	*Unstable*
Climb	6	0	0	5	1	3	1
Level	5	1	0	5	1	2	0
L/C	3	1	0	1	3	1	0
General	2	0	0	0	0	0	0

Remarks.—Unstable due to aileron boost—1. Did not like boost ailerons, especially at take-off; feeling of being in slip-stream—1.

Directional	*Good*	*Fair*	*Poor*	*Positive*	*Weak, Positive*	*Neutral*	*Unstable*
Climb	5	1	1	4	3	2	0
Level	5	1	0	4	2	2	0
L/C	3	2	0	1	1	0	1
General	2	2	0	0	0	0	0

Remarks.—Slow to damp out—1. Good roll inducer—1. Slow—1.

Controllability

Elevators	*Good*	*Fair*	*Poor*	*High*	*Moderate*	*Light*
Force	6	2	0	6	1	4
Effectiveness	16	0	0	0	1	0
General	5	0	0	0	0	0

Remarks.—Normal—3. Nice curve—1. Excellent—1. Sluggish—1.

P-38L—Continued

Ailerons	*Good*	*Fair*	*Poor*	*High*	*Moderate*	*Light*
Force	4	0	0	0	0	19
Effectiveness	14	1	0	0	0	2
General	3	0	0	0	0	0

Remarks.—No feel—1. Too light compared to other control forces—1. Too much related to other controls—1. Ailerons snatch—2.

Rudders	*Good*	*Fair*	*Poor*	*High*	*Moderate*	*Light*
Force	2	0	0	16	3	1
Effectiveness	10	5	0	0	0	0
General	1	0	0	0	0	0

Remarks.—Too heavy, especially for jinking at high speeds—1. Easy to overbalance—1. Very nice with change in speed—1.

Trimmability

Remarks

Good	12
Fair	4
Poor	0
Other	8
Blank	4
Total	28

Little trim change necessary for change in speed—3. Bad tab locations—3. Aileron trim would help—1. Balance of three controls not good—1. Can be trimmed for all flight conditions, inclusive of one-engine operation—1.

	A	*B*	*C*	*D*	*Other*	*Blank*
Maneuverability	2	10	13	1	0	2

Remarks.—None.

	A	*B*	*C*	*D*	*Other*	*Blank*
Night and Instrument Flying	1	7	6	1	1	12

Remarks.—With boost ailerons, bad; otherwise excellent—1.

	A	*B*	*C*	*D*	*Other*	*Blank*
Gun Platform Rating	2	13	6	2	0	5

Remarks.—None.

Diving Characteristics

	Good	*Fair*	*Poor*	*High*	*Moderate*	*Light*
Acceleration	6	2	0	3	0	1
Stick Force	8	0	0	4	2	2

Remarks.—Ailerons light, elevator moderate, rudder heavy—1.

	Good	*Fair*	*Poor*	*Fast*	*Moderate*	*Slow*
Recovery	8	0	0	0	0	0
General	0	0	0	0	0	0

Remarks.—Dive flaps make recovery effortless—4. Excellent at 410 IAS—1.

Stall Characteristics

	Clean		*L/C*		*Accelerated*
	Power Off	*Power On*	*Power Off*	*Power On*	*3G*
IAS (mph):					
Average	90	100	80	84	170
Range	80–95	95–115	70–98	75–100	170

Warning:

Good

Fair

Poor

Type Good buffet, shudder. Buffet. Loss of aileron control. Drops nose. Elevator shudder.

Recovery.—Normal all conditions—4. Very easy, except clean with power on—1.

P–47D

Army—1	British —4
Navy—14	Contractors—10

Remarks

Cockpit

Good	11
Fair	4
Poor	0
Other	11
Blank	3
Total	29

Roomy—7. Cluttered—7. Leg room short—5. Lay-out good—5. Canopy control good—4. Good visibility—2. Needs seat adjustment—2. Hot—2. Needs more ventilation—2. Needs rudder pedal adjustment—2. Attitude instruments should be in a plane for blind flying—1. Trimmer excellent—1. Would like electric primer—1. No landing gear position indicator; circuit-breaker panel rather inaccessible—1. Too many red, green, and yellow marks on instruments—1. Several controls hard to reach and operate—1. Visibility excellent except forward and on beams—1.

Comfort

Good	21
Fair	1
Poor	0
Other	4
Blank	3
Total	29

Quiet—3. Hot—2. Amazing draught—free properties with hood open—1.

Ground Handling

Good	19
Fair	1
Poor	0
Other	9
Blank	0
Total	29

Forward visibility poor—6; fair—5; good—1. Brakes good—10. Stability good—2. Rudder control fair—1. Poor—1. Tail wheel lock arrangement good—1.

Power Plant Operation

Good	18
Fair	0
Poor	0
Other	11
Blank	0
Total	29

Smooth—12. Turbo control good—1. Auto prop control good—1. Prop control awkward to use—1. Like the tie-up of controls on the quadrant—1. Too complicated—1. Needs auto oil cooler and intercooler controls—1. Coolant flaps hidden under hatch rim—1.

Take-Off

Good	6
Fair	2
Poor	0
Other	19
Blank	3
Total	29

Long run—14. Good control—8. Blind ahead—3.

Approach and Landing

Good	14
Fair	2
Poor	0
Other	12
Blank	1
Total	29

Control good—3. Braking good—1. Fast—9. Comfortable—9. Visibility good—2. In gradual let-down at 130 I. A. S. plenty of control except for wave-off—1. Unable get tail down when power reduced—1. Below 150 rudder and ailerons feel sloppy—1. Elevators get heavy in getting tail on ground—1.

P-47D—Continued

Wave-Off

Good	5
Fair	0
Poor	5
Other	6
Blank	12
Total	28

Sluggish—10. Control fair—2. Poor turbo response—1.

Bail-Out

Good	16
Fair	0
Poor	0
Other	0
Blank	13
Total	29

Combat Qualities

Good	3
Fair	1
Poor	1
Other	17
Blank	5
Total	27

Large turning circle—performance not too good at low altitude—1. Directionally too weak; visibility over nose too little; acceleration only fair; W/S too high; adds up to "only fair"—1. Hit-and-run tactics only: would need good long runways—1. Not a dogfighter, but should be good strafer and bomber interceptor, except for mediocre rate of climb. Bubble canopy is very good, and electric operation fine—1.

Visibility in all after angles from dead ahead is very good—maneuverability at high speeds very good. Looks as though gunnery runs would be difficult. High-wing loading makes acceleration easy on the pilot—1.

Need for large field, comparatively speaking; lack of maneuverability compared to Jap VF. A whole lot of airplane!—1.

Versus Jap VF at a great disadvantage. For ground support, appears to be very good. Visibility aft very good—1. Rate of climb is pretty low, but having achieved its best altitude advantage can be taken of good dive and zoom. Maneuverability is poor compared to some of its opponents. Armament is good—1. Very hot in cockpit low down. Not taken to altitude but would not like to fight with enemy aeroplanes below 25,000 feet. Would make a good steady dive bomber, etc.—1.

Good ground strafer, good high level escort, not an interceptor—1. A most excellent airplane. With the exception that its climb is poor, I could find nothing in this aircraft in one short flight which I didn't like—1.

Only fair, due to complexity and lack of maneuverability—1. Very low rate of climb, and low-altitude speed very low; best mission strafing—1.

Very good high altitude ship. Bubble canopy makes tremendous improvement. Firepower excellent. Range excellent—1. Feels big and heavy, but an excellent strafing airplane and good general VF for hit-and-run tactics. Excellent vision all around—beautiful bubble. Very steady gun platform. OK from big fields—1.

Stability

Longitudinal	*Good*	*Fair*	*Poor*	*Positive*	*Weak Positive*	*Neutral*	*Unstable*
Climb	7	2	0	1	4	0	0
Level	4	0	0	0	0	1	0
L/C	1	1	0	0	0	2	1
General	0	0	0	0	0	2	0

Remarks.—L/C: Unstable with auxiliary tank full—1.

Lateral	*Good*	*Fair*	*Poor*	*Positive*	*Weak Positive*	*Neutral*	*Unstable*
Climb	9	0	0	2	1	0	0
Level	5	0	0	3	0	0	0
L/C	1	0	1	0	2	1	0
General	2	0	0	0	0	1	0

Remarks.—Climb: Unstable after roll—1.
General: Not enough lateral control for CV—1.

P–47D—Continued

Directional	Good	Fair	Poor	Positive	Weak, Positive	Neutral	Unstable
Climb	7	1	0	0	3	0	1
Level	4	0	0	1	1	1	0
L/C	1	0	2	0	0	1	1
General	1	0	0	1	0	0	0

Remarks.—Level: Too weak—1.
L/C: Rudder reversal at high deflection—5. (O. K. up to 5° yaw—1; locks—3.)

CONTROLLABILITY

Elevators	Good	Fair	Poor	High	Moderate	Light
Force	13	0	0	4	6	2
Effectiveness	20	0	1	0	0	0
General	0	0	0	0	0	0

Remarks.—Unstable under accelerated condition—2.

Ailerons	Good	Fair	Poor	High	Moderate	Light
Force	15	0	0	8	0	6
Effectiveness	16	4	0	0	0	0
General	0	0	0	0	0	0

Remarks.—Heavy at high speed—8.

Rudder	Good	Fair	Poor	High	Moderate	Light
Force	13	0	0	5	0	2
Effectiveness	13	3	2	1	0	0
General	0	0	0	0	0	0

Remarks.—Reverses—1. Too heavy—2.

TRIMMABILITY

Good	14
Fair	0
Poor	0
Other	15
Blank	0
Total	29

Remarks

Touchy, and elevator slow—1. Good aileron and rudder tab, poor elevator tab—1. Requires rudder and elevator trimming as speed increases—1. Elevator trim tab requires excessive movement—1. Ideal except for rudder, which requires slight trimming in dive—1. Lot of movement needed to get elevator trim; rudder requires attention in dive—1. Satisfactory, except prefer aileron trim tab to be in vertical plane; too much trim change with power and speed—1. Calls for constant tab adjustment with 100 miles per hour speed change, especially rudder—1. Rudder trim inadequate in a nose-high power-on attitude—1. Requires exceptionally little trim; controls easy to reach—1. O. K. except rudder—not enough low speed high power; much too much change of trim with changes in speed—1.

	A	B	C	D	Other	Blank
Maneuverability	2	11	9	4	1	2

Remarks.—Too heavy—2. Stalls steep turn at 3g—1.

	A	B	C	D	Other	Blank
Night and Instrument Flying Rating	3	7	1	3	0	15
Gun Platform	9	12	2	0	0	6

DIVING CHARACTERISTICS

	Good	Fair	Poor	High	Moderate	Light
Acceleration	6	1	0	4	1	0
Stick Force	7	1	0	2	0	3

	Good	Fair	Poor	Fast	Moderate	Slow
Recovery	7	0	0	1	0	1
General	4	0	0	0	0	0

Remarks.—Recovery: Dive flaps function very well—3.

P-47M

Army—4	British —1
Navy —3	Contractors—3

Cockpit		Remarks
Good	6	Comfortable—3. Roomy—2. Too many instruments and controls—2. Controls, switches, and instruments all logically placed—1. Good view all around—2. Electric canopy is good—1. Gun-sight in pilot's face—1. Rudders have been shortened from D model—1. Not enough rudder adjustment for tall man—1. Too little leg room—1. Can't keep toes off brakes—1.
Fair	1	
Poor	0	
Other	3	
Blank	1	
Total	11	

Comfort		
Good	5	Leg adjustment too short; brake pedals should be adjustable or contain forward tilt—1.
Fair	0	
Poor	1	
Other	0	
Blank	5	
Total	11	

Ground Handling		
Good	8	Good brakes—2. Easy to taxi—2. View is bad—1. Vision in three-point position good—1.
Fair	0	
Poor	0	
Other	2	
Blank	1	
Total	11	

Power Plant Operation		
Good	6	Much improved. Switches for control of cowl flaps, oil cooler, and intercooler flaps are particularly well located on left side and near throttle. Same is true of turbo regulator—1. Rough—surges—1. Auto controls (intercooler, oil cooler, cowl flaps) cause radio interference when they operate, which is often very annoying—1. Considerable interference between different controls on throttle box. Pitch control is too stiff and jerky with sufficient friction to keep throttle open—1. Super charger control and throttle easy to operate—1. One of smoothest pilot has flown behind—excellent—no vibration—1.
Fair	0	
Poor	0	
Other	4	
Blank	1	
Total	11	

Take-Off		
Good	4	Long—3. No swing—1. Blind until tail is up—1.
Fair	0	
Poor	0	
Other	3	
Blank	3	
Total	11	

Approach and Landing		
Good	6	Stable—2. Easy landing—2. Good control—1. Visibility just fair—1. Ground roll is long, and fair-sized field is necessary—1. This aircraft could be landed on a carrier (but wouldn't get off again)—1.
Fair	0	
Poor	0	
Other	2	
Blank	3	
Total	11	

P–47D—Continued

STALL CHARACTERISTICS

	Clean		L/C		
	Power Off	*Power On*	*Power Off*	*Power On*	*Accelerated 3g*
IAS (mph):					
Average	114	109	96	94	170
Range	100–130	102–115	85–103	88–100	150–195
Warning:					
Good	6	4	4	4	2
Fair	1	0	1	0	1
Poor	0	0	0	0	0
Type	Buffet Aileron shake	Buffet	Buffet	Buffet	Buffet
Recovery:					
Good	10	7	10	7	5
Fair	1	1	2	1	0
Poor	0	0	0	0	0

Remarks.—Straightforward all conditions—4. Dropped wing—1. Normal—3.

P–47M—Continued

Wave-Off

Good	2	Elevator trim comes naturally to hand. Trim change not too high but climb response is slow—1.
Fair	0	
Poor	2	
Other	1	
Blank	6	
Total	11	

Bail-Out

Good	5
Fair	0
Poor	0
Other	0
Blank	6
Total	11

Combat Qualities

Good	4
Fair	0
Poor	0
Other	3
Blank	4
Total	11

Good—for high-altitude escort or fighter-bomber. The best P47 yet—1. Would rather fight F6F or F4U. This plane is too "heavy." Would like to see more immediate reaction from tab movement—1. Speed and long range make up for somewhat reduced comparative maneuverability. Pilot's comfort and natural feel (coordination, position, etc.) of cockpit layout and controls gives one a sense of "fighting confidence" in the aircraft—1. This airplane would be best in E. T. O. at 30,000–40,000 feet as long-range high cover. It is no mixer but could obviously do great damage at strafing—1. Excellent. Observe record of Three Hundred and Forty-eighth Fighter Group at New Guinea; 343 enemy aircraft for the loss of 1 P47—1.

Stability

Longitudinal	*Good*	*Fair*	*Poor*	*Positive*	*Weak Positive*	*Neutral*	*Unstable*
Climb	4	1	0	0	0	2	0
Level	5	0	0	0	1	0	0
L/C	5	0	0	0	0	0	0
General	0	0	0	0	0	0	0

Remarks.—None.

Lateral	*Good*	*Fair*	*Poor*	*Positive*	*Weak Positive*	*Neutral*	*Unstable*
Climb	4	1	0	1	0	1	0
Level	3	2	0	0	0	0	0
L/C	5	0	0	0	0	0	0
General	1	0	0	0	0	0	0

Remarks.—None.

Directional	*Good*	*Fair*	*Poor*	*Positive*	*Weak Positive*	*Neutral*	*Unstable*
Climb	4	0	0	0	1	1	0
Level	4	0	0	0	0	0	0
L/C	4	0	0	0	0	0	0
General	0	0	0	0	0	0	0

Remarks.—With new dorsal fin, stability has been improved considerably in climb, level flight, and dive, especially directionally. Recovery in compressibility dives has been improved considerably with dive recovery flaps—1.

Controllability

Elevators	*Good*	*Fair*	*Poor*	*High*	*Moderate*	*Light*
Force	4	0	0	1	1	2
Effectiveness	7	0	0	0	0	0
General	0	0	0	0	0	0

Remarks.—Forces seem to lighten with G—1. Too much inertia all around—1.

P-47M—Continued

Ailerons	*Good*	*Fair*	*Poor*	*High*	*Moderate*	*Light*
Force	4	0	0	1	0	3
Effectiveness	7	0	0	0	0	0
General	0	0	0	0	0	0

Remarks.—Force: High at high speed—1. Recommend 30-pound force—1. Nice force curve gradient—1. Effectiveness: Rate of roll too slow—1.

Rudders	*Good*	*Fair*	*Poor*	*High*	*Moderate*	*Light*
Force	2	0	0	5	0	1
Effectiveness	6	0	0	0	0	0
General	0	0	0	0	0	0

Remarks.—O. K. for small angles—1. Recommend boost—1. Too much inertia all around—1. O. K. for small angles—1.

TRIMMABILITY

		Remarks
Good	10	Elevator trim should have more response—2. Too much change in directional trim with change in velocity—1.
Fair	1	
Poor	3	
Other	6	
Blank	1	
Total	21	

	A	*B*	*C*	*D*	*Other*	*Blank*
Maneuverability	0	3	4	0	0	4

Remarks.—C at low altitude—probably B at 30,000–40,000 feet by comparison—1.

	A	*B*	*C*	*D*	*Other*	*Blank*
Night and Instrument Flying Rating	1	2	3	0	0	5
Gun Platform Rating	3	2	1	0	0	5

DIVING CHARACTERISTICS

	Good	*Fair*	*Poor*	*High*	*Moderate*	*Light*
Acceleration	2	0	0	0	0	0
Stick Force	1	0	0	0	0	0

Remarks.—Not too heavy—1.

	Good	*Fair*	*Poor*	*Fast*	*Moderate*	*Slow*
Recovery	2	0	0	0	0	0
General	3	0	0	0	0	0

Remarks.—Normal—dive flaps a help in pull-outs at 400 miles per hour—1.

STALL CHARACTERISTICS

IAS (KTS):

	Clean		*L/C*		*Accelerated*
	Power Off	*Power On*	*Power Off*	*Power On*	*3G*
Average	108	106	97	95	0
Range	95–118	0	90–103	0	0
Warning:					
Good	3	2	3	2	2
Fair	0	0	0	0	0
Poor	0	0	0	0	0
Recovery:					
Good	2	2	2	2	2
Fair	0	0	0	0	0
Poor	0	0	0	0	0

P–51D

Army— 1 British — 3
Navy —19 Contractors—15

Remarks

Cockpit

Good	28
Fair	2
Poor	1
Other	5
Blank	2
Total	38

Visibility good—1, except over nose—3. Gun sight obscures instruments and forward vision—15. Cramped—5. Long—3. Flap lever poor location—2. Trim tabs poor arrangement—4. Fuel gauges hard to see—2. Very pistol handle in way—1. Canopy crash arrangement poor—1. No rear vision mirror—1. Landing gear lever hard to reach—1. Switch boxes in way—1. Starting groups scattered—1. Would like panel nearer pilot—1.

Comfort

Good	24
Fair	8
Poor	1
Other	0
Blank	5
Total	38

Warm—3. Adjustment of seat and rudder adequate—1. Position comfortable, noise and vibration high—1. Cramped—2. Poor, seat at wrong angle—1.

Ground Handling

Good	28
Fair	1
Poor	3
Other	6
Blank	0
Total	38

Tail wheel lock: 11 for, 2 against. Specific criticisms: Tail wheel lock sticks, tail wheel locked when it should have unlocked, should lock in more off position, should lock nearer neutral. Visibility: Good—2, poor—6. General comment: Visibility poor ahead, good otherwise, brakes good but weak, cockpit hot—1, excessive CO—1, directional control good—1. Tends to ground-loop with tail wheel unlocked.

Power Plant Operation

Good	35
Fair	1
Poor	0
Other	0
Blank	2
Total	38

Automatic features: 6 for, 1 against. Specific points: Boost regulator poor. All for power plant in general. Specific comments: Noisy, rough between 1,600 and 1,800 revolutions per minute, quiet, requires minor changes in fuel system to prevent back flow to auxiliary tank, harmonic every 15 seconds, smooth except at 2,100 revolutions per minute. Like controls and small quadrant.

Take-Off

Good	26
Fair	6
Poor	4
Other	0
Blank	2
Total	38

Run: Long—7, normal—1, short—1. Rudder: Weak—2, strong—1. Specific comments: Instability with full rear tank disconcerting—1. Take-off not too good using full flap and stalling off—1. Easily controlled directionally—1. Tends to swing hard to left—1. Blind until tail is raised—1.

Approach and Landing

Good	23
Fair	7
Poor	1
Other	0
Blank	7
Total	38

Landing speed: High—5, normal—3. Stability and controllability bad near stall—1. Blind in landing condition—1. Carrier landing, against—2.

P–51D—Continued

WAVE-OFF

Good	5
Fair	5
Poor	13
Other	0
Blank	15
Total	38

Rudder: Sufficient—2, insufficient—6. Ailerons: Sufficient—1, insufficient—3. Acceleration: Sufficient—0, insufficient—2. Specific comments: Too much rudder trim required—1. Rudder tab not sufficient to trim—1. Quite good for Army ship—1.

BAIL-OUT

Good	23
Fair	0
Poor	0
Other	0
Blank	15
Total	38

Canopy release mechanism looks good—3. O. K. roll on back—1. O. K. on ditching of canopy—1. General impression favorable.

COMBAT QUALITIES

Good	19
Fair	6
Poor	4
Other	1
Blank	8
Total	38

Very fine airplane—1. Very good, best ailerons in show. Simple comfortable cockpit. Easy to fly. Six guns—long and short range—excellent performance—1. A fair fighter for shore base using hit and run tactics—1. Directional stiffness entirely unsatisfactory. Frankly, I wouldn't like it because of general instability—1. Like F4U or F6F better—1. Directional stability should be improved for both air to air and air to ground gunnery—1. Good except slow rate of climb and small lead angle for deflection shooting—1. Has good possibilities—1. Good for land based operations on medium and long fields—1. Aft fuel tank caused negative longitudinal stability. Needs more directional stability for gun platform—1. Good in Pacific except for strafing—maneuverability could be improved at expense of straight away speed—1. Appreciate automatic features—give pilot chance to look around. Visibility good. Control response makes for nimble aircraft—1. Visibility excellent with this canopy. Sacrifices stronger skin for deflecting 30 mm. Good at high speed for maneuverability. Climb is disappointing at normal power. Take off run long without flaps. Ranks high with all present planes—1. Rudder trim changes too much with speed and power—1. Good for shore based—1. Excellent. Visibility arcs are very good. Maneuverability and diving qualities good. Directional characteristics poor. Change of trim with speed and lack of directional stiffness undesirable. Consider suitable for carrier—1. Climb and high speed make it excellent for low altitude strafing and fighting. Dive attacks prejudiced by large change on most important gunnery control, namely the rudder—1. Very good but cockpit is very hot, noise and vibration too great—1. Very good. Eliminate rudder trim change for gunnery—1. Should have 8—50 cal.—1. Good because of range, speed and firepower. Do not like directional characteristics at high speeds—change in directional trim—and sudden tendency to yaw—1. Vision aft not as good as expected. Longerons aft wider than necessary obscuring visions. Combat qualities best in conference—1. O. K. for hit and run—1. Unstable with rear tank filled—1. Cannot sit so as to see out and also see instruments because of blocking by gunsight—1. Ailerons very poor at 100 mph. Visibility very good with this canopy. Can see past vertical tail—1. Should be excellent due to performance, excellent vision and high restricted speed in dives—1.

STABILITY

Longitudinal	*Good*	*Fair*	*Poor*	*Positive*	*Weak Positive*	*Neutral*	*Unstable*
Climb	3	1	2	1	6	6	8
Level	2	2	2	0	7	5	2
L/C	2	2	1	0	3	1	3
General	0	0	0	0	0	0	0

Remarks.—No good with fuselage tank—1. Unstable dynamically, neutral statically—1. Unstable, dives stick free—1.

Lateral	*Good*	*Fair*	*Poor*	*Positive*	*Weak Positive*	*Neutral*	*Unstable*
Climb	6	0	0	3	9	6	2
Level	3	1	0	2	7	6	2
L/C	3	1	1	0	4	6	2
General	0	0	0	0	0	4	0

Remarks.—Nose drops with wing when rudder is applied—1. Fair—ailerons weak at low speed—1.

P-51D—Continued

Directional	Good	Fair	Poor	Positive	Weak Positive	Neutral	Unstable
Climb	4	2	0	4	12	3	0
Level	2	2	1	2	6	2	0
L/C	2	3	1	1	6	2	0
General	0	0	0	0	0	0	0

Remarks.—Not enough directional stiffness—1. Not enough vertical fin—1. Bad as speed increases—1. Unstable in dive—1. Returned to heading if yawed with wings level—1.

Controllability

Elevators	Good	Fair	Poor	High	Moderate	Light
Force	5	2	1	1	2	16
Effectiveness	19	4	3	0	0	1
General	1	0	0	0	0	1

Remarks.—None.

Ailerons	Good	Fair	Poor	High	Moderate	Light
Force	7	5	0	1	3	13
Effectiveness	18	10	0	0	1	0
General	1	0	0	0	0	1

Remarks.—None.

Rudder	Good	Fair	Poor	High	Moderate	Light
Force	5	7	0	9	3	3
Effectiveness	13	8	0	4	2	0
General	0	0	0	0	0	1

Remarks.—Excessive trim change with change of speed and power—3.

Trimmability

		Remarks
Good	21	Trim tabs effective—rudder requires too frequent trim—10. Elevator requires too much attention—2. General—too much attention required—5.
Fair	3	
Poor	6	
Other	4	
Blank	2	
Total	36	

	A	B	C	D	Other	Blank
Maneuverability	7	18	8	0	0	5

Remarks.—Ship seems to decelerate too rapidly under accelerated conditions—1. Poor turning radius—1.

	A	B	C	D	Other	Blank
Night and Instrument Flying Rating		5	14	3	0	16
Gun Platform Rating	4	16	9	2	1	6

Diving Characteristics

	Good	Fair	Poor	High	Moderate	Light
Acceleration	21	0	0	0	0	0
Stick Force	5	0	0	0	1	10

	Good	Fair	Poor	Fast	Moderate	Slow
Recovery	11	0	0	0	1	0
General	1	0	0	0	0	0

Remarks.—High acceleration—1. Forward press necessary in dive—trim necessary—1. Recovery necessitates quite a lot of rudder trim—1. General—ship changes directional trim violently and rudder very heavy—1.

P–51D—Continued

STALL CHARACTERISTICS

IAS (mph):	Clean Power Off	Clean Power On	L/C Power Off	L/C Power On	Accelerated 3G
Average	99	94	90	86	159
Range	92–107	83–100	85–100	70–98	150–170

General	*Remarks*
Warning:	
Good 9	
Fair 8	Sharp instability with clean condition with power off—1. Shake—3. Buffet—5. Shudder—3. Left wing drops—2. Nose down—1.
Poor 0	
Type -	
Recovery:	
Good 17	Straight forward—2. Gentle, left wing down—1. More difficult than F6F or F4U-1—1.
Fair 7	
Poor 1	

YP-59A

Army—0 British — 0
Navy —2 Contractors—14

Remarks

Cockpit

Good	4
Fair	1
Poor	1
Other	7
Blank	2
Total	15

Visibility to rear and down very poor—should have bubble—1. Visibility poor—1. Blind aft—1. Comfortable—2. Roomy—1. Quiet—1. Smells of kerosene—1. Needs regrouping—1. Not even cluttered with the experimental gauges—1. Messy—1.

Comfort

Good	8
Fair	3
Poor	0
Other	1
Blank	3
Total	15

Quiet—2. No vibration—1. Messy cockpit—1.

Ground Handling

Good	10
Fair	0
Poor	0
Other	3
Blank	2
Total	15

Requires much braking because it tends to taxi fast—2. Brakes weak—3. Nice brakes—1. Can be done readily with one jet—1. Visibility ahead excellent—2. Slow to accelerate from stop—1. Lot of power to overcome nose wheel—1.

Power Plant Operation

Good	6
Fair	0
Poor	0
Other	7
Blank	2
Total	15

Smooth—3. Quiet—2. Simple—2. Engines howl when throttled back—1. Used one-half of fuel supply in 30 minutes' flight—1. Difficult to synchronize but O. K. after 20 minutes of familiarization—1. Kerosene odor—1.

Take-Off

Good	4
Fair	3
Poor	0
Other	5
Blank	3
Total	15

Good dirctional control—3. Long run—3. Slow acceleration—2. Poor acceleration at start, good at end of run—1. Smooth—1. Good visibility—1.

Approach and Landing

Good	6
Fair	0
Poor	0
Other	4
Blank	5
Total	15

Difficult to slow down because it's clean—1. Hard to lose speed to lower gear—1. Good, if speed is kept slow—1. Flat glide—2. Light wing loading provides good handling characteristics—1. Ailerons a little mushy in approach—1. Good vision—2. Slow effectiveness of throttle—1. Takes a lot of power for slight increase of speed in the landing condition—1.

YP-59A—Continued

Wave-Off

Good	0
Fair	2
Poor	4
Other	3
Blank	6
Total	15

Acceleration not adequate—1. Slow to take power—1. Acceleration as good as Mosquito or P-47—1. At 80 knot approach speed a fair wave-off can be made. It is recommended that one-half or one-fourth flaps be used—1.

Bail-Out

Good	5
Fair	0
Poor	0
Other	2
Blank	8
Total	15

Airspeed pilot on fin is an extra hazard—1.

Combat Qualities

Good	0
Fair	0
Poor	3
Other	8
Blank	4
Total	15

No good—limited endurance—no fire power—1. No good for combat in any theater due to limited range, limited armament, and not too excellent performance—1. Excellent for high altitude, high-speed combat—1. I consider airplane an excellent trainer—1. An excellent first design in a radical field. Much impressed with quietness, single-unit operation, simplicity of control, possibilities in angle of vision, etc.—1. Transition from conventional airplane apparently very easy—1. Not unless definite combat technique developed or acceleration improved or both—1. Exceptionally good except range and take-off characteristics—1. Remember to keep plenty of air speed at all times—1. Not enough performance for present-day combat—1. Great possibilities—1. Aircraft is purely a test platform for the units—1. This pilot prefers this unit to the Westinghouse in most respects—1. Visibility restricted due to formers—1. Inverted flight good—1.

Stability

Longitudinal	*Good*	*Fair*	*Poor*	*Positive*	*Weak Positive*	*Neutral*	*Unstable*
Climb	4	0	0	0	0	0	0
Level	3	0	0	1	0	0	0
L/C	1	0	0	0	0	0	0
General	0	0	0	0	0	0	1

Remarks.—General: Divergence—1.

Lateral	*Good*	*Fair*	*Poor*	*Positive*	*Weak Positive*	*Neutral*	*Unstable*
Climb	3	0	0	0	0	0	0
Level	2	0	0	1	0	0	0
L/C	1	0	0	0	0	0	0
General	0	0	0	0	0	0	1

Remarks.—General: Divergence—1.

Directional	*Good*	*Fair*	*Poor*	*Positive*	*Weak Positive*	*Neutral*	*Unstable*
Climb	3	0	0	0	0	0	0
Level	2	0	0	1	0	0	0
L/C	1	0	0	0	0	0	0
General	0	0	0	0	0	0	0

Remarks.—Level: Positive (oscillated)—1.
General: In dive not too good—1. Positive+++—1. Weak natural—1

Controllability

Elevators	*Good*	*Fair*	*Poor*	*High*	*Moderate*	*Light*
Force	4	0	0	2	1	1
Effectiveness	7	1	0	0	0	0
General	0	0	0	0	0	0

Remarks.—Force: Slightly high—4.
Effectiveness: Tends heavy—2.

YP-59A—Continued

Ailerons	*Good*	*Fair*	*Poor*	*High*	*Moderate*	*Light*
Force	3	0	0	1	1	1
Effectiveness	6	1	0	1	0	0
General	0	0	0	0	0	0

Remarks.—Force: Slightly high—1. Heavy at 200—1. Heavy at 275 knots—1.
Effectiveness: Rate of roll low—1.
General: Sloppy just over stalling speed clean—1.

Rudders	*Good*	*Fair*	*Poor*	*High*	*Moderate*	*Light*
Force	5	1	0	0	1	1
Effectiveness	5	0	0	0	0	1
General	0	0	0	0	0	0

Remarks.—Effectiveness: Weak—2.

TRIMMABILITY

		Remarks
Good	9	Only has elevator trim—very easy to trim—1.
Fair	0	
Poor	0	
Other	0	
Blank	6	
Total	15	

	A	*B*	*C*	*D*	*Other*	*Blank*
Maneuverability	2	5	1	1	0	6

Remarks.—Underpowered—1.

	A	*B*	*C*	*D*	*Other*	*Blank*
Night and Instrument Flying Rating	2	0	1	0	0	12
Gun Platform Rating	0	3	1	1	1	0

Remarks.—Directionally unstable—1.

DIVING CHARACTERISTICS

	Good	*Fair*	*Poor*	*High*	*Moderate*	*Light*
Acceleration	1	1	0	0	1	0
Stick Force	3	0	0	0	0	0
Recovery	1	0	0	0	0	0
General	0	1	0	0	0	0

Remarks.—Acceleration: Quite rapid—1.
Stick Force: Little high—1.
General: Directionally unstable—3. Steady on shallow dive up to 240 IAS—1.

STALL CHARACTERISTICS

IAS (MPH):	*Clean* *Power Off*	*Clean* *Power On*	*L. C.* *Power Off*	*L. C.* *Power On*	*Accelerated 3g*
Average	91.5	89.5	83.5	82	
Range	76–81	76–82	69–82	68–85	

Warning:

- Good ...... 4.
- Fair ...... Clean-power: Poor—1. Little—1.
- Poor ......
- Type ...... Shudder, shake, buffet—2.

Recovery:

- Good ...... 8.
- Fair ......
- Poor ......
- Remarks ......

YP-59A—Continued

Ailerons	*Good*	*Fair*	*Poor*	*High*	*Moderate*	*Light*
Force	3	0	0	1	1	1
Effectiveness	6	1	0	1	0	0
General	0	0	0	0	0	0

Remarks.—Force: Slightly high—1. Heavy at 200—1. Heavy at 275 knots—1.
Effectiveness: Rate of roll low—1.
General: Sloppy just over stalling speed clean—1.

Rudders	*Good*	*Fair*	*Poor*	*High*	*Moderate*	*Light*
Force	5	1	0	0	1	1
Effectiveness	5	0	0	0	0	1
General	0	0	0	0	0	0

Remarks.—Effectiveness: Weak—2.

TRIMMABILITY

Remarks

Good 9 — Only has elevator trim—very easy to trim—1.
Fair 0
Poor 0
Other 0
Blank 6
Total 15

	A	*B*	*C*	*D*	*Other*	*Blank*
Maneuverability	2	5	1	1	0	6

Remarks.—Underpowered—1.

	A	*B*	*C*	*D*	*Other*	*Blank*
Night and Instrument Flying Rating	2	0	1	0	0	12
Gun Platform Rating	0	3	1	1	1	0

Remarks.—Directionally unstable—1.

DIVING CHARACTERISTICS

	Good	*Fair*	*Poor*	*High*	*Moderate*	*Light*
Acceleration	1	1	0	0	1	0
Stick Force	3	0	0	0	0	0
Recovery	4	0	0	0	0	0
General	0	1	0	0	0	0

Remarks.—Acceleration: Quite rapid—1.
Stick Force: Little high—1.
General: Directionally unstable—3. Steady on shallow dive up to 240 IAS—1.

STALL CHARACTERISTICS

IAS (MPH):	*Clean* *Power Off*	*Clean* *Power On*	*L/C* *Power Off*	*L/C* *Power On*	*Accelerated 3g*
Average	91. 5	89. 5	83. 5	82	
Range	76-81	76-82	69 82	68-85	

Warning:
- Good 4.
- Fair Clean-power: Poor—1. Little—1.
- Poor
- Type Shudder, shake, buffet—2.

Recovery:
- Good 8.
- Fair
- Poor
- Remarks

P–61

Army—1 Contractors—11
Navy —7 British —2

Remarks

Cockpit

Good	2
Fair	2
Poor	1
Other	15
Blank	1
Total	21

Cluttered—8. Instrument panel too far from pilot—10. Cabin formers restrict visibility—5. Good grouping of instruments—4. Trim controls inconveniently located: rudder trim should be in horizontal plane—1. Flight instruments off center and not grouped around scope—1. U. V. lighting poor for night adaptation—1. Flaps hard to operate—1. Could simplify starting procedure—1.

Comfort

Good	14
Fair	3
Poor	0
Other	0
Blank	4
Total	21

Continuous longitudinal oscillation—1. Cold—1.

Ground Handling

Good	15
Fair	1
Poor	0
Other	5
Blank	0
Total	21

Brakes sensitive but powerful—2. Unusual brake pedal action—1. Don't like feel of brakes—1. Brakes appear defective—1. Takes too much brake and engine to turn—1. Excessive power required—3. Visibility forward excellent—1. Would like taxi lights to shine a little further out—1.

Power Plant Operation

Good	9
Fair	0
Poor	0
Other	11
Blank	1
Total	21

Prop and throttle controls should be together—3. Prop feathering switches awkwardly placed—1. Props smoothly controlled—1. No prop vernier control—1. Supercharger control too far from quadrant—1. Single supercharger lever would be advantageous—1. Head temperatures ran high in climb, low-pressure level flight, and approach with gear extended (cowl flaps closed)—1. At maximum cruise power heads run hot, but effective cowl flaps—1. Engines difficult to synchronize—1. Should have synchroscope—1. Auto oil flaps and auto intercooler flaps are good; would be better with auto cowl flaps—1.

Take-Off

Good	13
Fair	0
Poor	0
Other	8
Blank	0
Total	21

Take-off was long—8. Good directional control—4. Good acceleration and control—1. Nice elevator control—1. Elevator much too heavy—1.

Approach and Landing

Good	13
Fair	0
Poor	0
Other	4
Blank	4
Total	21

Comfortable except for large change at trim with flaps and for slow aileron response—elevator response good—1. Settles fast with power off—1. Visibility good. Nose wheel hits hard—1. Not enough lateral effectiveness—1. Elevator much too heavy—1. Control better than expected. Landing good—1. Visibility excellent. Rate of sink without power and with full flaps extremely high—good to prevent overshooting at night. Deceleration with full flaps very rapid. Contact characteristics are excellent. Slow landing speed. Stable—1. Easy to control, easy to land, good vision all the way—1. Elevator moderately heavy—directional control good—1. Flare-out good. Considerable power necessary for slow approach with full flap—1.

P-6J—Continued

Wave-Off

Good	1
Fair	3
Poor	0
Other	3
Blank	14
Total	21

Slow and sluggish—2. Requires a great deal of power, acceleration slow (with full flaps). Control good and change of trim not great—1. Awkward with full flap—1. Probably not too good, but far superior to F7F—1.

Bail-Out

Good	3
Fair	3
Poor	7
Other	0
Blank	8
Total	21

Combat Qualities

Good	0
Fair	0
Poor	1
Other	15
Blank	5
Total	21

Poor, inasmuch as it lacks necessary speed and rate of climb. The cockpit is too complicated. Visibility is not good. Poor airplane for a quick scramble take-off. Should have a stick instead of a wheel—1. For a night fighter against enemy in Pacific excellent until the time they speed up and learn to hit in force and take evasive action—1. Excellent night fighter against large bombers or nonmaneuvering targets. Low on visibility. Not good as auxiliary ship night bomber. Little changes would improve it, as change radio button, reflect light on screen, dim it even more, do away with luminous paint on switches, etc. The instrument lights are too bright—1. Not enough performance for a night fighter—1. Too heavy; pilot armor, good range, poor cockpit, poor gun platform—1. O. K. in most respects. A little big for an interceptor but excellent for an intruder. Needs more range, better climb, air brakes, better cockpit in both arrangement and lighting—1. Just too damn big for a night fighter—1. Would be satisfactory if enemy does not go above 30,000 feet—1. This aircraft is simple and quite pleasant to fly, except for the infernal wobble which only detracts from its gun-platform qualities. As a night fighter it's rather cumbersome. As a day fighter it would be quite hopeless—1. Probably good as a night fighter or medium bomber only—1. Ship needs performance, but is adequate for Pacific now—1. Poor for day, probably excellent for night—1. This airplane is a useful combat machine, but can stand much improvement in speed and cockpit—1. Too short on performance with present power plants—1. Excellent, should have more performance, however—1. Not maneuverable enough, too complicated—1.

Stability

Longitudinal	*Good*	*Fair*	*Poor*	*Positive*	*Weak Positive*	*Neutral*	*Unstable*
Climb	6	1	0	2	1	4	0
Level	7	1	0	3	2	1	0
L/C	5	0	0	2	0	0	0
General	1	0	0	0	0	0	0

Lateral	*Good*	*Fair*	*Poor*	*Positive*	*Weak Positive*	*Neutral*	*Unstable*
Climb	9	1	0	2	0	2	0
Level	6	2	0	4	3	1	0
L/C	7	0	0	0	2	0	0
General	2	0	0	0	0	0	0

Remarks.—Roll due to slight yaw—1.

Directional	*Good*	*Fair*	*Poor*	*Positive*	*Weak Positive*	*Neutral*	*Unstable*
Climb	8	1	0	3	1	0	0
Level	7	1	0	7	0	1	0
L/C	6	0	0	1	0	0	0
General	3	0	0	0	0	0	0

Remarks.—Moderate—1. Strong—1.

Level: Oscillation—1. Airplane directionally stable, but yawing moment would not damp in less than four oscillations when cruise trimmed. Rudder moved, so that I suspect rudder difficulty and not directional

P–61—Continued

cause—1. A rudder oscillation tends to aid yaw. This oscillation manifested itself, after moderate rudder displacements at 250 miles per hour and then release of rudder, in a movement of the rudder control cables on the order of ⅛ inch to ¼ inch, such movement having a phase angle of about 45° compared to the angle of sideslip. The self-induced yaw results in roll, hence poor gun platform. Would like to recheck, but would look for rudder camber or peculiarity of balance.

CONTROLLABILITY

Elevators	*Good*	*Fair*	*Poor*	*High*	*Moderate*	*Light*
Force	5	0	0	10	0	1
Effectiveness	14	2	0	0	0	0
General	0	0	0	0	0	0

Remarks.—Force: Light at normal speeds, too heavy to hold nose-high attitude—1. Too heavy at low speeds—1. Force high for a fighter—1.

Effectiveness: O. K. at above 250; below this reaction is slow—1.

Spoilers	*Good*	*Fair*	*Poor*	*High*	*Moderate*	*Light*
Force	14	0	0	1	0	6
Effectiveness	5	4	0	0	0	0
General	1	0	0	0	0	0

Remarks.—Force: Shallow gradient generally over speed range—1. Force heavy at speeds over 350, very light at low speeds—1. No appreciable increase with speed—1.

Effectiveness: Slow response—5. Low rate of roll—2. Very effective at high speeds; very effective in stall area, but large deflections required—1. Fair at cruising speed and above; poor to fair in landing condition and low speed—1.

General: There is considerable rather heavy buffet when spoilers are displaced to a large extent at speeds in neighborhood of 250 miles per hour—1.

Rudder	*Good*	*Fair*	*Poor*	*High*	*Moderate*	*Light*
Force	6	0	0	7	1	2
Effectiveness	11	1	0	0	0	0
General	1	0	0	0	0	0

Remarks.—Force high at high speeds. Can hold military power with one engine at stall—1. Fairly heavy forces, but rudder is not needed much—1.

TRIMMABILITY

Good	11
Fair	0
Poor	0
Other	10
Blank	0
Total	21

Remarks

Tabs are sensitive—5. Tabs are too sensitive—6. Tabs poorly situated—4. Like absence of aileron trim, and light aileron forces make it unnecessary—1. Should have aileron trim—1.

	A	*B*	*C*	*D*	*Other*	*Blank*
Maneuverability	1	3	10	3	0	4

Remarks.—Its maneuverability is good for such a large plane, but it may find it hard to stay behind a Jap night bomber. The P-61 is not easily controlled in the clean condition below 110 miles per hour. The Japs sometimes fly this slow just to bitch up night fighters—1. Turning radius is excellent considering wing loading. Acceleration too slow; not enough power—1.

	A	*B*	*C*	*D*	*Other*	*Blank*
Night and Instrument Flying Rating	11	2	1	1	0	6
Gun Platform Rating	5	8	1	3	0	4

DIVING CHARACTERISTICS

	Good	*Fair*	*Poor*	*High*	*Moderate*	*Light*
Acceleration	4	0	1	0	2	2
Stick Force	6	0	0	6	0	0

P–61—Continued

	Good	*Fair*	*Poor*	*Fast*	*Moderate*	*Slow*
Recovery	9	0	0	0	0	0
General	5	0	0	0	0	0

Remarks.—Stick Force: Aileron and elevator leads just right; rudder too heavy—1.
Recovery: Controls well harmonized—1. Nice directional stability in dive—1.
General: Controls, except rudder, much lighter than at low speeds—no sign of hunting—recovery good—1. Small change in directional trim, but rudder forces should be lighter—1.

STALL CHARACTERISTICS

	Clean		*L/C*		*Accelerated*
	Power Off	*Power On*	*Power Off*	*Power On*	*3g*
IAS (MPH):					
Average	112	101	92	85	137
Range	105–120	90–112	80–105	70–99	125–150
Warning:					
Good	8	5	10	4	6
Fair	0	0	0	0	0
Poor	0	0	0	0	0
Type	Buffet—2 Pitch—2 Shudder—1 Terrific—1	Buffet—2 Terrific—1	Buffet—2 Pitch—2 Shudder—1	Buffet—2 Vertical Oscillation—1	*See below

Warning:
*Continued acceleration results in buffeting and periodic dropping of nose. No fall-out—1.

	Clean		*L/C*		*Accelerated*
Recovery:	*Power Off*	*Power On*	*Power Off*	*Power On*	*3g*
Good	10	3	12	3	2
Fair	0	0	0	0	0
Poor	0	0	0	0	0
Remarks	Pitching oscillation	Shakes	Pitching oscillation	Rolls left	**

**Buffet, but stable—1. Buffeting downward (no wing drop). Pitching and recovery—1.
General Remark.—Every stall results in nose dropping forward. Never any tendency to fall off on one wing—1.

P-63

Army—1 British —1
Navy —7 Contractors—12

		Remarks
Cockpit		
Good	3	Small, cramped—13. Roof is too low—1. Visibility obstructed by door frames—5. Poor vision except ahead—1. Poor side visibility—1. Very blind—1. Good lay-out—6. Cluttered, confused—3. Too many gadgets—1. Many hazards to pilot in crash—1. Trim tabs hard to reach—2. Rudder tab is badly placed—1. Hand crank for wheel hits leg—1. Seat not adjustable—1. Flap and landing gear switch not easy to recognize—1. Automatic features for engine very good—2. Noisy—1.
Fair	2	
Poor	0	
Other	16	
Blank	0	
Total	21	
Comfort		
Good	3	Too small, cramped—10. Noise level high—2. Too much vibration—2. No seat adjustment—1. Ventilation good—1.
Fair	4	
Poor	3	
Other	10	
Blank	1	
Total	21	
Ground Handling		
Good	12	High revolutions per minute (1,000–1,300) necessary for taxying because engine idles very roughly—requires excessive use of brakes—5. Brakes poor—2. Too soft—3. Too stiff, heavy—3. Positive and effective—1. Visibility good—3. Control good—1.
Fair	1	
Poor	0	
Other	8	
Blank	0	
Total	21	
Power Plant Operation		
Good	2	Rough at idling speeds—8. Rough—6. Rough in fast throttle changes—1. Smooth—1. Noisy—2. Beat vibration at 2,240 revolutions per minute—1. Pulsating vibration below 2,350 revolutions per minute—seems like carburation—1. Like combination propeller and throttle control—4. Recommend optional manual control override—2. Combination propeller and throttle control undesirable because it prohibits small revolutions per minute changes and makes engine rough in fast throttle changes —1.
Fair	1	
Poor	1	
Other	17	
Blank	0	
Total	21	
Take-Off		
Good	7	Long, slow—6. Good rudder control—2. Not enough rudder control—3. Takes full right rudder—1. Tendency to swerve left with sudden application of power—1. Strong torque—1. Exceptional power—1. Mushes if nose wheel pulled off too soon—1.
Fair	0	
Poor	1	
Other	11	
Blank	2	
Total	21	
Approach and Landing		
Good	10	Good visibility and control—3. Very easy—typical tricycle—1. Long roll—2. Approach speed too high—1. Landing speed too high—1. Good control prior to landing. Good directional control after landing—1. Good roll. No tendency to swing—1. At landing, stick forces very heavy—1. Do not like the rudder - slow landing condition. Too weak laterally in landing condition—1.
Fair	0	
Poor	0	
Other	5	
Blank	6	
Total	21	

P–63—Continued

WAVE-OFF

Good	3
Fair	2
Poor	5
Other	0
Blank	11
Total	21

Light and plenty of power—1. Insufficient rudder—1. At 110 IAS in landing condition it takes 2 seconds for full power to develop. Could not be used as CV airplane—1.

BAIL-OUT

Good	4
Fair	2
Poor	7
Other	1
Blank	7
Total	21

Good chance of striking tail—3.

COMBAT QUALITIES

Good	1
Fair	0
Poor	4
Other	6
Blank	10
Total	21

Visibility to rear not too good. Cockpit heater insufficient above 12,000 feet—1. Not up to most VF, but much better than old P–39. Difficult to cool in hot climate. Visibility aft not too good. Engine very rough. Not a carrier airplane. I do not like the rudder, slow L/C, or fast strafing dive. Plane too weak directionally for good gunnery and too weak laterally in L/C. Need aileron trim tab. Engine roughness and vibrations—miserable—1. If visibility can be improved this would be a desirable plane from performance standpoint—1. Probably not enough performance. Maneuverability is limited because of heavy aileron forces. Not enough visibility to the sides and aft—1. This airplane is what the P–39 ought to have been. Too late now—1. Excellent at high speeds (above 230 miles per hour) for maneuverability. Visibility is good. Stick force light. Small target for strafing runs. Climb is good at normal power, excellent at powers above this. Zoom is excellent from cruise power—1. Slow level and acceleration. It would take superior pilot technique to use in Pacific—1. Not as good as P–51 or F6F as all-round fighter—1. Climb too slow possibly. Better rearward and all-around visibility needed—1. Would be best Army fighter with a bubble canopy—1. Not suitable for combat. Heavy controls and instability. Poor vision—1.

STABILITY

Longitudinal	*Good*	*Fair*	*Poor*	*Positive*	*Weak Positive*	*Neutral*	*Unstable*
Climb	7	1	0	0	1	1	0
Level	8	0	0	1	1	0	0
L/C	1	0	0	0	0	0	0
General	1	0	0	1	0	0	0

Remarks.—Climb: Dynamically positive, statically weak—1.
Level: Weak—1.
L/C: Rudder reverses—1.

Lateral	*Good*	*Fair*	*Poor*	*Positive*	*Weak Positive*	*Neutral*	*Unstable*
Climb	6	0	1	1	1	2	0
Level	7	0	0	2	1	1	0
L/C	1	0	0	1	0	0	1
General	0	2	0	1	0	0	0

Remarks.—L/C: Much change of longitudinal trim with rudder drift—1.
General: Lack of aileron trim tab makes steady flight difficult—1.
Right wing heavy—1. Poor during power changes—1.

Directional	*Good*	*Fair*	*Poor*	*Positive*	*Weak Positive*	*Neutral*	*Unstable*
Climb	4	1	2	2	2	1	0
Level	4	1	1	2	2	0	0
L/C	1	0	0	1	0	0	0
General	2	0	0	1	0	0	0

Remarks.—Climb: Rudder gets soft in climb—1. Rudder reversal—1.
General: Too weak directionally for good gunnery—1. Satisfactory for gun platform—1.

P–63—Continued

Controllability

Elevators	Good	Fair	Poor	High	Moderate	Light
Force	5	0	0	1	2	2
Effectiveness	10	1	0	0	0	0
General	4	0	0	0	0	0

Remarks.—Force: Low (too little movement per G)—1. Heavy with speed—1.

Ailerons	Good	Fair	Poor	High	Moderate	Light
Force	2	0	0	7	0	1
Effectiveness	8	1	0	1	0	0
General	3	0	0	0	0	0

Remarks.—Force: Heavy at high speeds (200–400 IAS and higher)—4.
Effectiveness: Roll fairly good—1. Rate of roll at high speeds satisfactory, could be improved at low speeds—1.
General: Needs aileron tab—1.

Rudder	Good	Fair	Poor	High	Moderate	Light
Force	5	0	0	0	1	2
Effectiveness	6	1	0	0	0	0
General	5	0	0	0	0	0

Remarks.—Force: Force gradient low—1. Not enough control—1. Too heavy—1.
Effectiveness: Strong—1. No good for take-off—1. Marginal to deficient for take-off—1.
General: Mushy—soft at high deflection—1. Tend to over balance—1. Moderate amount of trim needed—1.

Trimmability

Remarks

Good	10
Fair	1
Poor	3
Other	6
Blank	1
Total	21

Need aileron trim tabs—4. Need aileron trim tab with landing gear down—1. Rudder tab inaccessible—1. Cannot reach tabs—1. Too much change of trim with speed and power—2. Changes directional trim rapidly with speed—2. Change in speed (directional) good—1. Does not require much trimming—1. Rudder requires constant attention—1. Sluggish—1. Elevator tab not sensitive enough—1.

	A	B	C	D	Other	Blank
Maneuverability	3	8	8	1	0	1

Remarks.—Excellent ailerons for rate of roll—1. Heavy ailerons—1. Large turning circle—1. Control difficulties Rudder reversal—1.

	A	B	C	D	Other	Blank
Night and Instrument Flying Rating	0	2	3	1	0	15

Remarks.—None.

	A	B	C	D	Other	Blank
Gun Platform Rating	0	9	4	1	0	7

Remarks.—Too much trim change—2. Poor visibility—1.

Diving Characteristics

	Good	Fair	Poor	High	Moderate	Light
Acceleration	5	0	0	0	0	0
Stick Force	2	0	0	2	1	2

	Good	Fair	Poor	Fast	Moderate	Slow
Recovery	7	0	0	0	0	0

Remarks.—Acceleration: Rapid—3. Normal—1. Slow—1.
Stick Force: A little stiff fore and aft—1. Controllable—1.
Recovery: Recovers without force—1. Normal—1.
Straight forward—1.
General: Ailerons too heavy—1. Trim changes from cruise too high—1.

STALL CHARACTERISTICS

	Clean *Power Off*	*Power On*	*L/C* *Power Off*	*Power On*	*Accelerated* *3g*
IAS (mpe)					
Average	93	86	90	82	132
Range	78–100	75–95	79–130	70–110	105–170

Warning:

Good 8
Fair 0
Poor 0
Type .... Buffet shake—8.
Remarks.—Slight warning in landing condition—4. Rudder effectiveness is weak—1.

Recovery:

Good 4

FIREFLY

Army—0 British —0
Navy —1 Contractors—9

Remarks

Cockpit		Remarks
Good	3	Poor arrangement—2. Control arrangement fair—1. Well set up; especially tabs nicely grouped—1. Accessibility good except for radio receiver. Harness particularly convenient—1. Typical RAF, not bad if you are used to it—1. Canopy distortion—1.
Fair	0	
Poor	2	
Other	2	
Blank	3	
Total	10	

Comfort		Remarks
Good	5	Too upright sitting position—1. Cockpit noisy—vibration from power plant uncomfortable—1.
Fair	1	
Poor	0	
Other	1	
Blank	3	
Total	10	

Ground Handling		Remarks
Good	5	Brake system good—3. Blind over nose—2. Rudder uncomfortable—1. Don't like brake system—1.
Fair	0	
Poor	0	
Other	2	
Blank	3	
Total	10	

Take-Off		Remarks
Good	7	Directional control on take-off marginal. This may have been due to opposite rudder torque—1.
Fair	0	
Poor	0	
Other	1	
Blank	2	
Total	10	

Approach and Landing		Remarks
Good	7	Ground handling excellent during landing roll—1. Doesn't need tail wheel lock, easy to hold straight—1.
Fair	0	
Poor	0	
Other	1	
Blank	2	
Total	10	

Wave-Off		Remarks
Good	5	None.
Fair	0	
Poor	1	
Other	0	
Blank	4	
Total	10	

Bail-Out		Remarks
Good	2	Light fit—1. O. K. if canopy can be ditched—1. Cookpit tight. Hit head on antenna getting in—1 Questionable—1.
Fair	0	
Poor	1	
Other	3	
Blank	4	
Total	10	

FIREFLY—Continued

Combat Qualities

Good	0
Fair	0
Poor	0
Other	6
Blank	4
Total	10

Do not see a spot for it either in Pacific or elsewhere. Japs could out-fight it and the Firefly could not run away. Do not consider it adaptable to a night fighter—1. Very similar in all respects to the Fairey "Battle" although many improvements add to its suitability—1. Performance low and rearward visibility bad, but stability controls and general handling qualities good for night fighting. Performance believed too low for modern warfare—1. Too slow and clumsy. Would make fair dive bomber if fitted with dive brakes—1. Vision aft is not good enough, performance not good enough, too much changing of lateral and directional trim required—1.

Stability

Longitudinal	*Good*	*Fair*	*Poor*	*Positive*	*Weak, Positive*	*Neutral*	*Unstable*
Climb	3	0	0	1	0	1	0
Level	2	0	0	2	0	0	0
L/C	2	0	0	0	0	0	0
General	2	0	0	0	0	0	0

Lateral	*Good*	*Fair*	*Poor*	*Positive*	*Weak, Positive*	*Neutral*	*Unstable*
Climb	3	0	0	1	0	1	0
Level	2	0	0	0	0	0	0
L/C	2	0	0	0	0	0	0
General	2	1	0	0	0	0	0

Remarks.—Not stable all ranges, but satisfactory—1.

Directional	*Good*	*Fair*	*Poor*	*Positive*	*Weak, Positive*	*Neutral*	*Unstable*
Climb	2	1	0	1	1	0	0
Level	1	1	0	0	0	0	0
L/C	1	1	0	0	0	0	0
General	2	1	0	0	0	0	0

Remarks.—Climb: Positive; too much so for a fighter—1.

Controllability

Elevators	*Good*	*Fair*	*Poor*	*High*	*Moderate*	*Light*
Force	4	0	0	0	1	1
Effectiveness	5	0	0	0	0	0
General	3	0	0	0	0	0

Remarks.—Stiff at high speeds—1.

Ailerons	*Good*	*Fair*	*Poor*	*High*	*Moderate*	*Light*
Force	3	1	0	0	1	1
Effectiveness	5	0	0	0	0	0
General	3	0	0	0	0	0

Remarks.—Stiff at high speeds—1. Effectiveness good except with full flaps—1.

Rudder	*Good*	*Fair*	*Poor*	*High*	*Moderate*	*Light*
Force	3	1	0	0	2	0
Effectiveness	5	0	0	0	0	0
General	3	0	0	0	0	0

Remarks.—Force becomes heavy with speed—2.

Trimmability

Good	6
Fair	0
Poor	0
Other	1
Blank	3
Total	10

	A	*B*	*C*	*D*	*Other*	*Blank*
Maneuverability	1	4	3	0	0	2
Night and Instrument Flying Rating	1	2	1	0	0	6
Gun Platform Rating	0	6	0	1	0	3

FIREFLY—Continued

Diving Characteristics

	Good	*Fair*	*Poor*	*High*	*Moderate*	*Light*
Acceleration	0	1	0	0	1	0
Stick Force	0	0	0	2	1	0
	Good	*Fair*	*Poor*	*Fast*	*Moderate*	*Slow*
Recovery	3	0	0	0	0	0
General	1	1	0	0	0	0

Remarks.—Acceleration: Fast—1.
Stick Force: Too much change in directional and longitudinal trim—1.

Stall Characteristics

	Clean		*L/C*		*Accelerated*
	Power Off	*Power On*	*Power Off*	*Power On*	*3g*
IAS (mph):					
Average	77	72	66	65	140
Range	75-80	65-80	62-69	65-65	
Warning:					
Good	2	2	2	1	3
Fair	0	0	0	0	0
Poor	0	0	0	0	0
Little	1	1	1	0	1
None	1	2	2	1	0
Recovery:					
Good	6	4	4	2	2
Fair	0	1	0	0	0
Poor	0	0	1	0	0

Remarks.— Lots of directional control—1. Drops right—1. Pitch—1. Right wing drop in split S—1.

SEAFIRE

Army—2 British —0
Navy —0 Contractors—9

Remarks

Cockpit		Remarks
Good	1	Poor arrangement—5. Unfamiliar—naturally am prone to like American arrangement—1. Typically British—seems odd at first but really is quite simple—1. Small; very close to all controls. Visibility in air good—1. Cramped, but O. K.—1. Like seat adjustment—1.
Fair	2	
Poor	0	
Other	8	
Blank	0	
Total	11	

Comfort		Remarks
Good	5	Cramped—3.
Fair	2	
Poor	3	
Other	0	
Blank	1	
Total	11	

Ground Handling		Remarks
Good	5	Poor visibility over nose—3. Brake system very good except caution necessary or brakes can be easily burned out as they cannot be felt out—1. Air brake hard to get used to—1. Definitely dislike hand brakes—1.
Fair	3	
Poor	0	
Other	3	
Blank	0	
Total	11	

Power Plant Operation		Remarks
Good	8	
Fair	1	
Poor	0	
Other	2	
Blank	0	
Total	11	

Take-Off		Remarks
Good	10	Rapid—2. Directional control good—1.
Fair	0	
Poor	0	
Other	1	
Blank	0	
Total	11	

Approach and Landing		Remarks
Good	7	Good control in ailerons down to the stall; good flare out—1. Handles nicely on approach; visibility good considering long nose. Not enough time on landings for a good opinion. Handled well—some tendency to float—1. Excellent, low landing speed and good control—1. Good. Elevators very touchy—1. Very slow landing—1.
Fair	1	
Poor	0	
Other	2	
Blank	1	
Total	11	

SEAFIRE—Continued

WAVE-OFF

Good	3
Fair	2
Poor	0
Other	0
Blank	6
Total	11

BAIL-OUT

Good	4
Fair	1
Poor	1
Other	1
Blank	4
Total	11

Would appear difficult to bail out of side or out of top due to small clearance and sharp protrusions—1.

COMBAT QUALITIES

Good	1
Fair	1
Poor	0
Other	4
Blank	5
Total	11

Just fair; it is surpassed by many others—1. An outstanding plane at the time it was designed. Still a good fighter, but naturally not equal to latest types (these are first opinions only, based on 1 hour's flight)—1. Good fighter if performance compares favorably with other modern fighters. Should have bubble canopy—1. Would be an excellent airplane because of its very slow landing speed and fast rated climb. War experience has proved it to be a good gun platform—1. Believe low V max. performance and poor rate of climb would not be compensated for in the good maneuverability of this ship. Japs would knock it hard, I believe—1. Very maneuverable—vision to rear poor—1.

STABILITY

Longitudinal	*Good*	*Fair*	*Poor*	*Positive*	*Weak Positive*	*Neutral*	*Unstable*
Climb	3	1	0	0	0	2	0
Level	3	2	0	0	0	0	0
L/C	1	0	0	0	0	0	0
General	0	0	0	0	0	0	0

Remarks.—Level: Sensitive—1. Dynamic instability—1.
General: Neutral dynamically, O. K. statically—1. At high speeds ship is longitudinally unstable; has a moderate tendency to dig when g is pulled in any plane of flight—1.

Lateral	*Good*	*Fair*	*Poor*	*Positive*	*Weak Positive*	*Neutral*	*Unstable*
Climb	3	1	0	0	0	2	0
Level	3	2	0	0	0	0	0
L/C	1	0	0	0	0	0	0
General	1	0	0	0	0	0	0

Remarks.—Climb: Sensitive—1.
General: Bad Dutch-roll effect—1.

Directional	*Good*	*Fair*	*Poor*	*Positive*	*Weak Positive*	*Neutral*	*Unstable*
Climb	2	1	0	0	0	1	0
Level	4	1	0	0	0	0	0
L/C	1	0	0	0	0	0	0
General	1	0	0	0	0	0	0

Remarks.—Climb: Sensitive—1. Noticeable yawing—1. Statically stable; too much so for a fighter—1.

CONTROLLABILITY

Elevators	*Good*	*Fair*	*Poor*	*High*	*Moderate*	*Light*
Force	3	0	0	0	1	5
Effectiveness	8	0	0	0	0	0
General	2	0	0	0	0	0

Remarks.—Force: Stick forces a bit light in accelerated turns—1.
Effectiveness: Too effective—1.

SEAFIRE—Continued

Ailerons	*Good*	*Fair*	*Poor*	*High*	*Moderate*	*Light*
Force	3	0	0	1	0	5
Effectiveness	9	0	0	0	0	0
General	1	1	0	0	0	0

Remarks.—Stiffen excessively at high speeds—1.

Rudder	*Good*	*Fair*	*Poor*	*High*	*Moderate*	*Light*
Force	3	0	0	0	0	6
Effectiveness	7	0	0	0	0	0
General	0	0	0	0	0	0

Remarks: Effectiveness: Very sensitive—1. Too effective—1.

Trimmability

Remarks

Good	7
Fair	2
Poor	1
Other	1
Blank	0
Total	11

O. K. Rudder control operates in wrong plane—1. Can be trimmed well hands off—1. Light—no high power used for take-off so no check on torque made, otherwise O. K.—1. Usual light and effective British controls—1.

	A	*B*	*C*	*D*	*Other*	*Blank*
Maneuverability	8	3	0	0	0	0
Night and Instrument Flying Rating	1	1	3	1	0	5
Gun Platform Rating	1	4	3	0	0	3

Diving Characteristics

	Good	*Fair*	*Poor*	*High*	*Moderate*	*Light*
Acceleration	4	0	0	0	0	0
Stick Force	0	0	0	2	0	4

	Good	*Fair*	*Poor*	*Fast*	*Moderate*	*Slow*
Recovery	5	1	0	0	0	0
General	1	0	0	0	0	0

Remarks.—Acceleration: Rapid—1. Slow—1.
Stick Force: Too heavy—1. Elevator force too high—1. Decrease in stick force with g—1. Reverse—1.
Recovery: Wants to snap out—1. Mushes moderately—1.

Stall Characteristics

	Clean		*L/C*		*Accelerated*
	Power Off	*Power On*	*Power Off*	*Power On*	*3g*
IAS (MPH):					
Average	70	63	58	51	140
Range	63-79	55-76	55-67	50-52	
Warning:					
Good	4	2	4	2	4
Fair	1	0	1	0	1
Poor	0	0	0	0	0
Type	Moderate shaking—1.				Normal—1 Buffet—1.
Recovery:					
Good	4	3	5	3	2
Fair	0	0	0	0	0
Poor	0	0	0	0	0
Remarks.—					Abrupt—1. Tends to dig with back stick pressure—1.

MOSQUITO

Army—0 Navy—2 British—0 Contractors—10

		Remarks
Cockpit		
Good	1	Great space for bubble. Position is upright and uncomfortable. Radio box interferes with elbow. Cramped. Rudder tab near gun sight. Opportunistic—1. Comfortable, but ingress and egress terrible; arrangement of controls not satisfactory; radio set-up awful—1. Confused—1. Horrible. Cluttered, crowded, uncomfortable, unsystematic arrangement. However, visibility is excellent—1.
Fair	0	
Poor	1	
Other	9	
Blank	1	
Total	12	
Comfort		
Good	4	No. Position too upright—1. Noisy—otherwise good—1. Bad—too cramped—1. Seemed cramped—1. Most comfortable British airplane at show—1.
Fair	0	
Poor	3	
Other	3	
Blank	2	
Total	12	
Ground Handling		
Good	7	Normal after getting accustomed to hand braking. Nose over—1. Good visibility, good brake control—1. Fairly good. Not used to brakes and believe them harder to use than toe brakes. No difficulty experienced however—1. Good English brakes are very easy to use but I do not prefer them. In fact the English method seems more complicated because it involves the use of both feet and hands—1.
Fair	1	
Poor	0	
Other	3	
Blank	1	
Total	12	
Power Plant Operation		
Good	4	Simple enough—auto blower shift seems very useful—1. Easy and simple—1. Tendency to overheat on ground—1. Easy-good shapes on handles—would prefer feather buttons on left—1. Quite like an American plane. Surprised to see manifold pressure gauge—1. O. K. if dependable. (2 mags went out in 2 days)—1. Normal—1. Slightly rough but good—1. Smooth—1.
Fair	0	
Poor	0	
Other	7	
Blank	1	
Total	12	
Take-Off		
Good	2	Long—4. Poor directional control at beginning of run—4. Gear pull-up too slow—1.
Fair	0	
Poor	0	
Other	7	
Blank	3	
Total	12	
Approach and Landing		
Good	2	Flaps down requires nose down trim. Ailerons overbalance completely. O. K. except for sloppy lateral control—1. Very poor. Sloppy control in approach condition and landing speed too high. Difficult to make smooth landing—1. Requires great power with wheels and flaps down—1. Good visibility. Needed quite a lot of power to drag in. Not too good directionally after landing. Needs tail wheel lock—1. O. K. except for rather high approach speed—1. Vision good, approach sloppy—1. Fast glide-landing run would be easier if airplane had tail wheel lock. Very hard gear—1.
Fair	0	
Poor	1	
Other	5	
Blank	4	
Total	12	

MOSQUITO—Continued

Wave-Off

Good	0
Fair	0
Poor	3
Other	3
Blank	6
Total	12

Acceleration slow—2. Ailerons overbalance, rudder weak—1.

Bail-Out

Good	3
Fair	1
Poor	5
Other	0
Blank	3
Total	12

Combat Qualities

Good	2
Fair	0
Poor	0
Other	6
Blank	4
Total	12

Seems useful because of speed and bomb load and ease of construction. Side by side good for VFN. Needs better flap for slower landing—1. Probably satisfactory from fighter standpoint, but maintenance difficulties along with attendant problems of plywood construction might rule against their use—1. Controls are too light. Climb is not good enough—1. Unsuitable for night operations because of landing and take-off characteristics and bad field and weather encountered in Pacific. Excellent for low level attack bombing—1. Very good machine but could use more speed now and lighter elevators. This plane is still one of the best fighter bombers in existence. A development program to increase speed and lighten the elevators to 10#/G would put it at the top of the list. It has the best rudder I have flown on a twin engine airplane—1. War record speaks for itself—1. Hit and run OK—1. Specialized purpose only—1.

Stability

Longitudinal	*Good*	*Fair*	*Poor*	*Positive*	*Weak Positive*	*Neutral*	*Unstable*
Climb	3	1	0	3	0	1	1
Level	2	1	0	2	1	1	1
L/C	0	0	0	0	1	0	1
General	0	0	0	0	0	0	0

Lateral	*Good*	*Fair*	*Poor*	*Positive*	*Weak Positive*	*Neutral*	*Unstable*
Climb	1	0	0	1	1	2	0
Level	4	0	0	2	0	1	0
L/C	0	0	0	0	0	1	0
General	0	0	0	0	0	0	0

Directional	*Good*	*Fair*	*Poor*	*Positive*	*Weak Positive*	*Neutral*	*Unstable*
Climb	2	0	1	2	2	0	0
Level	2	0	0	1	3	0	0
L/C	0	0	0	0	0	0	1
General	0	1	0	0	0	0	0

Remarks.—General: Adequate except for slow single engine operation—1. Too much pitch with yaw—1.

Controllability

Elevators	*Good*	*Fair*	*Poor*	*High*	*Moderate*	*Light*
Force	3	0	0	2	0	2
Effectiveness	3	1	0	0	0	0
General	0	0	0	0	0	0

Remarks.—Force: Sloppy at low speeds—1.
Effectiveness: Poor at low speed—2. Good above 200 mph—1. Lag in elevator—1.
General: Good feel but too heavy—1.

MOSQUITO—Continued

Ailerons	*Good*	*Fair*	*Poor*	*High*	*Moderate*	*Light*
Force	3	1	0	0	0	2
Effectiveness	3	1	0	0	0	1
General	0	0	0	0	0	0

Remarks.—Force: Sloppy at low speeds—1.
Effectiveness: Poor at low speeds—4.
General: Don't like at approach speeds—1. Nice feel and good rate of roll—1.

Rudder	*Good*	*Fair*	*Poor*	*High*	*Moderate*	*Light*
Force	2	0	0	0	0	3
Effectiveness	3	0	0	0	0	1
General	1	0	0	0	0	0

Remarks.—Effectiveness: Poor at low speed—3. General: Good balance of forces—2.

TRIMMABILITY

		Remarks
Good	5	Elevator tab wheel is inaccessible—2. Change of trim with radiator flap not good—1. Requires little trim change with change of configuration or speed—1. Must use considerable amount of elevator trim—1.
Fair	1	
Poor	0	
Other	4	
Blank	2	
Total	12	

	A	*B*	*C*	*D*	*Other*	*Blank*
Maneuverability	2	5	3	1	0	1
Night and Instrument Flying Rating	2	2	2	0	0	6
Gun Platform Rating	2	5	3	0	0	2

DIVING CHARACTERISTICS

	Good	*Fair*	*Poor*	*High*	*Moderate*	*Light*
Acceleration	1	0	1	0	0	0
Stick Force	2	0	0	2	1	0
Recovery	3	0	0	0	0	0
General	0	0	1	0	0	0

Remarks.—Acceleration: Rapid—2. Normal—1.
Stick Force: High when pulled quickly—1. 6–8 lbs/g—1.

STALL CHARACTERISTICS

	Clean		*L/C*		*Accelerated*
IAS (kts):	*Power Off*	*Power On*	*Power Off*	*Power On*	*3g*
Average	133	122	107	97	0
Range	125–155	120–125	85–119	90–102	0
No. of Est	7	4	7	3	0
Warning:					
Good	2	0	3	0	3
Fair	0	0	0	0	0
Poor	1	0	0	0	0

Remarks.—Clean Power Off: Slight shake—1. Good shake and buffet—1. Buffet—2. Good buffet—1.
Clean Power On: Slight shake—1. Good shake and buffet—1. Buffet—1. Good buffet—1.
L/C Power Off: Very slight shake—1. Elevators shake—1. Shake and buffet—1. Buffet—2. Good and very rough buffet—1. Good but sharp—1.
L/C Power On: Shake and buffet—1. Elevator shake—1. Buffet—1. Good buffet—1.
Accelerated 3/g: Aileron overbalanced, rudder weak, elevator OK at stall—1. No wingdropping—1. Some rudder buffet—1. Reversal of aileron forces—1.

Recovery:

	Clean		*L/C*		*Accelerated*
IAS (kts):	*Power Off*	*Power On*	*Power Off*	*Power On*	*3g*
Good	6	3	7	2	1
Fair	0	0	0	0	0
Poor	0	0	0	0	0

ZEKE 52

Army—0 British —0
Navy —1 Contractors—9

COCKPIT

Good	0
Fair	3
Poor	1
Other	6
Blank	0
Total	10

Good vision—2. Poor arrangement—2. Small but all right for small person—2. Too many gadgets for guns—1. Best seat adjustment in any ship—1.

COMFORT

Good	4
Fair	1
Poor	1
Other	2
Blank	2
Total	10

Too cramped—3. Good for small pilot—3.

GROUND HANDLING

Good	6
Fair	1
Poor	0
Other	3
Blank	0
Total	10

Good brakes—2. Poor brakes—5. Poor rudder operation—2. Needs tail-wheel lock—1. Needs self-centering tail-wheel—1.

POWER PLANT OPERATION

Good	4
Fair	1
Poor	0
Other	5
Blank	0
Total	10

Operates fairly smoothly, similar to any small plane like an SNJ—1. Fairly smooth, and cooling does not seem critical. Prop governing stinks. Easy to overboost. Don't like cooling flap controls—1. Normal to United States airplanes—1. No auto-boost regulator—1. Quick acting prop—1. Noisy—1.

TAKE-OFF

Good	7
Fair	1
Poor	0
Other	2
Blank	0
Total	10

Very little torque, easy to take off—1. Swerves left—1. Short initial climb good—1. Short, directional control O. K.—1. Tab settings unimportant—1.

APPROACH AND LANDING

Good	5
Fair	0
Poor	0
Other	3
Blank	2
Total	10

Very easy to land. Flies like a light trainer—1. Excellent visibility, good control, poor directional control after landing—1. Very slow landing. Control good—1. Plenty of vision, ease of landing, no difficulty—1. Approach was comfortable but rudder action on ground was uncomfortable. Did not trust brakes—1.

ZEKE 52—Continued

WAVE-OFF

Good	6
Fair	0
Poor	0
Other	0
Blank	4
Total	10

Excellent acceleration—1. Plenty of excess power available—1.

BAIL-OUT

Good	5
Fair	1
Poor	0
Other	1
Blank	3
Total	10

No emergency egress—1.

COMBAT QUALITIES

Good	0
Fair	0
Poor	2
Other	4
Blank	4
Total	10

It is a dangerous airplane to dog fight at slow speeds—4. Fighting qualities good at low speed, poor at high speed—1. Excellent for low altitude offensive combat or any turning fight where radius of turn or maneuverability is required as prime—1. Maneuverability is best feature, but such items as poor pilot protection, extremely poor ailerons and only fair performance detract from its usefulness—1. Very poor in relation to present American fighters because of low performance, no armoring and stiffness of controls at high speeds—1. Record stands for self—1.

STABILITY

Longitudinal	*Good*	*Gair*	*Poor*	*Positive*	*Weak Positive*	*Neutral*	*Unstable*
Climb	2	0	0	2	1	1	0
Level	1	0	0	1	2	0	0
L/C	0	0	0	0	1	0	0
General	1	0	0	2	0	0	0

Lateral	*Good*	*Fair*	*Poor*	*Positive*	*Weak Positive*	*Neutral*	*Unstable*
Climb	3	0	0	0	0	1	1
Level	2	1	0	0	0	1	1
L/C	1	0	0	0	0	0	0
General	0	1	0	2	0	0	0

Remarks.—Climb: Normal dihedral effect—1.

Directional	*Good*	*Fair*	*Poor*	*Positive*	*Weak Positive*	*Neutral*	*Unstable*
Climb	3	0	0	1	1	0	0
Level	2	0	0	1	0	0	0
L/C	1	0	0	0	0	0	0
General	1	0	0	2	0	0	0

Remarks.—Climb: Yaw due to yaw normal, damping slightly weak—1.

CONTROLLABILITY

Elevators	*Good*	*Fair*	*Poor*	*High*	*Moderate*	*Light*
Force	0	0	0	1	2	1
Effectiveness	5	0	0	0	0	0
General	4	0	0	0	0	0

Remarks.—Force: Light at low speeds, freeze at high speeds—1. Lighter than rest of controls.
General: Forces and effectiveness good at slow speeds—1.

ZEKE 52—Continued

Ailerons	*Good*	*Fair*	*Poor*	*High*	*Moderate*	*Light*
Force	0	0	0	6	0	0
Effectiveness	0	1	1	0	0	0
General	1	1	0	0	0	0

Remarks.—Force: Light at low speed—5. Freeze at high speed—8.
Effectiveness: Good at low speed—3.

Rudders	*Good*	*Fair*	*Poor*	*High*	*Moderate*	*Light*
Force	1	0	0	1	0	3
Effectiveness	3	2	0	0	0	0
General	2	1	0	0	0	0

Remarks.—Force: Light at low speeds, freeze at high speeds.
General: Forces and effectiveness very good at slow speeds—1.

TRIMMABILITY

Good	2
Fair	1
Poor	3
Other	3
Blank	1
Total	10

Rudder trim: inconvenient—1. Ineffective—1. Insufficient—2. None—1. Elevator trim: ineffective—1, fair—1, too much change with change in speed—1. Aileron trim: None.

	A	*B*	*C*	*D*	*Other*	*Blank*
Maneuverability	5	1	0	0	4	0

Remarks.—A at low speeds—6. D at high speeds—3.

	A	*B*	*C*	*D*	*Other*	*Blank*
Night and Instrument Flying Rating	2	2	0	0	0	6
Gun Platform Rating	1	3	3	0	0	3

Remarks.—A at slow speeds—1.

DIVING CHARACTERISTICS

	Good	*Fair*	*Poor*	*High*	*Moderate*	*Light*
Acceleration	0	1	2	0	0	0
Stick Force	0	0	1	4	0	0
Recovery	2	0	2	0	0	0
General	1	0	1	0	0	0

Remarks.—Acceleration: Slow—2. Moderately fast—1. Fast to 250 knots—1.
Stick Force: Ailerons very heavy—3. Directional and longitudinal trim fair—1. Rudder force change with increase in speed is excessive—1.
Recovery: Slow—2.
General: Not very steady.

STALL CHARACTERISTICS

IAS (mph):	*Clean*		*L/C*		*Accelerated 3g*
	Power Off	*Power On*	*Power Off*	*Power On*	
Average	72	64	67	63	
Range	68-75	60-70	60-73	62-63	

ZEKE 52—Continued

Warning:	Clean Power Off	Clean Power On	L/C Power Off	L/C Power On	Accelerated 3g
Good	5	4	4	3	4
Fair	0	0	0	0	0
Poor	0	0	0	0	1

Description

Remarks.—Clean Power Off: Tail Buffet—2.
Clean Power On: Tail Buffet—2. Slight shudder—1. Sloppy controls—1.
L/C Power Off: Tail Buffet—2.
L/C Power On: Tail Buffet—2.
Accelerated: Stick Buffet—1.

Recovery:	Clean Power Off	Clean Power On	L/C Power Off	L/C Power On	Accelerated 3g
Good	5	5	1	3	0
Fair	0	0	0	0	0
Poor	0	0	0	0	0

Remarks.—Clean Power Off: Bad drop to right.
Accelerated: Roll either way—1. Quite gentle—1.

HOOVER HORIZON

Army—1 Navy—3 British—1 Contractors—6

Good	4
Fair	0
Poor	0
Other	7
Blank	0
Total	11

Remarks

The gyro gives a true picture of flight and is easy to fly from moment of first use. Would probably be of little use to a pilot of accrued instrument time but would undoubtedly be a boon to the beginner or individual running into his first instrument conditions. The unit itself is a little large but could be reduced, and if the limitations of this unit are comparable to the present gyros it would be a desirable installation.

This instrument is a very interesting variation of the gyro horizon. Its two main advantages over the present standard instrument are:

1. It gives a clearer and simpler picture of the plane's attitude in relation to the earth.
2. It combines the gyro horizon and gyro compass in one instrument.

Its disadvantages, as seen so far:

1. Large package.
2. Complicated design.
3. Cannot adjust "plane" in relation to horizon.
4. Questionable as suitable for night work—can it be dim enough and illuminated by red light?
5. Can it be as reliable as the present gyro horizon?

If all the above items can be solved satisfactorily the Hoover Horizon is a definite improvement over the present instrument now in use.

A bit confusing at first because of reversed action on gyro. After familiarization this instrument should provide a very restful means of flying instrument. Very easy to maintain altitude in level flight or turns without reference to other instruments.

Coordination of rudder in turns not good—attribute this to unfamiliarity with plane rather than instrument. Constant rate of descent or climb in spirals easy to maintain. Altogether consider this a step forward for instrument flight aids.

This instrument has the advantage of allowing the pilot to more closely simulate contact flying. If the installation could be cut down in size and could be properly lighted it would be a welcome addition to the instrument panel of any night fighter.

The gyro horizon is very easy to get used to and exceptionally easy to interpret. There appears to be a considerable lag on the instrument (this may be due to the number of gyros on one pump). The main disadvantages seem to me to be:

1. No standardization of which way the horizon would move might cause a number of accidents with people who had been used to flying the normal instruments.
2. All the airplanes now built would have to be modified, which would be quite a job (civil airplanes).

Some method of dimming the lights for night flying will have to be provided. Suggest an ultraviolet lighted airplane with an ultraviolet lighted horizon.

The horizon is most interesting—particularly in its possible applications to night fighting. Unfortunately the undersigned has been unable to learn its limits of operation. It appears to have the following disadvantages:

1. Any instrument with light as a background is unsatisfactory. I suggest a horizon line with diagonals on one side, for example:

—–O—–

―――――――――――― ←horizon
////////////////// ← earth

These lines to be in luminous paint, activated by red lights.

2. A device for setting miniature airplane on horizon is needed for different trim conditions.
3. The fact the heading readings on the screen are in the opposite direction from the normal compass or directional gyro readings causes some initial difficulty—a pilot, however, can easily become accustomed.
4. If the gyros ever "spill" under any condition of flight, should be able to cage.
5. Complete instrument much too large.

This instrument is excellent from the point of view of giving the pilot a realistic picture of the airplane's attitude. The incorporation of the gyro compass in this instrument eliminates the necessity for the pilot to constantly glance back and forth from the artificial horizon to the gyro compass. The direction in which the gyro compass operates in this installation is also more natural than the standard gyro compass. I believe that if there is no appreciable weight or maintenance penalty to pay, that this instrument is worthy of installation if it can be installed in the space formerly occupied by both the artificial horizon and the gyro compass. It would be desirable to incorporate the full swiveling horizon in this design.

Much interested in gyro horizon. Apparently makes instrument flying simpler and less fatiguing. Should think it would be of great value to inexperienced instrument pilots. Suggest trying more clearly marked terrain on film (roads, section lines, etc.).

This horizon is a great development. The use of this instrument will simplify instrument flying materially. It is especially advantageous to the pilot who does not practice a lot.

This gadget seems to me to be the ideal method for instrument flying and should be investigated and refined for production. Instrument flying is a pleasure with this instrument.

A marvelous development.

ANTI-G SUIT

Remarks

The suit is a definite aid to pull-outs. The pressure is not too noticeable in the stomach regions but is very high on the upper legs. There would be a tendency to tighten up a steep bank, or a vertical, unknowingly and do a high speed stall. Suit is light and worked O. K. even though it was worn over my flying clothes. I would like one in combat.

Suit works beautifully. Pulled 7 G with no gray-out at all. Only difficulty in its wide use might be ability of pilots to exceed strength of airplanes.

Very good and positive in action. Hose connection sticks (weak point of equipment). Suit very light and cool. Each pilot will have to be fitted by flight test of suit. Wearing one of these suits will call for more indoctrination of pilots in their planes' limitations.

Not more than steady 5 G. Good for short high G turns but should be very carefully considered before putting in production airplanes.

The G-suit is one of the finest pieces of equipment yet tried by this pilot. The suit is comfortable and not bulky and adds to pilot comfort in all light turns and pull-outs. The suit should be furnished to all service pilots to be used at their discretion.

G-suit appears to materially increase the G which can be withstood before gray-out or black-out. Did not find it uncomfortable to wear. Made pull-outs up to 7.9 G. Should think it would be a decided asset in any flying where high G is encountered.

Suit functioned O. K. I built up G slowly so that I would have gone black after approximately 4 seconds at 6 G. Suit comes unplugged inadvertently too easily.

G-suit well liked.

This device with its lack of cumbersome equipment is excellent for test work, and I was pleased with its operation. A 4 G turn was held for 720° with no sign of fade or black-out. I am susceptible to fade with about 360° of turn usually. I consider this one of the highlights of the conference.

O. K. Like for high G pull-outs. Don't find too much help for long-time 4–5 G pull-outs. Not too comfortable on long acceleration periods.

G-suit could fit tighter. 4½ G, good; 6 G, good; 7½ G, good.

G-suit will definitely reduce fatigue due to constant pulling of G.

G-suit excellent.

Very good. Can fly indefinitely at 5 G, but lags a little for abrupt pull-outs above 6 G.

ANALYSIS OF DATA FROM CONFERENCE MEMBER'S QUESTIONNAIRES

A questionnaire (sample shown in Appendix D) was submitted to the Conference. Fifty-one completed questionnaires were received; tabulation of the answers is given in this section.

In the summary which follows, the figure shown opposite the question asked is the percentage of completed questionnaires on which some answer to the question was given. This is called the "total vote" on that question. The votes received by each aircraft are listed opposite the aircraft designation. They are expressed as a percentage of total votes cast on that question. The aircraft, furthermore, are listed in order of preference.

Several questions were answered in such a way that tabulation by the above method was unsuitable. These have been summarized at the end of the section. Reasons given for choice of best and worst cockpit were duplicates of the comments on flight reports and have not been included here.

One question in particular—"Which installation had the most impressive characteristics?"—was interpreted differently on almost every questionnaire. The answers were so varied in purpose that it was not thought profitable to include the tabulation.

General comments on the Conference expressed appreciation of the content and execution of the program. Several pilots made specific suggestions, and these have either been summarized or quoted in full.

SUMMARY OF QUESTIONNAIRES

Division of Votes

Army—9 Navy—15 British—7 Contractors—20 Total—51

Best all-around cockpit: Total vote—91 percent

	Percent
F8F	36
F7F	20
F6F	16
F4U4	12
P-51	7
F2G	5
P-47	4

Worst cockpit: Total vote—95 percent

	Percent
P-38	55
Mosquito	11
P-61	10
F4U1	9
P-63	4
P-47	3
F6F	2
P-51	2
Seafire	2
P-59	1
FM	1

Nicest arrangement of engine controls: Total vote—91 percent

	Percent
P-51	20
F7F	17
P-47	13
P-63	12
F8F	10
F4U4	9
P-61	5
F6F	3
Mosquito	3
F4U1	2
FM	2
Mock-up	2
P-38	1
F2G	1

Do you believe single power control is practicable? Total vote—100 percent

	Percent
Yes	60
No	21
Yes; with override	16
Indecisive	3

Are automatic cooler flaps practicable? Total vote—100 percent

	Percent
Yes	84
Yes; with override	13
Indecisive	3

Most convenient gear and flap controls: Total vote—91 percent

	Percent
F8F	19
F6F	18
F7F	14
P-51	11
F4U1	8
F4U4	7
P-63	5
Mosquito	3
P-47	3
P-61	3
Mock-up	3
F2G	2
FM	2
P-38	2

Best cockpit canopy: Total vote—96 percent

	Percent
P-47	13
P-51	21
F8F	16
F4U1	5
F2G	4
F6F	3
F7F	2
Any good bubble	3

Most comfortable cockpit: Total vote—92 percent

Aircraft	Percent
P-47	24
F8F	18
F6F	15
F7F	11
F4U4	9
P-61	8
P-51	7
F4U1	3
FM	2
Seafire	2
F2G	1

Best all-around visibility: Total vote—93 percent

Aircraft	Percent
P-51	35
P-47	27
F8F	22
F6F	6
F7F	5
F2G	2
F4U1	2
P-63	1

Best all-around armor: Total vote—59 percent

Aircraft	Percent
P-47	27
F4U1	22
F4U4	14
F8F	9
F7F	8
F6F	7
F2G	7
P-51	6

Best for overload take-off from a small area: Total vote—90 percent

Aircraft	Percent
F6F	28
F8F	21
F4U1	10
F7F	9
P-38	7
FM	5
F4U4	5
P-61	5
Seafire	5
F2G	3
P-51	2

Nicest harmonization of control forces: Total vote—89 percent

Aircraft	Percent
F4U1	26
P-51	20
F6F	14
F8F	10
P-47	8
Mosquito	6
F7F	5
Seafire	5
P-63	4
FM	2

Best ailerons at 350 MPH: Total vote—83 percent

Aircraft	Percent
P-51	33
F4U1	20
P-38	19
F6F	9
F8F	6
P-47	4
P-61	3
F7F	2
Seafire	2
Mosquito	2

Best ailerons at 100 MPH, landing condition: Total vote—88 percent

Aircraft	Percent
F6F	36
F4U1	18
Seafire	12
P-47	6
FM	5
P-51	5
F8F	5
Zeke	4
P-38	3
F2G	2
F7F	2
P-61	2

Best elevator: Total vote—73 percent

Aircraft	Percent
F4U1	20
F8F	13
F7F	13
F6F	13
P-51	13
Seafire	6
P-47	5
P-59	3
P-61	3
Zeke	3
FM	2
P-38	2
P-63	2
Mosquito	2

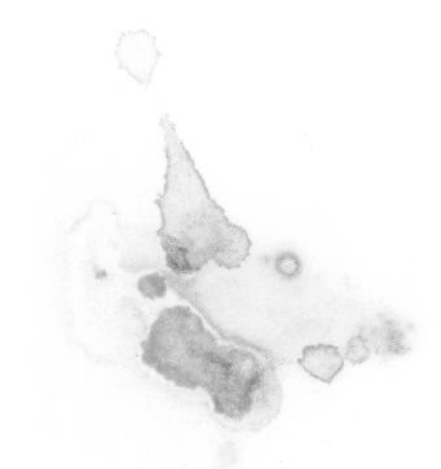

Best rudder: Total vote—76 percent

	Percent		Percent		Percent
F7F	20	Seafire	11	P-61	3
F6F	18	P-38	6	F8F	2
F4U1	14	P-51	6	P-59	2
Mosquito	13	P-47	3	P-63	2

Fighter exhibiting nicest all-around stability: Total vote—76 percent

	Percent		Percent		Percent
F6F	33	P-47	11	P-63	3
F4U1	23	F8F	7	Mosquito	2
P-61	14	P-51	6	P-38	1

Fighter appearing to have the best stability and control in a dive: Total vote 75 percent

	Percent		Percent		Percent
F4U1	25	P-51	10	Mosquito	3
P-47	23	F8F	4	FM	2
F6F	13	P-63	4	P-38	2
F7F	11	P-61	3		

Are dive recovery flaps "the answer"? Total vote—73 percent

	Percent		Percent		Percent
Yes	14	As a stop-gap	61	Indecisive	25

Best characteristics at 5 MPH above stall: Total vote—87 percent

	Percent		Percent		Percent
F6F	47	F7F	4	F4U1	3
P-61	17	FM	4	Firefly	3
P-38	7	P-51	3	Zeke	2
Seafire	7	F8F	3		

Best instrument and night flying qualities: Total vote—56 percent

	Percent		Percent		Percent
F6F	35	F4U1	6	Firefly	3
P-61	32	P-47	6	Mosquito	3
F7F	10	P-38	5		

Which powerplant operation inspired the most confidence? Total vote—84 percent

	Percent		Percent
Pratt and Whitney R-2800	79	Jet	5
Merlin	7	Rolls Royce	3
Packard	5	Allison	1

Best solely night fighter among P-61A, F7F, P-38J, F6F-5N, F4U-1N, Mosquito: Total vote—73 percent

	Percent		Percent		Percent
F7F	42	Mosquito	21	P-38	2
P-61	26	F6F	9		

Best combination day and night fighter among P-61A, F7F-1, P-38J, F6F-5N, F4U-1N, Mosquito. Total vote—82 percent

	Percent		Percent		Percent
F7F	43	F6F	15	P-38	9
F4U1	21	Mosquito	12		

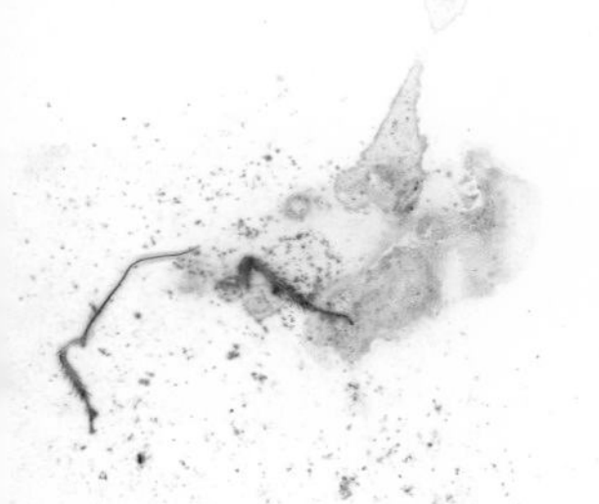

NAS Patuxent River, MD - 16-23 October 1944

Best all-around fighter above 25,000 feet: Total vote—82 percent.

	Percent		Percent		Percent
P-47	45	F6F	3	P-38	1
P-51	39	F4U4	3		
F4U1	7	Seafire	2		

Best all-around fighter below 25,000 feet: Total vote—89 percent

	Percent		Percent		Percent
F8F	30	F7F	6	F4U4	2
P-51	29	F6F	2	F2G	2
F4U1	27	Mosquito	2		

Best carrier based fighter in production: Total vote—87 percent.

	Percent		Percent
F4U1	61	F7F	5
F6F	31	FM	3

Potentially the best carrier fighter (modified for carrier operation if necessary): Total vote—74 percent

	Percent		Percent		Percent
F8F	50	F7F	10	F2G	3
F4U1	15	P-51	6		
F4U4	13	F6F	3		

Which set of radio controls do you prefer? Total vote—70 percent.

	Percent
Individual control boxes—VHF push button radio control	47
Individual control boxes—VHF selector switch radio control with internal dial and windows	3
Console arrangement of all radio and electrical controls—VHF selector switch radio control	26
Unclassified	24

Best fighter bomber: Total vote—72 percent.

	Percent		Percent		Percent
F4U1	32	F6F	12	P-38	5
P-47	19	F7F	11		
Mosquito	14	P-51	7		

Best strafer: Total vote—75 percent.

	Percent		Percent		Percent
P-47	41	P-51	9	F8F	2
F4U1	18	F6F	7	P-63	2
F7F	17	P-38	3	F2G	2

Best fighter torpedo plane: Total vote—56 percent.

	Percent		Percent
F7F	78	F4U1	10
F6F	10	Mosquito	2

Cockpit Mock-up

Figures in parentheses after comment show number of pilots, according to service, making comment in question: thus—(A1, B2, C3) means 1 Army, 2 British, and 3 Contractors.

Modifications Recommended

Add—

Fore-and-aft and up-and-down seat adjustment (A1, B2, C1).
Center panel between pilot's legs (N1, C1).
Magnetic Compass (B1).
Hand brake system (B2).
Ring grip on stick (B1).

Remove—

Desk (A1). Free air temperature gauge (C1)—use carburetor air temperature gauge with calibration.
Hook (A1).

Move—

Pedals closer together (A1).
Flight instruments together (A1, N2, C3) and center them (A1, N2).
Feathering switches (N3, B3, C5).
Pitch controls (B2, C3).
Hand pump (B2, C3) left forward side of seat (B1) right side of seat (B1, C1) right side of seat, forward (C1) back (N1).
Gunsight nearer face (B1, C1).
Gunsight switches near gunsight (A1).
Harness release (A1, B1, C4) forward (A1, C1) put on left of seat itself (C1) change shoulder strap release to standard position on seat (C1).
Shoulder strap forward (A1).
Interchange landing gear and flap controls (B1).
Tank pressuring controls away from winglock control, putting them near gas-cock (B1).
Shutter controls to more convenient position (C1).
Fuel cocks forward (C1).
Seat raising latch lever to right of seat (C1).
Auto oil and cylinder-head temperature switches next to their respective gauges (C1).
Ignition switch off main panel (C1).
Oxygen and radio outlets to position under seat (A1).
Interchange voltmeter and free air temperature gauge (C1).
Stick forward (C1).
Oxygen regulator to present position of prop feathering switches (C1)—diluter interferes with landing gear knob and would not interfere with hook control.

Other Changes—

Provide more hand-grip on elevator trim (A1).
Slant instrument panel forward so pilot can better see instruments (N1).
Have armor slide with canopy (N1).
Install a center panel between pilot's legs, shrink cockpit, use a higher foot position, use a lower level for the rear end of the consoles on each side to make it easier to reach (N1).

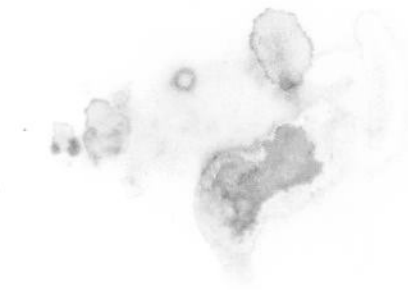

Other Changes—Continued.

Prefer less bulky throttle knobs. Incorporate F4U type tab handles (N1).
Make flap control larger for more positive control (B1).
Substitute as soon as possible Hoover horizon for gyro horizon (B1).
Make the engine instruments a different shape and color, and standardize as in all British airplanes (B2).
Rudder trimmer needs deeper notches in the wheel. Aileron trim wheel is not enough enclosed (B1).
A new type of harness—such as the British 0 type—might be tried (B1).
Decrease cumbersome size of throttle levers (B1).
Stick too short (C1).
Elevator trim tab wheel should be more exposed—with stiff tab and cold fingers could not operate (C1).
Why not have one master circuit breaker button which contacts and resets all buttons instead of the plate idea? This would make more room available (C1).
The seat will probably need modifying as the angles involved by the tilting are not compensated by an equal back tilt (C1).
Believe seat should be a compromise between the present type and transport type, moving fore and aft (C1).
I believe a major project should be started for pilot comfort to the extent of requiring more space and a couple hundred pounds in weight. There is a definite need for navigation aids and auto-pilot for long range fighters (C1).
Why have separate switches for guns? One switch would be simpler (C1).
Give pilot a little more room between calves of legs and instrument panel.
Believe the oxygen regulator will be difficult to install and remove readily (C1).
Throttle propeller control that can be used singly or combined (F4U4 type) (C1).
Not enough room for grip on L. G. handle. (C1).
Flap control obviously afterthought—no room for connecting the handle to anything (C1).
Windshield and cabin should be added to mock-up—looks as if windshield may come out to be too far ahead when gunsight is that far forward (C1).
Some question in my mind about position of hand pump handle—may be hard to build up full pressure and may interfere somewhat with proper stick handling in the air (C1).
Believe short travel of stick in both directions is very optimistic in any but the smallest airplanes from a force standpoint and even if forces can be made satisfactory most pilots will not like it. The 8F had originally the same thing and was increased for both reasons (C1)
My personal preference is for the F4U4 type of aileron trim tab control. The type shown is similar to F4F which was not ideal by any means (C1).
Cabin control similar to 8F would be better—cutting away longeron as was done in mock-up simply can't be done (C1).

Other Changes—Continued.

I would prefer fuel cock controls on a low front center panel with gauges and fuel pump and drop tank controls along side. The handles shown are not big enough to turn the large valves in 7F. However, the clear space between pedals is very desirable for changing foot positions on long flights (C1).

It will be difficult to read std. hyd. gauge in position shown (C1).

Epithets—

Very good (A4, N10, C2).

Good (A2, N1, C8).

On the whole good (A1, B2, C5).

Step in the right direction (A1, N2, B1, C2).

Fairly good for twin engine (B1).

Not as good as F2G or F4U4 (B1) F7F (B2) F8F (B1).

Too large (N1); too large for liquid cooled engine (C1).

Too cramped for anything but short-range fighter (C1).

Perhaps ideal for Navy, not for Army (A1).

Very clean and as a whole quite comfortable (C1).

Special Features—

Night instrument panel is a fine idea (A1).

Instrument setup open to question (C1).

Brake tilt-back feature for pilot rest is especially nice (C1).

Volt and ammeter with engine instruments, where it belongs (C1).

Oxygen easy to check; barber chair rudder feature should add to pilot comfort; gunsight out of way, but may be too far away and too high; armament and gun switches in good location and easy to read; control stick back where it should be and correct size—this feature can be appreciated after flying P-38; grouping of engine and flight controls good (C1).

Chart board arrangement and the plate over the circuit breakers are particularly attractive (C1).

Personally I don't like instrument lay-out—prefer Army standard flight panel (C1).

Will oxygen hose interfere with throttles (C1)?

Instrument panel is too near one's knees (B1).

Treatment of gun switches excellent (C1).

Treatment of power plant controls excellent—especially in covering up all rods, etc. (C1).

I like idea of "barber chair" brake pedals (C1).

Oxygen arrangement excellent except for connection (C1).

Single reset plate for breaker switches good (C1).

Remarks on Idea of Standardizing—

Think a standard arrangement will be ultimate answer (A1).

Wish Army, Navy, and British could get together on a standardized cockpit, but don't think they ever will (A1).

Modify it only enough to meet Army requirements (A3).

Doubt wisdom of standardization in every particular (C1).

AAF answer will probably be different, but Navy has very logical arrangement (A1).

What do you think of the "anti-g" suit? Total vote—47 percent.

	Percent		Percent
Very good	74	Fair	9
Good	17		

Remarks

Wouldn't want to go back to combat without one. Essential for combat; I would not want to be without it. Would want one in a dog fight, but she's a bear on tightening vertical turns.

Worked nicely, but do not think it practical for use in tropical climates.

Am enthusiastic about its combat possibilities, especially for fatigue reactions. This means that, while there is a limit to operating in combat, pilots will deliver a better job on the average throughout the period. Of benefit as a fatigue reliever, if nothing else. Should be used by test pilots to relieve fatigue from constant pulling of "g."

Best piece of pilot equipment out recently. So good that it would have to be used with care. Some indoctrination as to proper use absolutely necessary. Enables pilot to exceed structural limits of aircraft. A good scheme as long as pilot's resistance to G doesn't overtake stress limits of airplane. Believe there should be no more apprehension over straining the plane structure with the "g" suit than there would be in diving to critical velocity through poor discretion.

Best device for test work since the invention of the wheel. Suit is rather cumbersome.

For test work we keep suit inflated at start of dive so there will be no lag in a snap pull-out.

Idea and principle seems sound, and in general suit works to advantage. I did not get enough aid at long-time low "g" turns (4—4.5 g). It seemed to work better for short time interval—high acceleration. Was disturbing at times to be pushed against on mild maneuvers of short duration.

So good that I wonder what would ultimately happen to the organism of a pilot who frequently used it to the full.

Suit is comfortable, and I believe it will be cool enough for summer wear.

One of the most interesting things at the conference.

Best cockpit illumination

F6F-5	4	F6F-5N	1
P-61	2	Ultra-violet	1
Navy indirect red lighting	1		

Remarks

No night flying completed. Suggest indirect infra-red lights exciting (not illuminating) the figures, or at least the important figures, on the flight instruments. A direct red floodlight of the engine instruments on a separate rheostat; an emergency system run-off a small battery in case of power failure; direct flood lighting of the flight and engine instruments with a switch; foregoing are ideas on subject.

Only P 38 and P 61 flown at night. Of these two, P 61 far better. Only objection to P 61 is some reflection off the small radiant dots on switch panel on left side of cockpit.

Remarks—Continued.

None observed. Believe indirect red lighting is best answer. Reflections from instrument faces must be reduced.

Like fluorescent lighting. Do not like red lights as installed in F6F-5.

Best oxygen installation

XF8F	4	XF2G	1
F7F	2	P-47	1
P-51	1		
Mock-up	2		

Remarks

Suggest single terminal for radio and oxygen connections located between pilot's legs and attached to pilot's seat.

F4U4 or F8F appears best, in that the leader tube is under pilot's arm and out of the way. Very easy to work oxygen in F4U4.

Prefer demand type.

F4U-4 system seems to overcome most objections. Supply pressure easy to read, tube to mask handy and out of way, and supply easily turned on.

Comments on Spring Tabs: Total vote—65 percent

	Percent
Good	8
Poor	3
Good idea for lightening stick forces	31
Good if no flutter trouble develops	18
Good on F6F	9
Other remarks	15

Remarks

Good if working like elevator of F7F, not like elevator of P-61.

Preferable to hydraulic boost; still don't like artificial helps.

Do not like the lag it gives in rolling from right to left or vice versa.

I believe this is a great improvement in executing combat maneuvers.

Probably best solution for modifying existing ailerons; however, will destroy feel if not properly designed.

How do you like the P-38 hydraulic boost ailerons?

Good	3	Too sensitive	1
Fair	1	Too light	5
Poor	7	No feel	5
No opinion	22	Poor harmonization	3
Needs improvement	4		

Remarks

Excellent at high speed, but somewhat disconcerting during landing and take-off and when raising or lowering gear and flaps.

Nuisance during approach and landing.

Poor control in take-off and landing.

Excellent at high speeds, ineffective at low speeds.

Overbalance at low speeds.

Required constant correction.

How do you like the F7F hydraulic boost rudders

Good	17
Fair	1
Poor	2
No opinion	14
Poor harmonization with other controls	5
Boost is too powerful	13

Remarks

A jury rig; should concentrate on aerodynamic balance or spring tabs.

Prefer principle of Mosquito spring assisted rudder.

The spring tab idea is particularly adapted to rudders.

Good, although it appears too sensitive because of poor directional stability.

Do not like lack of centering forces.

Forces should be linear with displacement.

To what degree should a land-based fighter be dynamically stable:

Total vote 60 percent

	Strong Percent	*Moderate* Percent	*Weak* Percent	*Neutral* Percent
Laterally	10	13	60	17
Directionally	16	10	65	9
Longitudinally	36	12	45	7

The following remarks are quoted as representative of the explanations given:

Laterally, should be weak positive, so that Dutch roll is inhibited; at the same time, too much damping in roll prevents high rolling velocities.

Longitudinally, the damping of the short period oscillation with controls free should be strong (should disappear after one cycle, otherwise the aircraft will exhibit unsatisfactory rough air characteristics). It is not considered important that the long period longitudinal oscillation (phugoid) exhibit any particular damping characteristics.

Directionally, the damping should be strong, since otherwise the airplane will provide a poor gun platform in disturbed air.

Weak positive or neutral longitudinally and laterally. Should have enough stability to prevent pilot's becoming tired on long flights (the F4U is good in this respect) and should have good instrument flying qualities. But—it must retain good maneuvering qualities. Directionally, should be positively stable.

A certain amount of positive stability is needed directionally to prevent undue skidding when firing. For lateral and longitudinal stability, just enough should be provided to prevent inadvertent slipping, or tightening up in turns.

Directionally and longitudinally it should exhibit strong stability, but it should be easily controllable with light forces. These characteristics are most desirable for long flights, instrument flights, and/or flights in turbulent air.

Weak positive, but not at the expense of maneuverability. I would prefer to accept neutral stability if the demand for weak positive stability means sacrificing maneuverability.

Laterally weakly positive and directionally fairly strong, to relieve hunting in gunnery attacks. Longitudinal immaterial as long as range is between neutral and strong positive, and elevator forces do not become high.

Remarks—Continued.

Adequate stability should be provided—then the control force and effectiveness should be made satisfactory. Stability should be weak positive to strong.

Positive in all conditions, but near neutral since the present method of getting force required on the controls is by leaning on unstable conditions.

I believe most of our fighters here have hit a good compromise on stability, and that changing this balance would be a detriment.

Should a carrier based fighter exhibit dynamic stability about all axes: Total vote 72 percent

	Percent		Percent
Yes	55	Indecisive	17
No	14	Weak positive	14

Remarks

The only mode that need not necessarily be damped is the long period longitudinal oscillation. The meaning and importance of dynamical lateral stability is somewhat obscure, and in general the static stability about all axes is so much more important for both land- and carrier-based fighters.

I would say that, of all three axes, it should be most stable directionally, primarily at low speeds, with a trend toward neutral stability at high speeds if possible.

Which gunsight and armament controls impressed you most? Total vote 45 percent

	Percent		Percent
Mk 23	39	F4U-1	9
F8F	17	FG1	4
F7F	13	P-47	4
Mk 8	4	P-51	4

Remarks

The F7F gunsight for day work. The Mosquito gun controls.

P-47M controls well located on left side to rear of throttle.

Mosquito armament controls best.

F4U-1 armament controls.

Personally like the combined cannon and machine gun button on the Seafire. None of the gunsight controls impressed me favorably, all being not readily accessible.

Like the Army armament controls with a single master switch. Mosquito control on stick is good. Mk 23 sight controls still do not appear too controllable.

General Remarks Concerning All Aspects of the Conference

The general opinion was highly complimentary, both as to content and execution of the program. It was felt that the conference had served a valuable purpose, and that similar conferences should be held at 6-month intervals. Both service pilots and contractors' representatives felt that they had benefited greatly by the opportunity to fly all types of fighter aircraft, and to exchange views and information on the spot.

There were several specific suggestions as to execution of the program of future conferences:

A number of members would have liked more time in the air—either by lengthening the conference, decreasing the attendance, or increasing the number of aircraft.

Several would have preferred a more rigid program of discussion, while others expressed a preference for less formalization:

and still others suggested that in future conferences all microphones be kept "hot" to allow greater freedom for questions and answers from the floor, and to permit more impromptu discussion.

On the content of the program:

One member proposed having more enemy types available;

one proposed having fewer ships on the restricted list;

and several suggested that the next conference be attended by four or five pilots recently returned from combat theaters, to represent the latest combat experience with own and enemy types, and to "speak for the ordinary combat pilot."

The following suggestion is quoted from a letter accompanying one of the questionnaires:

"To treat upon the business end of things for a moment, I would like to offer a suggestion for future meets. It is my belief, which I think is shared by others, that the benefits of such conferences are shared by the nonflying personnel to an extent which, while now perhaps intangible, could be made measurable by the inclusion of a questionnaire for them as well as the pilots.

"I have reference particularly to the fact that those of us who are field engineers and installation engineers benefit considerably through the opportunity to examine numerous installations, thereby gaining valuable knowledge which will help us and, in turn, the Army and Navy. Comments based on such examination concerning exhaust systems, ducts, cowling, controls, accessibility, etc., could be recorded just as well as comments on stability and control, and I think should be.

"This idea might be furthered even more if one or two evenings—or bad weather days—were set aside, during which the cowling of one of each type plane involved were removed to permit examination of installation details by installation engineers present. With as many people present as there were this time, it would obviously be necessary to limit the examining personnel to those with a direct interest in the installations, as many more would just be getting in each other's hair.

"I am sure it could be suitably arranged, and would be interested in knowing the general reaction to it. The "Comment Card" for this phase of activities could be patterned very much like the flight comment cards, listing such things as Best Control System, Best Installation from Standpoint of Accessibility, Ease of Maintenance, Cowl Removal, etc."

GENERAL DATA

Model FM-2. airplane.

Description Single seat, single engine, shipboard fighter by Eastern Aircraft.

Engine Wright R-1820-56.

TAKE-OFF POWER

Blower	BHP	RPM	MAP	Altitude
Low	1,300	2,600	46.5	S. L.

NORMAL POWER

Blower	BHP	RPM	MAP	Altitude
Low	1,200	2,500	43.5	5,000
High	900	2,500	39.0	19,000

MILITARY POWER

Blower	BHP	RPM	MAP	Altitude
Low	1,300	2,600	46.5	4,000
High	1,000	2,600	44.0	17,000

COMBAT POWER

I N O P E R A T I V E

MAXIMUM CRUISE—AUTO LEAN

Blower	BHP	RPM	MAP	Altitude
High	750	2,200	31.5	20,300

Propeller Curtiss Electric.

Dia. 10'-0''. Blades 3.

Loading: Full internal fuel and ammo.

Wt. Empty—Lbs. 5,325.

Gross wt.—lbs. 7,420.

C. G.—Percent Mac (gear up) 28.6.

Power loading (T. O.) 5.7 lb./Hr.

Wing loading 28.5 lb./sq. ft.

Fuel—Gal. Main 126.

Oil—Gal. 9.

Armament 4-.50 cal. guns. 1,600 rds. ammo.

Radio

General features: Lower flaps below 130 knots.

No other restrictions specified.

GENERAL DATA

Model F6F-5. Airplane

Description Single seat, single engine, Ship-board fighter by Grumman.

Engine Pratt & Whitney R-2800-10W.

TAKE-OFF POWER

Blower	BHP	RPM	Map	Altitude
Neut.	2,000	2,700	54.0	S. L.
NORMAL POWER				
Neut.	1,675	2,550	45.0	5,500
Low	1,625	2,550	49.5	18,000
High	1,550	2,550	49.5	24,000
MILITARY POWER				
Neut.	2,000	2,700	52.5	1,700
Low	1,800	2,700	53.5	17,000
High	1,650	2,700	52.5	23,000
COMBAT POWER				
Neut.	2,250	2,700	57.5	S. L.
Low	2,135	2,700	59.0	15,000
High	1,975	2,700	59.5	20,000
MAXIMUM CRUISE—AUTO LEAN				
High	950	2,050	34.0	26,000

Propeller Hamilton Standard Hydromatic.

Dia. 13'-1". Blades 3.

Loading	Full internal fuel and ammo.
Wt. empty—lbs.	9,000.
Gross wt.—lbs.	12,500.
C. G.—Percent Mac (gear up)	29.4.
Power loading (T. O.)	6.25 lb./BHP.
Wing loading	37.4 lb./sq. ft.
Fuel—Gal. Main	175.
Res.	75.
Oil—Gal.	16.
Armament	6v .50 cal. guns.
	2,400 rds. per gun.
Radio	AN/APX-1 (IFF).
	AN/ARC-5 (Radio).
	AN/ARR-2 (Homing).

General features: Lower gear below 135 knots; Spring flaps will ''blow up'' at 170 knots. Max. dive speeds; less than 5 g, 390 knots below 15,000 ft., 370 knots above 15,000 ft. From 5-7g, 320 knots at all altitudes.

GENERAL DATA

Model F7F 1. Airplane

Description 2-engine, single seat, tricycle gear, shipboard fighter by Grumman.

Engine Pratt and Whitney R 2800 22W.

TAKE-OFF POWER

Blower	BHP	RPM	MAP	Altitude
Low	2,100	2,800	53.0	S.L.

NORMAL POWER

Blower	BHP	RPM	MAP	Altitude
Low	1,700	2,600	41.5	12,500
High	1,450	2,600	42.0	22,500

MILITARY POWER

Blower	BHP	RPM	MAP	Altitude
Low	2,100	2,800	53.0	10,000
High	1,600	2,800	48.5	22,000

COMBAT POWER

I N O P E R A T I V E

MAXIMUM CRUISE—AUTO LEAN

Blower	BHP	RPM	MAP	Altitude
High	975	2,250	30.8	24,000

Propeller Hamilton Standard Hydromatic.

Dia. 13' 2" Blades 3.

Loading	Full internal fuel and ammo.
Wt. empty—Lbs.	15,750.
Gross wt.—Lbs.	21,450.
C. G.—Percent Mac (gear up)	26 percent.
Power loading (T. O.)	5.1 lb./HP.
Wing loading	47 lb./sq. ft.
Fuel—Gal. Main	210.
Res.	156.
Aux.	60.
Oil—Gal.	24.
Armament	4 20 mm. cannon.
	4 .50 cal. mach. guns.
	800 rds. 20 mm. 120 rds. 50 cal.
Radio	ARC 5 (Radio).
	AN/ARR 2 (Homing).
	AN/AP5 6 (Radar).
	AN/APN 1 (Radio altimeter). AN/APS 2 (IFF).

General features: Lower gear below 140 knots.

Lower flaps 15° below 170 knots.

Lower flaps 30° below 135 knots.

Lower flaps 45° below 125 knots.

Max. dive speeds: 430 knots @ 10,000 ft., 350 knots @ 20,000 ft.

Do not exceed 3 g at these speeds.

GENERAL DATA

Model XF8F. Airplane

Description Single-engine, single-seat, mono-plane landplane fighter for use aboard aircraft carriers.

Engine Pratt and Whitney R-2800-22W.

TAKE-OFF POWER

Blower	BHP	RPM	MAP	Engine Rating Only Altitude
Low	2,100	2,800	53	S. L.

NORMAL POWER

Blower	BHP	RPM	MAP	Engine Rating Only Altitude
Low	1,700	2,600	41.5	S. L. 7,000
High	1,450	2,600	42.0	18,500

MILITARY POWER

Blower	BHP	RPM	MAP	Engine Rating Only Altitude
Low	2,100	2,800	53.0	1,000
High	1,600	2,800	48.5	16,000

COMBAT POWER

I N O P E R A T I V E

MAXIMUM CRUISE—AUTO LEAN

Propeller Aero Products.

Dia. 12'-7". Blades 4.

Loading	Full internal fuel and ammo.
Wt. empty—Lbs.	6,779.
Gross wt.—Lbs.	8,810.
C. G.—Percent Mac (gear up)	24.64.
Power loading (T. O.)	4.2
Wing loading	40.0.
Fuel—Gal. Main	150.
Belly (Drop Tank)	(100-not installed).
Oil—Gal.	9.5.
Armament	4 wing—.50 cal. guns and 1,200 rds. of ammunition.
Radio	AN/ARC-1.
	AN/ARR-2A.
	AN/APX-1.
	BC-1206-C Range finder.
General features:	External brake plate.

GENERAL DATA

Model F4U-1C. Airplane 57977

Description single-seat, single-engine, shipboard fighter by Chance Vought.

Engine Pratt and Whitney R-2800-8W

Blower	BHP	RPM	MAP	Altitude
TAKE-OFF POWER				
Neut.	2,000	2,700	54.0	S. L.
NORMAL POWER				
Neut.	1,675	2,550	45.0	5,500
Low	1,625	2,550	49.5	20,100
High	1,550	2,550	49.5	24,000
MILITARY POWER				
Neut.	2,000	2,700	52.5	1,700
Low	1,800	2,700	53.5	17,000
High	1,650	2,700	52.5	23,000
COMBAT POWER				
Neut.	2,250	2,700	57.5	S. L.
Low	2,135	2,700	59.0	15,000
High	1,975	2,700	59.5	20,000
MAXIMUM CRUISE—AUTO LEAN				
High	950	2,050	34.0	26,000

Propeller Hamilton Standard Hydromatic.

Dia. 13'-4" Blades 3.

Loading	Full internal fuel and ammo.
Wt. empty—Lbs.	9,100.
Gross wt—Lbs.	12,095.
C G.—Percent Mac (gear up)	32.5.
Power loading (T. O.)	6.05 lb./HP.
Wing loading	38.6 lb./sq. ft.
Fuel—Gal.	237.
Oil—Gal	20.
Armament	4 20-mm. cannon.
	800 rds. ammo.
Radio	ATA-1/ARA-2 (radio).
	ZB-1 (radio).
	ABK-1 (recognition).

General features: Speed restrictions: Lower gear below 200 knots. 20° flap below 200 knots, full flap below 130 knots.

Max. diving speeds: 10,000 ft. 3.5 g—385 knots; 6.5g—350 knots., 20,000 ft. 320 knots, 30,000 ft. 260 knots.

GENERAL DATA

Model FG-1. Airplane

Description Single-seat, single-engine, shipboard fighter by Goodyear (same as F4U-1).

Engine Pratt & Whitney R-2800-8W.

TAKE-OFF POWER

Blower	BHP	RPM	Map	Altitude
Neut.	2,000	2,700	54.0	S. L.
NORMAL POWER				
Neut.	1,675	2,550	45.0	5,500
Low	1,625	2,500	49.5	20,100
High	1,550	2,550	49.5	24,000
MILITARY POWER				
Neut.	2,000	2,700	52.5	1,700
Low	1,800	2,700	53.5	17,000
High	1,650	2,700	52.5	23,000
COMBAT POWER				
Neut.	2,250	2,700	57.5	S. L.
Low	2,135	2,700	59.0	15,000
High	1,975	2,700	59.5	20,000
MAXIMUM CRUISE—AUTO LEAN				
High	950	2,050	34	26,000

Propeller Hamilton Standard Hydromatic.

Dia. 13'-4" Blades 3.

Loading	Full internal fuel and ammo.
Wt. empty—Lbs.	9,000.
Gross wt.—Lbs.	11,803.
C. G.—Percent Mac (gear up)	
Power loading (T. O.)	5.90 lb./HP.
Wing loading	37.6 lb./sq. ft.
Fuel—Gal.	237.
Oil—Gal.	20.
Armament	6-.50 cal. guns.
	2400 rds. ammo.
Radio	ATA-1/ARA-2 (radio).
	ZB-1 (radio).
	ABK-1 (recognition).

General features: Speed restrictions: Lower gear below 200 knots; 20° flap below 200 knots, full flap below 130 knots.

Max. diving speeds: 10,000 ft.: 3.5g-385 knots; 6.5g-350 knots; 20,000 ft.: 320 knots; 30,000 ft.: 260 knots.

GENERAL DATA

Model F2G. Airplane

Description Single-engine, Single-seat, monoplane landplane fighter for use aboard aircraft carriers.

Engine Pratt & Whitney R-4360-4 single stage—variable speed supercharger.

Blower	BHP	RPM	MAP	Altitude
TAKE-OFF POWER				
Low	3,000	2,700	52.0	S. L.
NORMAL POWER				
Low	2,500	2,500	44.0	5,000
High	2,200	2,550		14,500
MILITARY POWER				
Low	3,000	2,700	51.0	1,500
High	2,400	2,700		13,500
COMBAT POWER				
MAXIMUM CRUISE—AUTO LEAN				
High	1,675	2,230	31.5	11,500

Propeller 14'-0'' Hamilton Standard hydromatic.

Dia. 14'-0''. Blades 4.

Loading	Full internal fuel and ammo.
Wt. empty—Lbs.	10034.
Gross wt.—Lbs.	12670.
C. G.—Percent Mac (gear up)	28.7.
Power loading (T. O.)	4.2.
Wing loading	40.3.
Fuel—Gal. Main	234.
Oil—Gal.	24.
Armament	4-wing—.50 cal. guns and 1600 rds. of ammunition.
Radio	ARC-5. ABA (wiring only). ABX (wiring only).

General features: Full blown bubble canopy-"Hi-performance" brakes Restricted Vmax-430 kts. IAS from S. L.-10,000 ft.; Do not exceed 6 g between 260 kts. IAS to 330 kts. IAS.

GENERAL DATA

Model: P-38L-5-LO. Airplane: 44-25077

Description: Single seat, twin engine, turbo supercharged, land-based fighter by Lockheed.

Engine: Allison V-1710-89 (R. H. rotation).
V-1710-91 (L. H. rotation).

Blower	BHP	RPM	MAP	Altitude
TAKE-OFF POWER				
Turbo	1,425	3,000	54	S. L.
NORMAL POWER				
Turbo	1,100	2,600	44	32,500
MILITARY POWER				
Turbo	1,425	3,000	54	26,500
COMBAT POWER				
Turbo	1,600	3,000	60	26,500
MAXIMUM CRUISE—AUTO LEAN				
Turbo	795	2,300	35	36,000

Propeller: Curtiss Electric.

Dia: 11'-6". Blades: 3.

Loading	Full internal fuel and ammo.
Wt. empty—Lbs.	
Gross wt.—Lbs.	17,488.
C. G.—Percent Mac (gear up)	
Power loading (T. O.)	6.15 lb./HP.
Wing loading	53.5 lb./sq. ft.
Fuel—Gal.	426.
Oil—Gal.	13.
Armament	1—20 mm. cannon.
	4—.50 cal. guns.
	150 rds. 20 mm. ammo.
	1,200 rds. .50 cal. ammo.
Radio	SCR-522.
	SCR-695 A.
General features:	Speed restrictions posted in cockpit.

GENERAL DATA

Model P-47D-30-RE. Airplane.

Description Single seat, single engine, turbo-super-charged, land-based fighter by Republic.

Engine Pratt & Whitney R-2800-59.

TAKE-OFF POWER

Blower	BHP	RPM	MAP	Altitude
Turbo	2,000	2,700	52	S. L.

NORMAL POWER

Blower	BHP	RPM	MAP	Altitude
Turbo	1,625	2,550	42.5	29,000

MILITARY POWER

Blower	BHP	RPM	MAP	Altitude
Turbo	2,000	2,700	52	27,000
22,000 RPM	(Max)	Determines	Critical	Altitude

COMBAT POWER

Blower	BHP	RPM	MAP	Altitude
Turbo	2,535	2,700	64	25,000
22,000 RPM				

MAXIMUM CRUISE—AUTO LEAN

Blower	BHP	RPM	MAP	Altitude
Turbo	740	1,400	35.5	25,000

Propeller Curtiss Electric.

Dia. 13'. Blades 4.

Loading	Full internal fuel and ammo.
Wt. empty—Lbs.	10,100.
Gross wt.—Lbs.	14,300.
C. G.—Percent Mac (gear up)	27.5.
Power loading (T. O.)	7.15 lb./HP.
Wing loading	47.7 lb./sq. ft.
Fuel—Gal. Main	270.
Aux.	100.
Oil—Gal.	28.7.
Armament	8-.50 cal. guns.
	2136 rds. ammo.
Radio	SCR-522.
	SCR-695.
General features:	Speed restrictions:

Lower gear below 200 MPH; flaps below 195 MPH. Max. Dive speeds: 10,000 ft.—500 MPH., 20,000 ft.—400 MPH., 30,000 ft.—300 MPH.*

*Speeds in excess of the above will result in compressibility. If compressibility is encountered use compressibility flaps for recovery.

GENERAL DATA

Model P-47M. Airplane

Description Single Place Single Engine Fighter.

Engine R-2800-C. General Electric CH-5 Turbo-Supercharger.

TAKE-OFF POWER

Blower	BHP	RPM	MAP	Altitude
Connected 22,000 Mas.	2,100 Turbo RPM	2,800	54.0	Sea level to 39,000 ft.

NORMAL POWER

Blower	BHP	RPM	MAP	Altitude
Connected	1,700	2,600	43.0	Sea level Critical Turbo RPM

MILITARY POWER

Blower	BHP	RPM	MAP	Altitude
Connected	2,100	2,800	54.0	Sea level to 37,000

COMBAT POWER

Blower	BHP	RPM	MAP	Altitude
Connected	2,800	2,800	72.0	32,600 level 28,000 climb

MAXIMUM CRUISE—AUTO LEAN

Blower	BHP	RPM	MAP	Altitude
Connected	740	1,400	35.0	0-3,000

Propeller Curtiss Electric 836.

Dia. 13.0. Blades Paddle.

Loading	
Wt. empty—Lbs.	11,330.
Gross wt.—Lbs.	14,450 combat.
C. G.—Percent Mac (gear up)	27.97 percent.
Power loading (T. O.)	6.9.
Wing loading	48.
Fuel—Gal. Main	270.
Aux.	100.
Oil—Gal.	21 Normal, 26 Overload.
Armament	8-.50 cal. MG.
Radio	VHF.
	Beam Rec.

General features: Electric Turbo Regulator. Automatic Manifold Pressure Regulator.

GENERAL DATA

Model P-51D-15-NA. Airplane

Description Single seat, single engine, land-based fighter by North American.

Engine Packard V-1650-7.

TAKE-OFF POWER

Blower	BHP	RPM	Map	Altitude
Low	1,490	3,000	61	S. L.

NORMAL POWER

Blower	BHP	RPM	Map	Altitude
Low	1,120	2,700	46	20,500
High	940	2,700	46	34,400

MILITARY POWER

Blower	BHP	RPM	Map	Altitude
Low	1,450	3,000	61	19,800
High	1,190	3,000	61	34,400

COMBAT POWER

Blower	BHP	RPM	Map	Altitude
Low	1,595	3,000	67	17,000
High	1,295	3,000	67	28,800

MAXIMUM CRUISE—AUTO LEAN

Blower	BHP	RPM	Map	Altitude
High	700	2,400	36	32,300

Propeller Hamilton Standard Hydromatic.

Dia. 11'-2". Blades 4.

Loading	Full internal fuel and ammo.
Wt. empty—Lbs.	
Gross wt.—Lbs.	9,500.
G. G.—Percent Mac (gear up)	26.3.
Power loading (T. O.)	6.4 lb./HP.
Wing loading	41 lb./sq. ft.
Fuel—Gal. Wing	184.
Oil—Gal.	12.5
Armament	6-.50 cal. guns.
	1,800 rds. ammo.
Radio	SCR-522.
General features:	Speed restrictions posted in cockpit.

GENERAL DATA

Model YP-59A. Airplane

Description Single seat, twin engine, land-based, jet propelled interceptor-fighter by Bell.

Engine General Electric 1-16.

TAKE-OFF POWER

Blower	Thrust	RPM	Map	Altitude
	1610(ea)	16,500		S. L.

NORMAL POWER

Blower	Thrust	RPM	Map	Altitude
	580(ea)	16,000		30,000

MILITARY POWER

Blower	Thrust	RPM	Map	Altitude
	670(ea)	16,500		30,000

COMBAT POWER

N O T A P P L I C A B L E

MAXIMUM CRUISE—AUTO LEAN

Blower	Thrust	RPM	Map	Altitude
	425(ea)	15,000		30,000

Propeller None.

Dia. Blades

GENERAL DATA

Model P-61B-1. Airplane

Description 3 place—2 engine night interceptor fighter by Northrop Aircraft.

Engine Pratt & Whitney R-2800-63.

Blower	BHP	RPM	MAP	Altitude
TAKE-OFF POWER				
Neut.	2,000	2,700	54.0	S. L.
NORMAL POWER				
Neut.	1,675	2,550	45.0	5,500
Low	1,625	2,550	49.5	18,000
High	1,550	2,550	49.5	24,000
MILITARY POWER				
Neut.	2,000	2,700	52.5	1,700
Low	1,800	2,700	53.5	17,000
High	1,650	2,700	52.5	23,000
COMBAT POWER—WATER INJECTION				
Neut.	2,250	2,700	57.5	S. L.
Low	2,135	2,700	59.0	15,000
High	1,975	2,700	59.5	20,000
MAXIMUM CRUISE—AUTO LEAN				
High	950	2,050	34.0	25,000

Propeller Curtiss Electric.

Dia. 12'-2". Blades 4 with cuffs.

Loading	Full internal fuel and ammo.
Wt. empty—Lbs.	No ballast for turret*. 19,500.
Gross wt.—Lbs.	27,000.
C. G.—Percent Mac (gear up)	25.6.
Power loading (T. O.)	6.75 lb./HP.
Wing loading	42.8.
Fuel—Gal. L. H. outbd.	205 R. H. outbd. 205.
L. H. inbd.	115 R. H. inbd. 115.
Total	640 gal.
Oil—Gal.	22 gal./Eng.
Armament	4-20 mm. 800 rds. ammo. *4-.50 cal. in G. E. remote turret- 2,000 rds. ammo.
Radio	VHF-SCR-522-2 sets. Detrola range recr. SCR-720 radar. SCR-729 radar. IFF.

General features: Lower gear 175 ind. max. Lower flaps 175 ind. max. Max. dive 430 ind.

*No turret or guns installed.

GENERAL DATA

Model P-63A-9-BE. Airplane

Description Single-seat, single-engine land-based fighter by Bell.

Engine Allison V-1710-93.

TAKE-OFF POWER

Blower	BHP	RPM	MAP	Altitude
Automatic	1,325	3,000	54.0	S. L.

NORMAL POWER

Blower	BHP	RPM	MAP	Altitude
Automatic	1,000	2,600	42.5	20,000

MILITARY POWER

Blower	BHP	RPM	MAP	Altitude
Automatic	1,150	3,000	50.0	25,000

COMBAT POWER

Blower	BHP	RPM	MAP	Altitude
Automatic	1,500	3,000	60.0	S. L.
	1,800	3,000	75"	10,000

MAXIMUM CRUISE—AUTO LEAN

Blower	BHP	RPM	MAP	Altitude
Automatic	670	2,280	31.2	20,000

Propeller Aeroprop Hydromatic.

Dia. 11'-0". Blades 4.

Loading	Full internal fuel and ammo.
Wt. empty—Lbs.	6,925 (basic).
Gross wt.—Lbs.	8,780.
C. G.—Percent Mac (gear up)	25.7.
Power loading (T. O.)	6.62 lb./HP.
Wing loading	35.4 lb./sq. ft.
Fuel—Gal. Wing	136.
Oil—Gal.	
Armament	1-37 mm. cannon.
	2-.50 cal. guns.
Radio	SCR-522.
	SCR-695.

General features: Lower flaps and gear below 155 MPH. RPM and Boost Controls Interconnected.

GENERAL DATA

Model: Firefly. Airplane.

Description: All metal two seater low wing monoplane. Equipped for arrested landing and catapulting.

Engine: Griffon II (12 cyl. liquid cooled).

Blower	BHP	RPM	Map Boost	Altitude
TAKE-OFF POWER (5 minutes)				
				S. L. to
M(low)	1,720	2,750	+12	1,000 ft.
Max. Climbing (1 hour limit)				
M	1,520	2,600	+ 9	3,250
S(high)	1,330	2,600	+ 9	15,750
Max. Cruise (Rich)				
M	1,320	2,400	+ 7	3,500
S	1,190	2,400	+ 7	15,250
COMBAT POWER (5 min. limit)				
M	1,720	2,750	+12	750
S	1,490	2,750	+12	14,000
ECONOMICAL CRUISE—AUTO LEAN				
M		2,400	+ 6	4,000
S		2,400	+ 6	16,000

Propeller: ROTOL—variable pitch.

Dia.: 13 feet. Blades: 3.

Loading		Full internal fuel and ammo.
Wt. empty—Lbs.		8929.
Gross wt.—Lbs.		11685.
C. G.—Percent Mac (gear up)		31.85" forward of datum.
Power loading (T. O.)		6.79.
Wing loading		35.6.
Fuel—Gal.	Main	145½ imp. gal. 171 U. S. gal.
	Wing	46 imp. gal. 55 U. S. gal.
	total	191½ gal. 226 U. S. gal.
Oil—Gal.		11½ imp. gal. 13.8 U. S. gal.
Armament		4-20 mm. Cannon.
Radio		ATA/ARA plus British version of IFF.

General features: Max. diving speed—370 kts. Lower landing gear below 150 kts. Lower flaps below 125 kts.

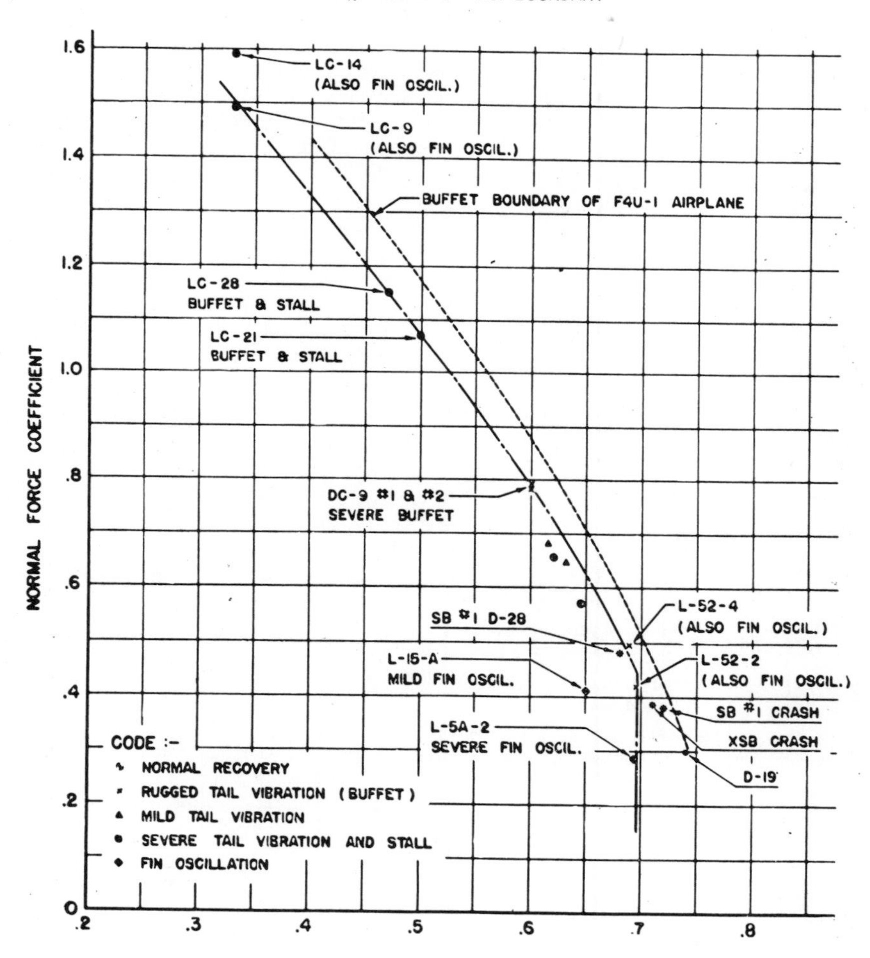

SUMMARY OF FLIGHT INVESTIGATION OF SB2C AIRPLANE
LIMITING C_N - VS. MACH NO. BOUNDARY
NORMAL FORCE COEFFICIENT
1.6
1.4
1.2
1.0
.8
.6
.4
.2
0
.2
.3
.4
.5
.6
.7
.8
LC-14 (ALSO FIN OSCIL.)
LC-9 (ALSO FIN OSCIL.)
BUFFET BOUNDARY OF F4U-1 AIRPLANE
LC-28 BUFFET & STALL
LC-21 BUFFET & STALL
DC-9 #1 & #2 SEVERE BUFFET
SB #1 D-28
L-52-4 (ALSO FIN OSCIL.)
L-15-A MILD FIN OSCIL.
L-52-2 (ALSO FIN OSCIL.)
SB #1 CRASH
XSB CRASH
L-5A-2 SEVERE FIN OSCIL.
D-19
CODE :-
NORMAL RECOVERY
RUGGED TAIL VIBRATION (BUFFET)
MILD TAIL VIBRATION
SEVERE TAIL VIBRATION AND STALL
FIN OSCILLATION

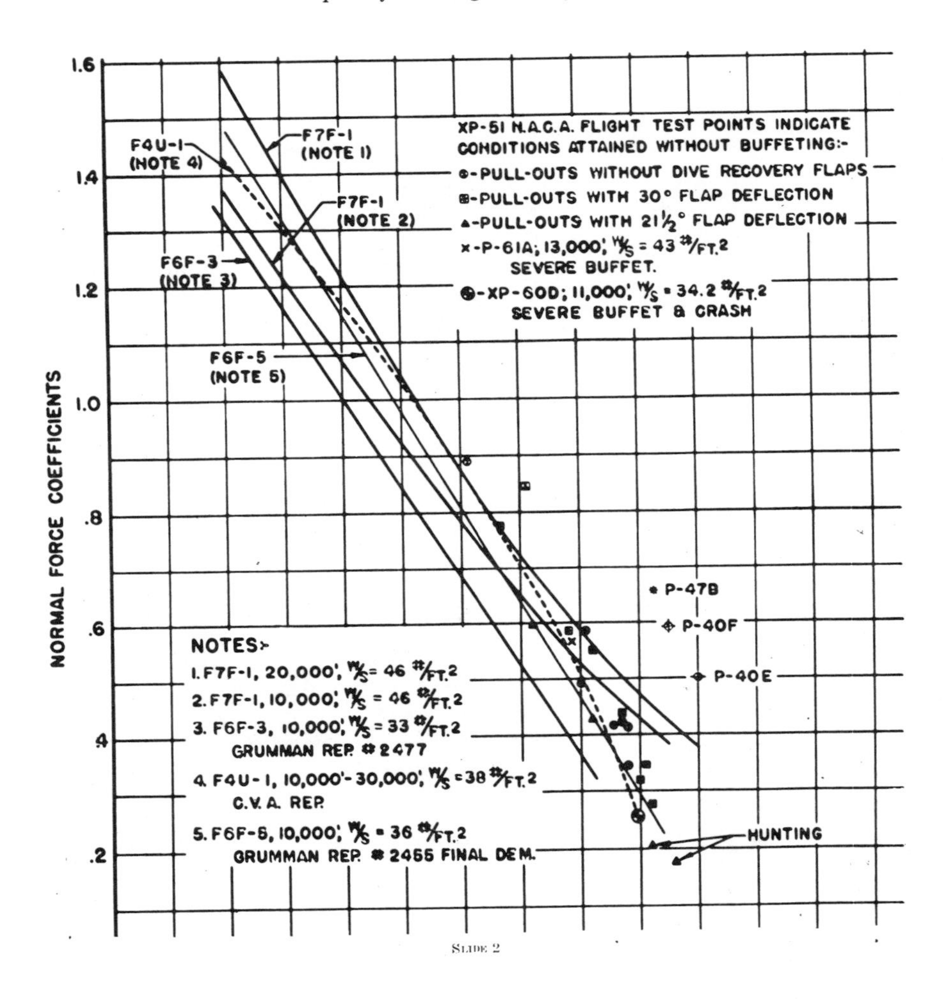
NORMAL FORCE COEFFICIENTS
1.6
1.4
1.2
1.0
.8
.6
.4
.2
F4U-1 (NOTE 4)
F7F-1 (NOTE 1)
F7F-1 (NOTE 2)
F6F-3 (NOTE 3)
F6F-5 (NOTE 5)
XP-51 N.A.C.A. FLIGHT TEST POINTS INDICATE CONDITIONS ATTAINED WITHOUT BUFFETING:-
⊙-PULL-OUTS WITHOUT DIVE RECOVERY FLAPS
⊞-PULL-OUTS WITH 30° FLAP DEFLECTION
▲-PULL-OUTS WITH 21½° FLAP DEFLECTION
×-P-61A; 13,000', W/S = 43 #/FT.2 SEVERE BUFFET.
⊕-XP-60D; 11,000', W/S = 34.2 #/FT.2 SEVERE BUFFET & CRASH
P-47B
P-40F
P-40E
HUNTING
NOTES:-
1. F7F-1, 20,000', W/S = 46 #/FT.2
2. F7F-1, 10,000', W/S = 46 #/FT.2
3. F6F-3, 10,000', W/S = 33 #/FT.2 GRUMMAN REP. # 2477
4. F4U-1, 10,000'-30,000', W/S = 38 #/FT.2 C.V.A. REP.
5. F6F-5, 10,000', W/S = 36 #/FT.2 GRUMMAN REP. # 2455 FINAL DEM.

SLIDE 2

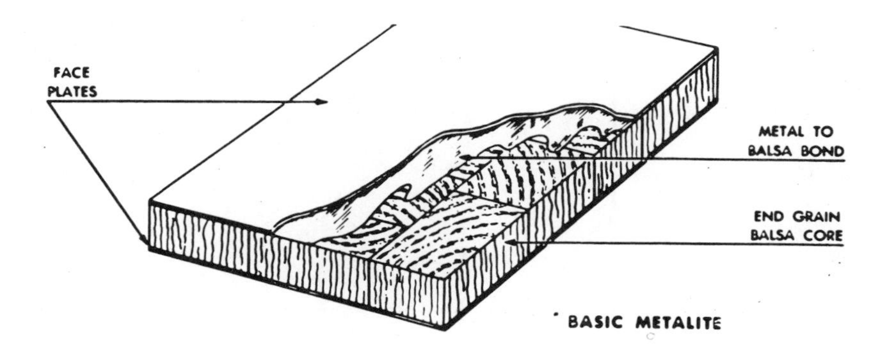
FACE PLATES
METAL TO BALSA BOND
END GRAIN BALSA CORE
BASIC METALITE

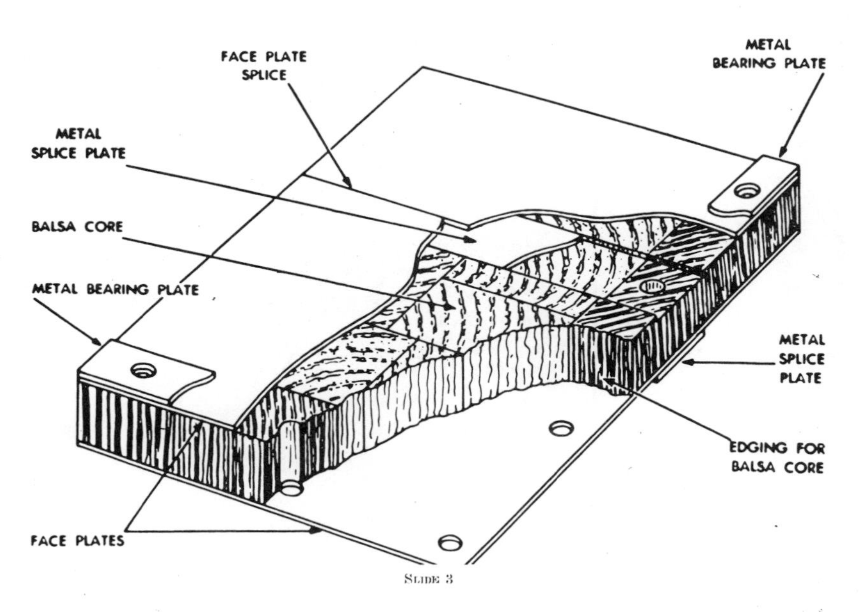
FACE PLATE SPLICE
METAL BEARING PLATE
METAL SPLICE PLATE
BALSA CORE
METAL BEARING PLATE
METAL SPLICE PLATE
EDGING FOR BALSA CORE
FACE PLATES

SLIDE 3

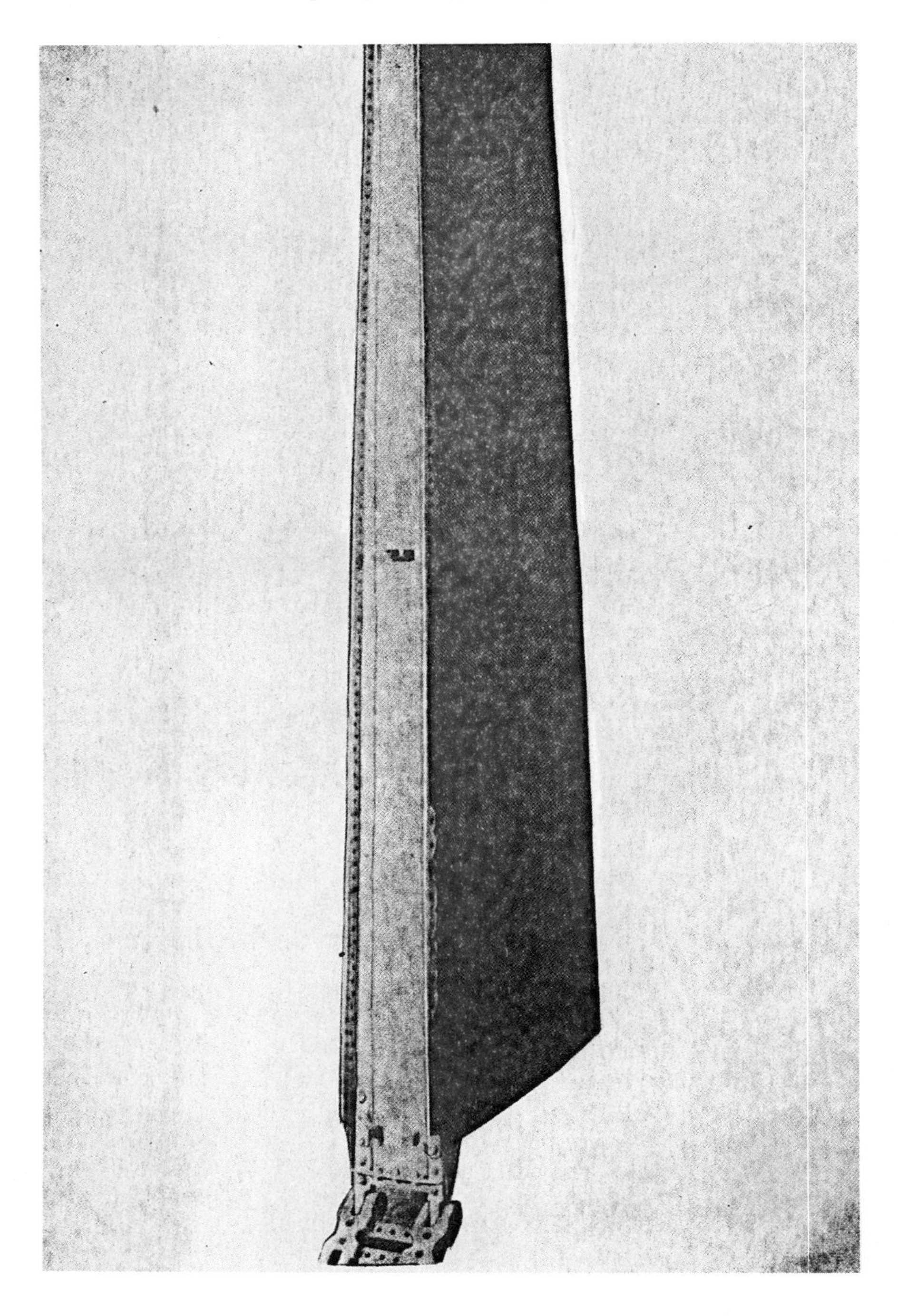

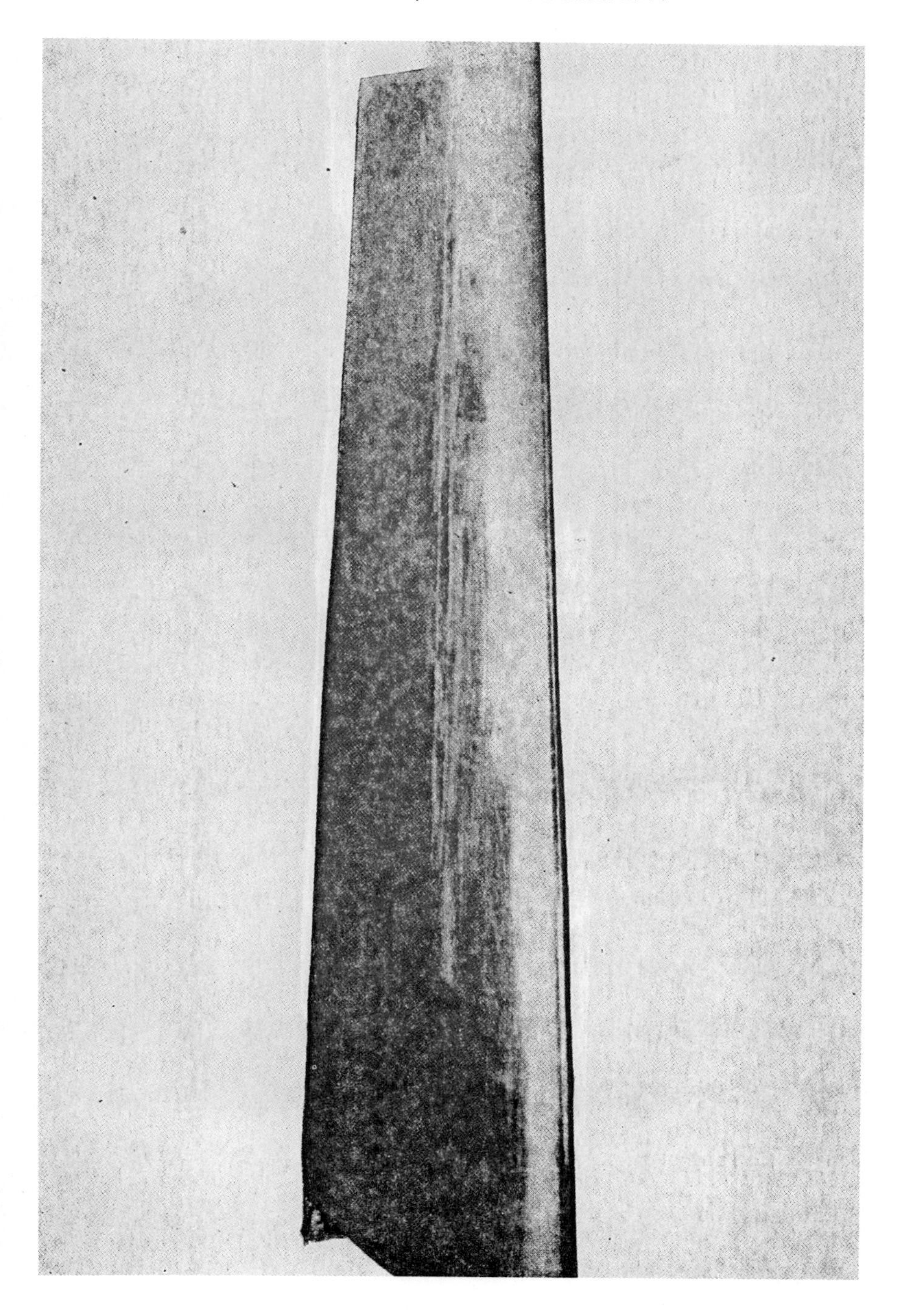

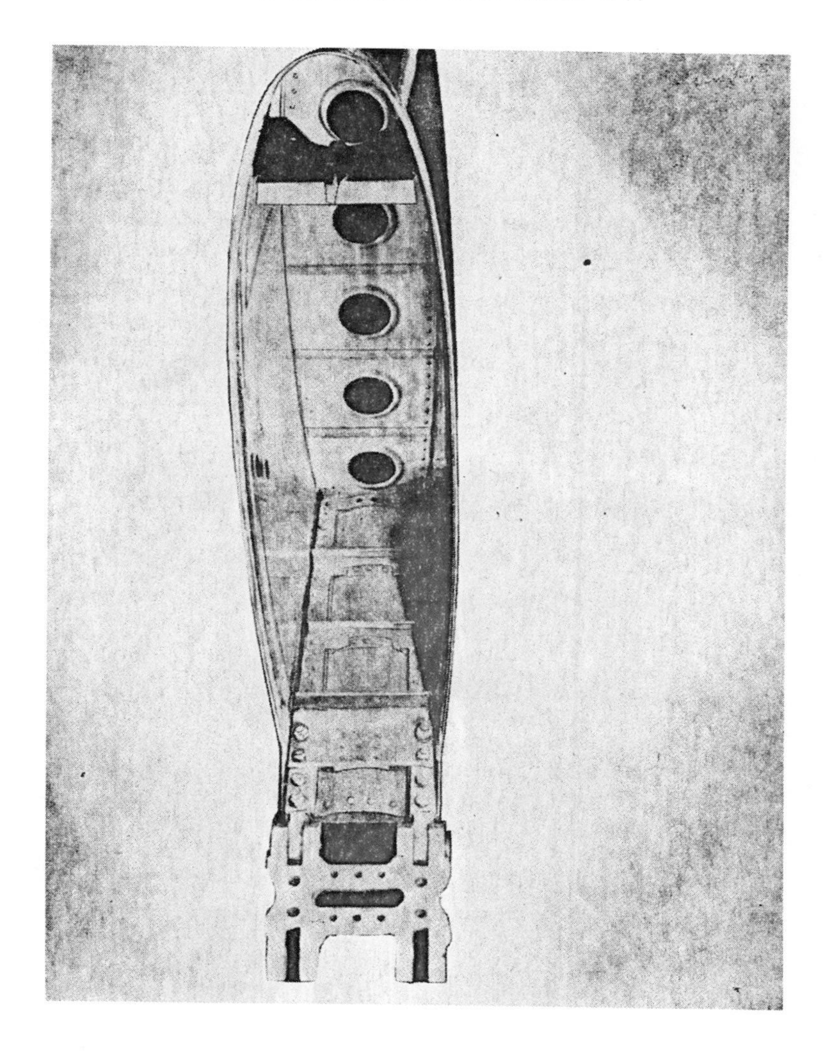

COMPARISON OF XF8F-1 AND F4F-4 BREAKABLE WING TIPS

SCALE $\frac{1}{80}$

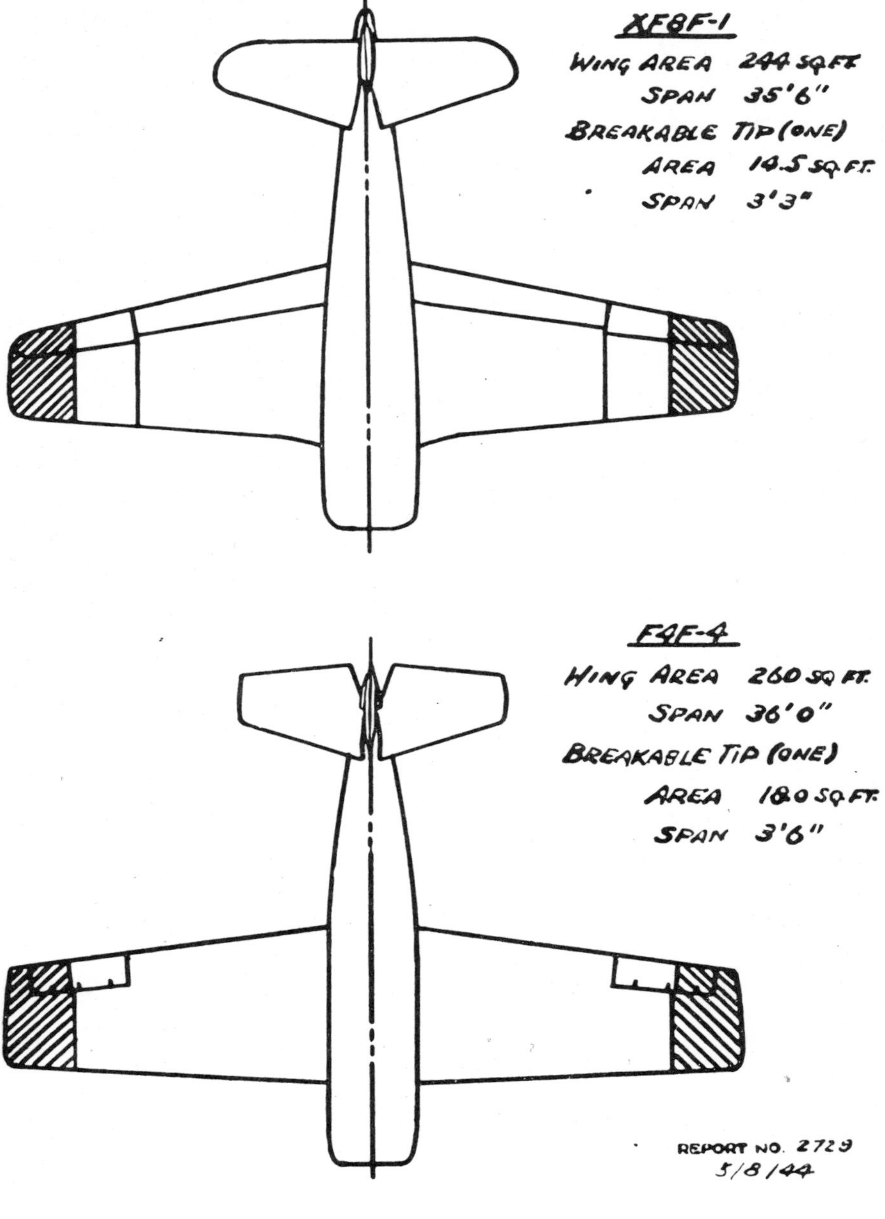

XF8F-1

WING AREA 244 SQ.FT.
SPAN 35'6"
BREAKABLE TIP (ONE)
AREA 14.5 SQ.FT.
SPAN 3'3"

F4F-4

WING AREA 260 SQ.FT.
SPAN 36'0"
BREAKABLE TIP (ONE)
AREA 18.0 SQ.FT.
SPAN 3'6"

REPORT NO. 2729
5/8/44

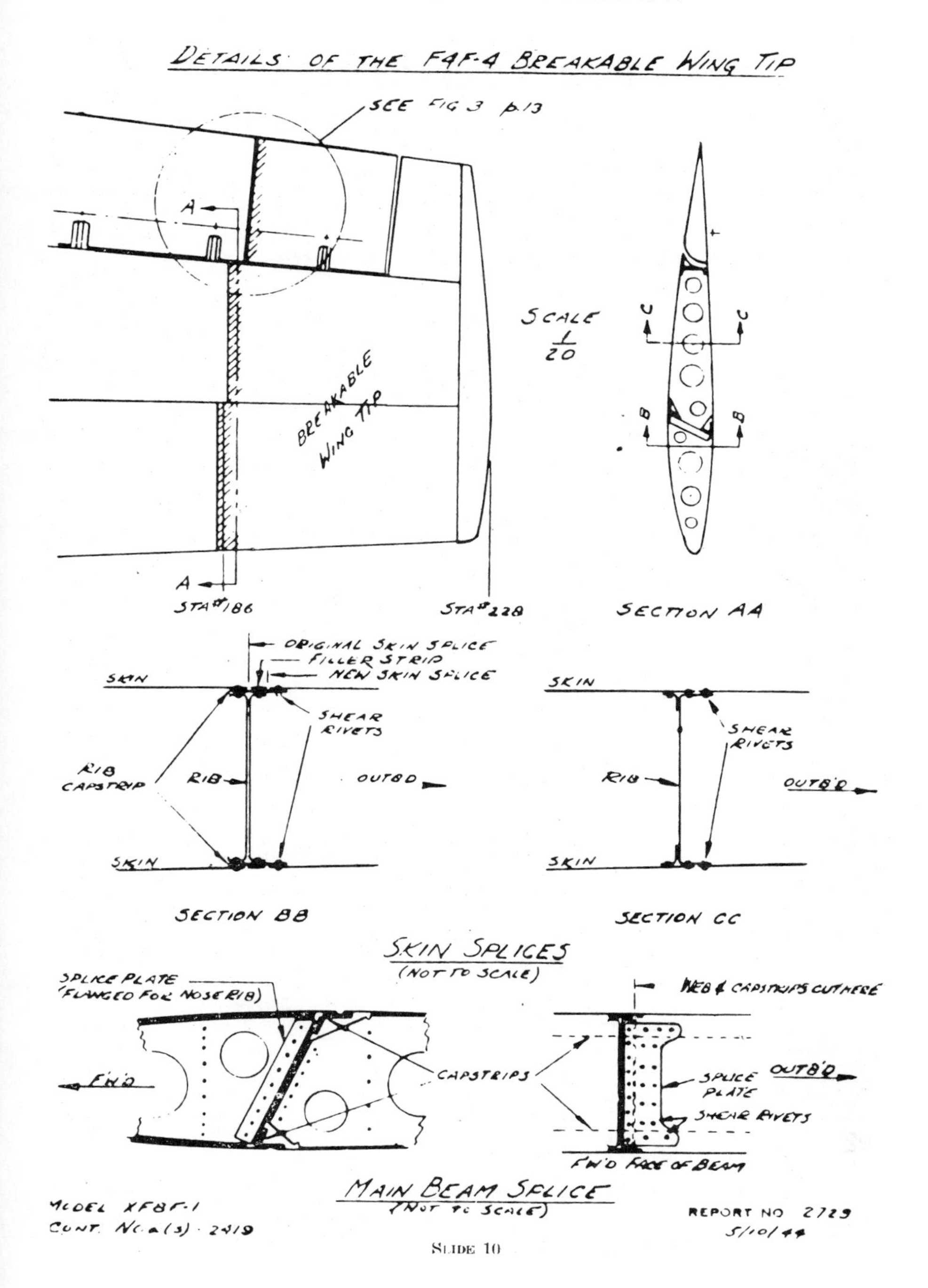
DETAILS OF THE F4F-4 BREAKABLE WING TIP
SEE FIG 3 p.13
A
BREAKABLE WING TIP
SCALE 1/20
C
B
A
STA#186
STA#228
SECTION AA
ORIGINAL SKIN SPLICE
FILLER STRIP
NEW SKIN SPLICE
SKIN
SHEAR RIVETS
RIB CAPSTRIP
RIB
OUTBD
SKIN
SECTION BB
SKIN
SHEAR RIVETS
RIB
OUTB'D
SKIN
SECTION CC
SKIN SPLICES
(NOT TO SCALE)
SPLICE PLATE (FLANGED FOR NOSE RIB)
FWD
CAPSTRIPS
WEB & CAPSTRIPS CUT HERE
SPLICE PLATE
OUTB'D
SHEAR RIVETS
FWD FACE OF BEAM
MAIN BEAM SPLICE
(NOT TO SCALE)
MODEL XF8F-1
CONT. NO a(s) - 2419
REPORT NO 2729
5/10/44

SLIDE 10

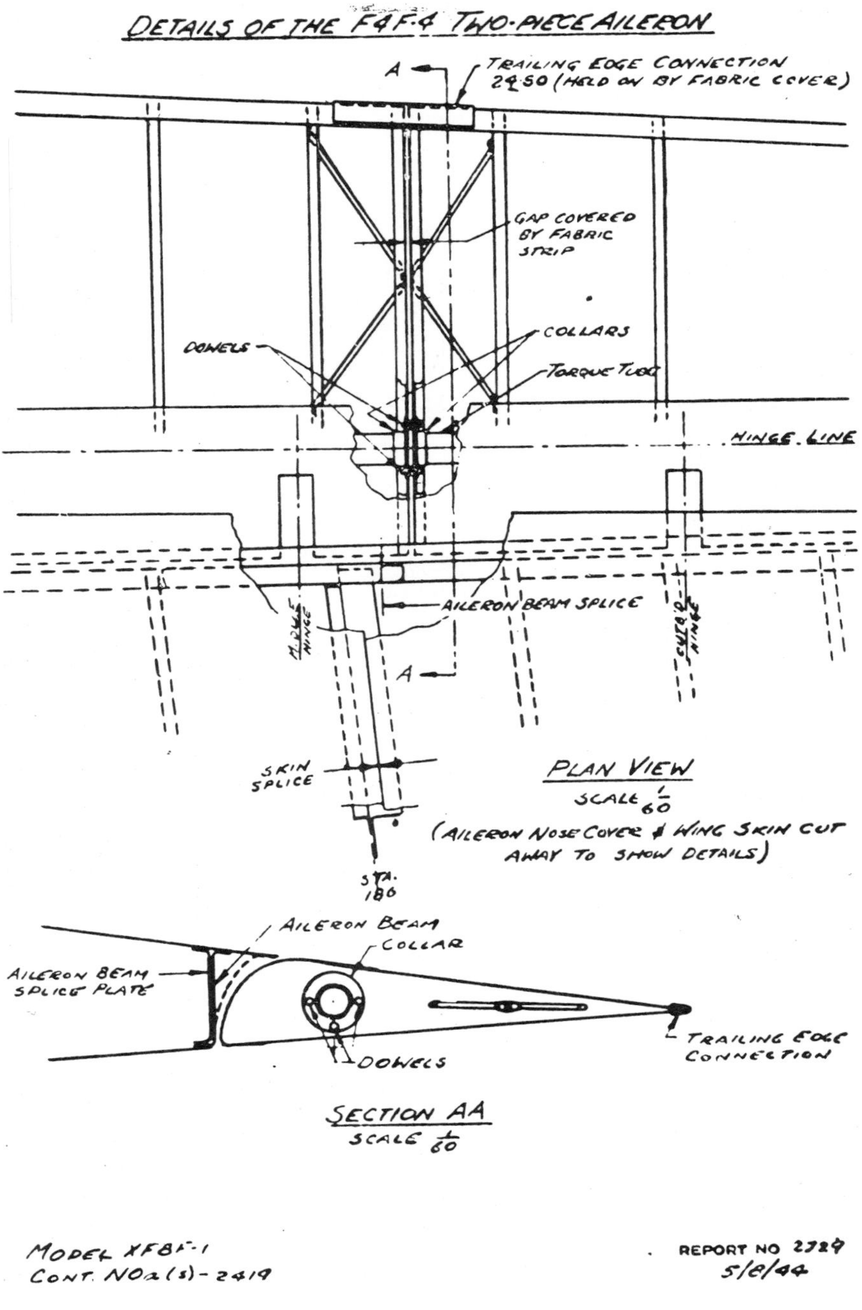

DETAILS OF THE F4F-4 TWO-PIECE AILERON

A
TRAILING EDGE CONNECTION
24-50 (HELD ON BY FABRIC COVER)
GAP COVERED
BY FABRIC
STRIP
COLLARS
DOWELS
TORQUE TUBE
HINGE LINE
AILERON BEAM SPLICE
MIDDLE HINGE
OUTBOARD HINGE
A
SKIN
SPLICE
STA.
186

PLAN VIEW
SCALE 1/60
(AILERON NOSE COVER & WING SKIN CUT
AWAY TO SHOW DETAILS)

AILERON BEAM
COLLAR
AILERON BEAM
SPLICE PLATE
DOWELS
TRAILING EDGE
CONNECTION

SECTION AA
SCALE 1/60

MODEL XF8F-1
CONT. NOa(s)-2419

REPORT NO 2329
5/8/44

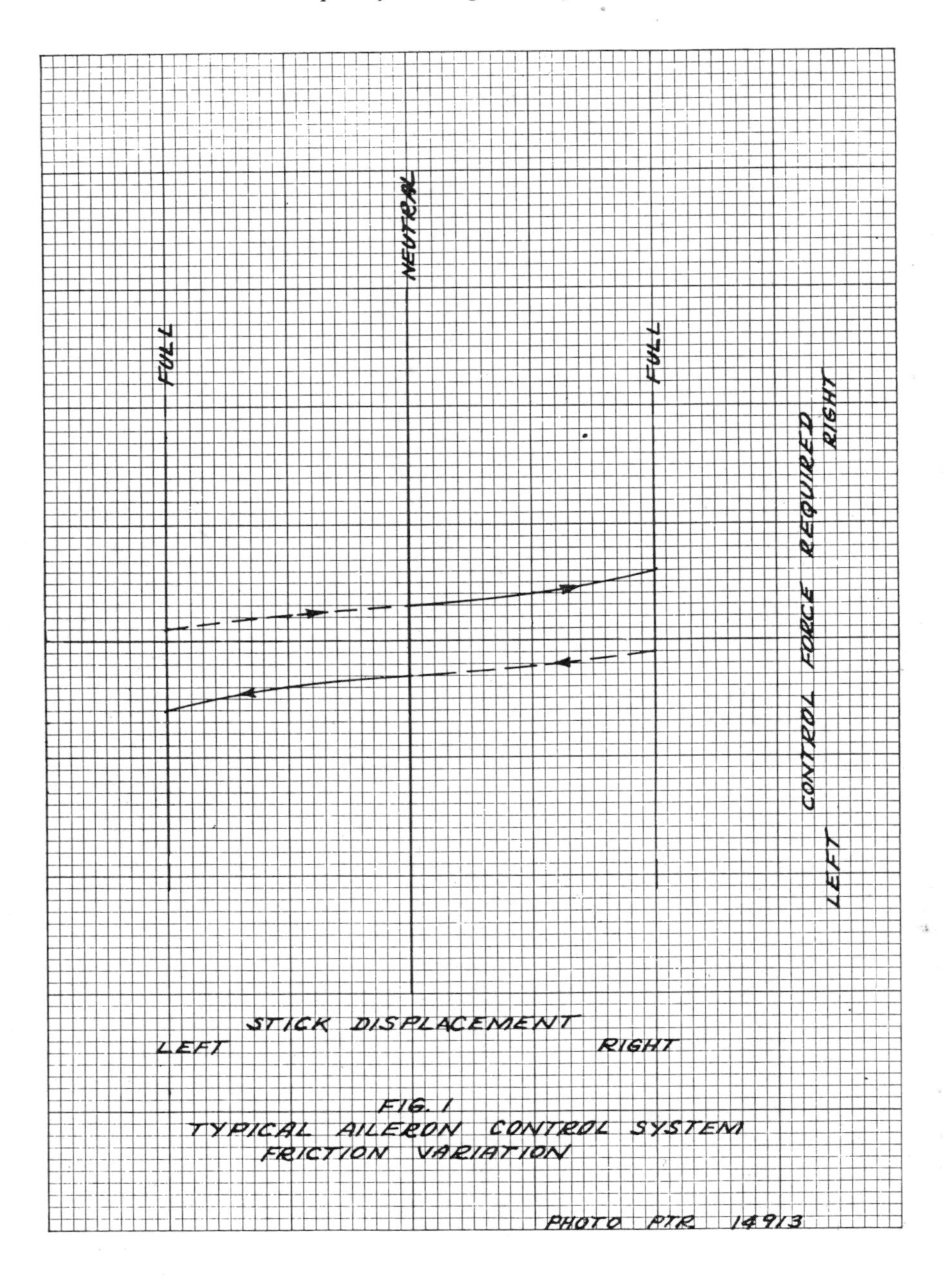

FIG. 1
TYPICAL AILERON CONTROL SYSTEM
FRICTION VARIATION

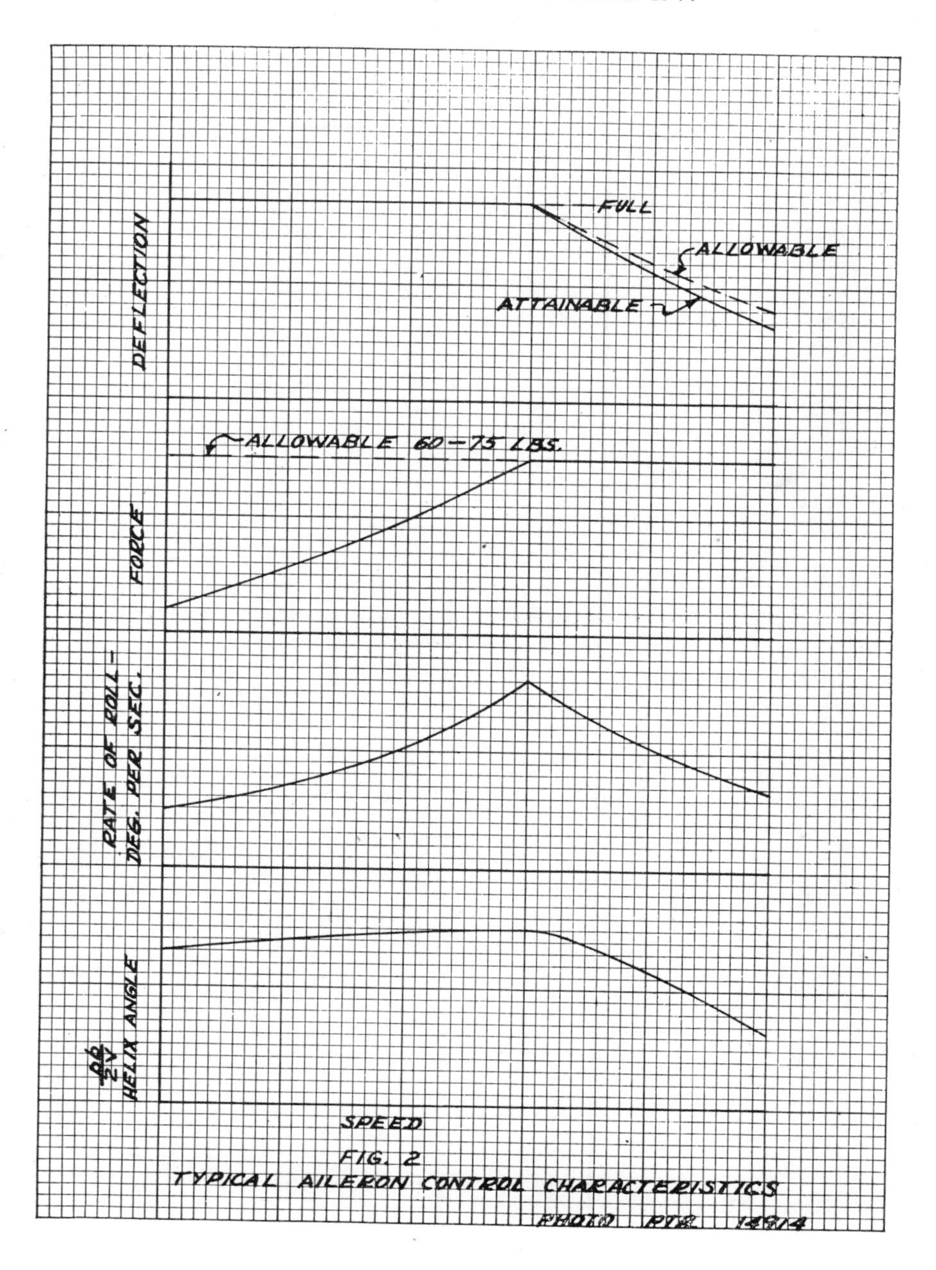

FIG. 2
TYPICAL AILERON CONTROL CHARACTERISTICS

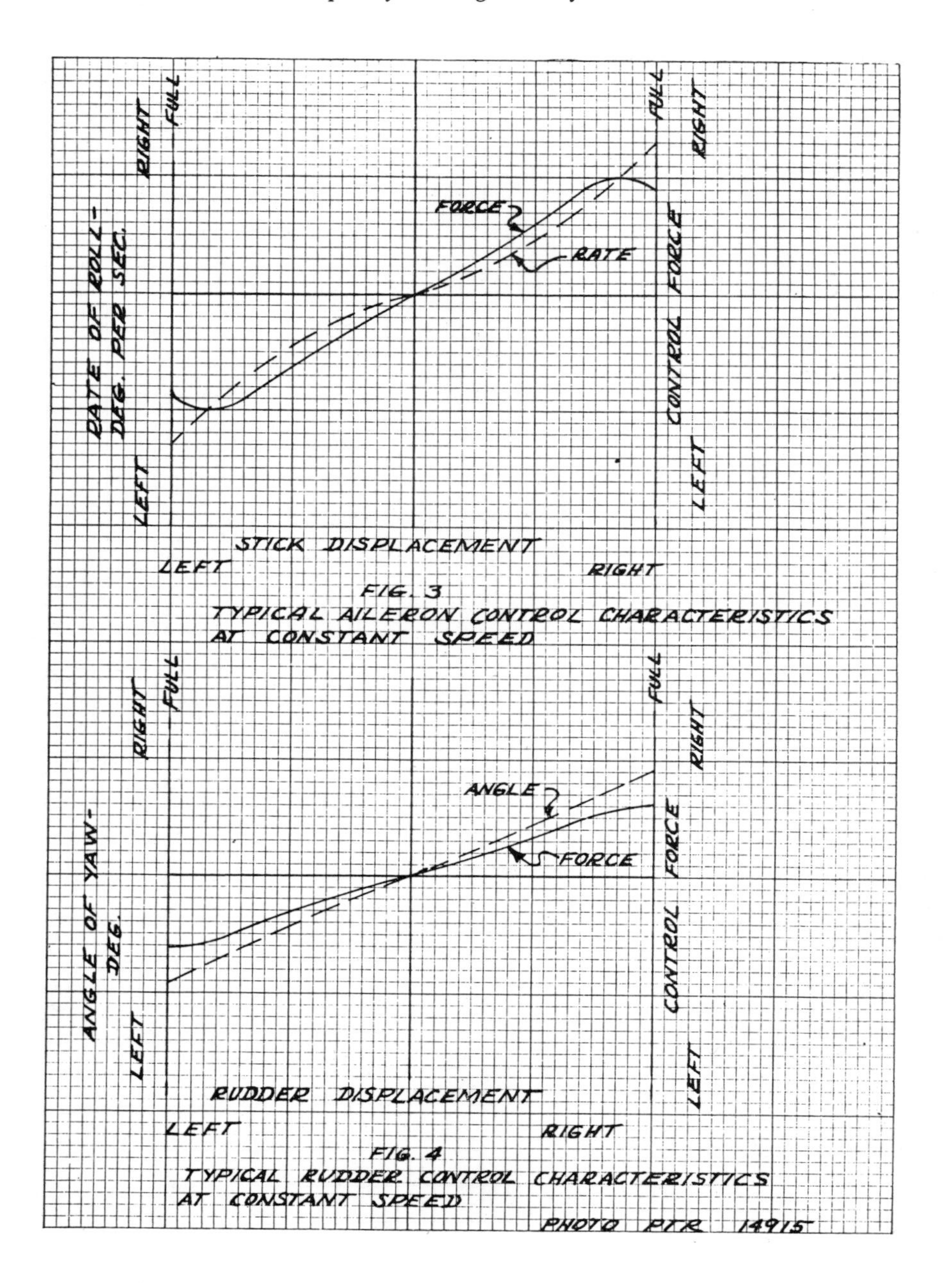

FIG. 3
TYPICAL AILERON CONTROL CHARACTERISTICS
AT CONSTANT SPEED

FIG. 4
TYPICAL RUDDER CONTROL CHARACTERISTICS
AT CONSTANT SPEED

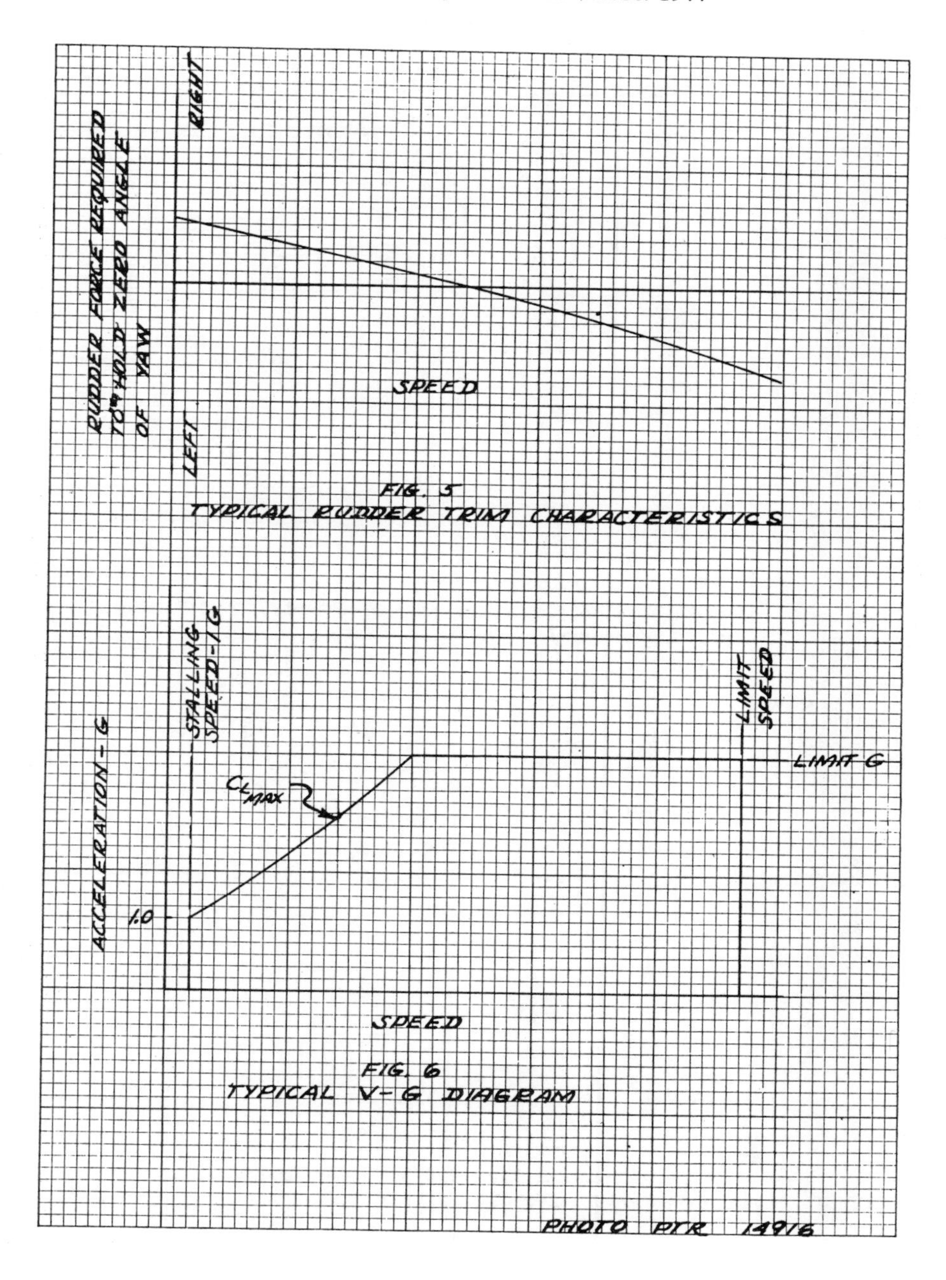

FIG. 5
TYPICAL RUDDER TRIM CHARACTERISTICS

FIG. 6
TYPICAL V-G DIAGRAM

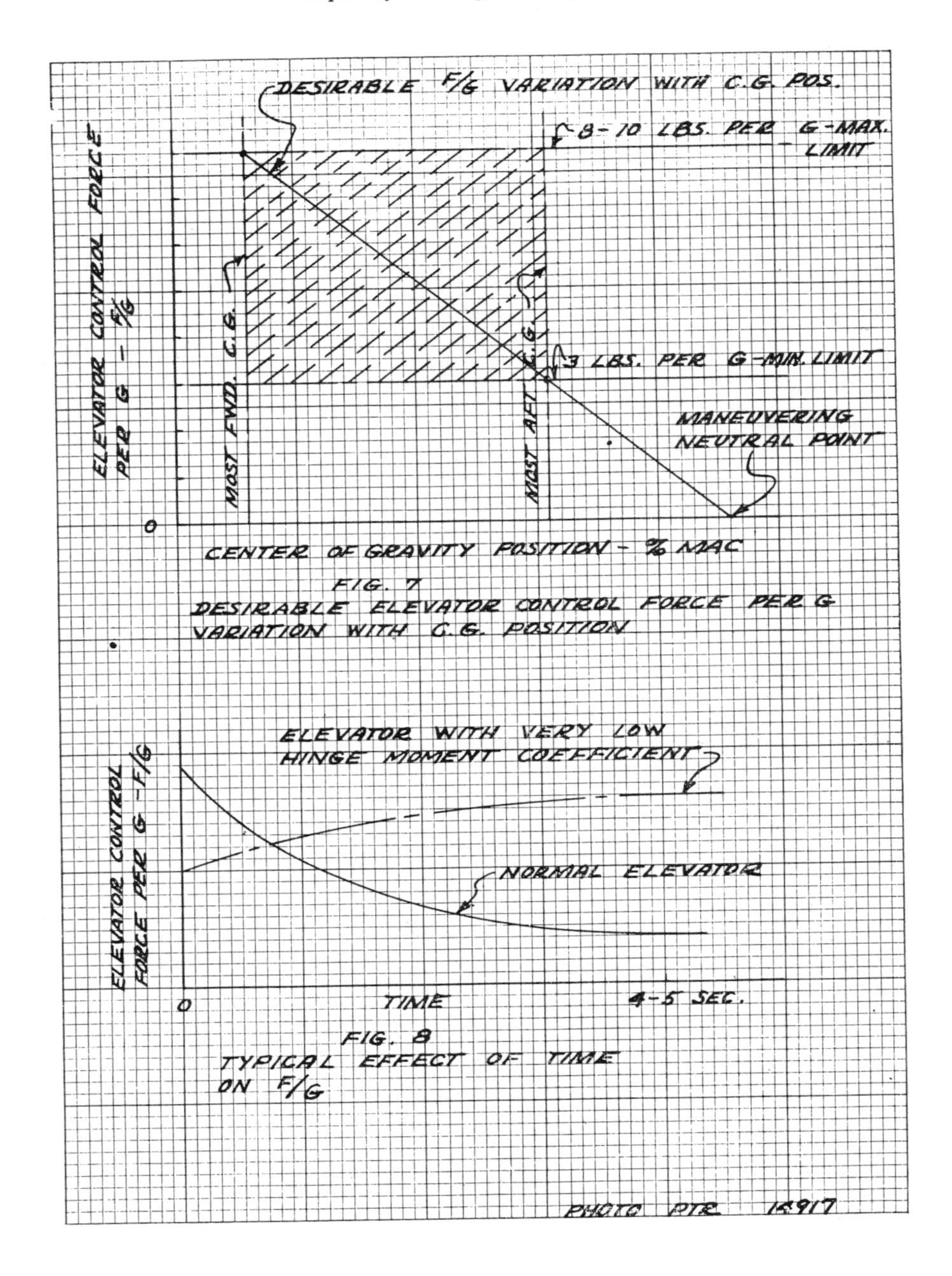
DESIRABLE F/G VARIATION WITH C.G. POS.
8-10 LBS. PER G-MAX. LIMIT
3 LBS. PER G-MIN. LIMIT
ELEVATOR CONTROL FORCE PER G - F/G
MOST FWD. C.G.
MOST AFT C.G.
MANEUVERING NEUTRAL POINT
0
CENTER OF GRAVITY POSITION - % MAC
FIG. 7
DESIRABLE ELEVATOR CONTROL FORCE PER G VARIATION WITH C.G. POSITION
ELEVATOR WITH VERY LOW HINGE MOMENT COEFFICIENT
NORMAL ELEVATOR
ELEVATOR CONTROL FORCE PER G - F/G
0
TIME
4-5 SEC.
FIG. 8
TYPICAL EFFECT OF TIME ON F/G
PHOTO PTR 18917

NOTES

NOTES

NOTES

NOTES

NOTES

NOTES

NOTES

Also from the Publisher

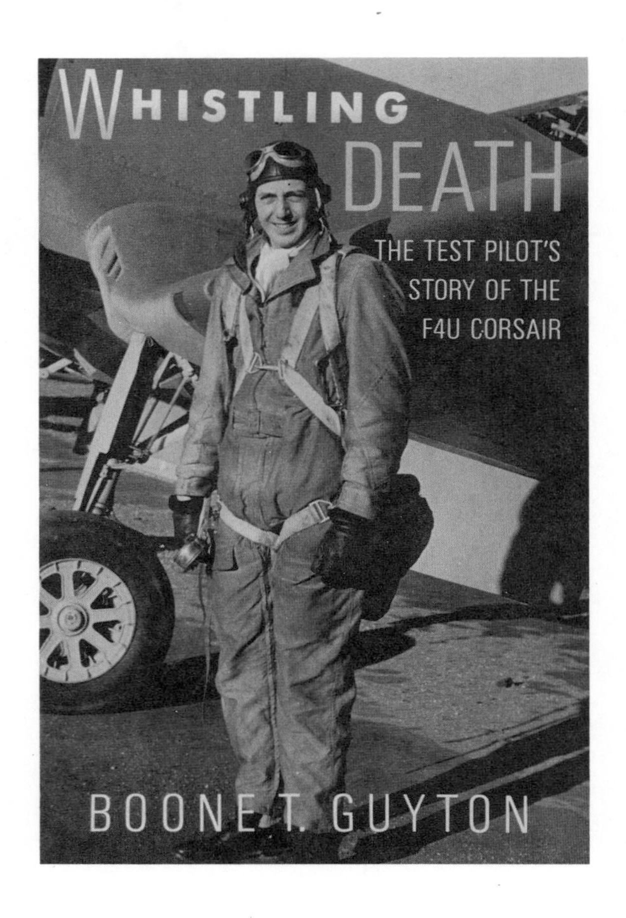
WHISTLING
DEATH
THE TEST PILOT'S
STORY OF THE
F4U CORSAIR
BOONE T. GUYTON